S0-DUW-247

Biogas from Waste and Renewable Resources

Edited by
Dieter Deublein and
Angelika Steinhauser

Related Titles

Soetaert, W

Biofuels

2008
ISBN 978-0-470-02674-8

Wiesmann, U., Choi, I., Dombrowski, E.-M.

Biological Wastewater Treatment

Fundamentals, Microbiology, Industrial Process Integration

2006
ISBN 978-3-527-31219-1

Dewulf, J., Van Langenhove, H. (eds.)

Renewables-Based Technology

Sustainability Assessment

2006
ISBN 978-0-470-02241-2

Olah, G. A., Goeppert, A., Prakash, G. K. S.

Beyond Oil and Gas: The Methanol Economy

2006
ISBN 978-3-527-31275-7

Kamm, B., Gruber, P. R., Kamm, M. (eds.)

Biorefineries – Industrial Processes and Products

Status Quo and Future Directions

2006
ISBN 978-3-527-31027-2

Collings, A. F., Critchley, C. (eds.)

Artificial Photosynthesis

From Basic Biology to Industrial Application

2005
ISBN 978-3-527-31090-6

Clark, C. W.

Mathematical Bioeconomics

The Optimal Management of Renewable Resources

2005
ISBN 978-0-471-75152-6

Gerardi, M. H.

The Microbiology of Anaerobic Digesters

2003
ISBN 978-0-471-20693-4

Biogas from Waste and Renewable Resources

An Introduction

Edited by
Dieter Deublein and Angelika Steinhauser

WILEY-VCH Verlag GmbH & Co. KGaA

The Authors

Prof. Dr.-Ing. Dieter Deublein
Deublein Consulting
International Management
Ritzingerstr. 19
94469 Deggendorf
Germany

Dipl.-Ing. Angelika Steinhauser
8, Dover Rise Heritage View
Tower A #11-08
Singapore 138679
Singapore

Library of Congress Card No.:
applied for

British Library Cataloguing-in-Publication Data
A catalogue record for this book is available from the British Library.

Bibliographic information published by the Deutsche Nationalbibliothek
The Deutsche Nationalbibliothek lists this publication in the Deutsche Nationalbibliografie; detailed bibliographic data are available on the Internet at <http://dnb.d-nb.de>.

Composition SNP Best-set Typesetter Ltd., Hong Kong

Printing Strauss GmbH, Mörlenbach

Bookbinding Litges & Dopf GmbH, Heppenheim

Cover Design WMX Design, Heidelberg

Printed in the Federal Republic of Germany
Printed on acid-free paper

ISBN 978-3-527-31841-4

Contents

Biogas from Waste and Renewable Resources. An Introduction.
Dieter Deublein and Angelika Steinhauser

ISBN: 978-3-527-31841-4

Preface

Rising crude oil prices force us to think about alternative energy sources. Of the different technologies, solar energy is considered the most effective, and can even afford the environmental protection of plants. Many visionaries think that rather biomass will probably convert the solar energy best and will replace all fossil energy resources in the future.

In the last decades, many companies have erected biogas plants worldwide. A lot of experience was gained, leading to a continuous process optimization of anaerobic fermentation and the development of new and more efficient applications. Overall, the basic knowledge of biogas production, the microorganisms involved, and the biochemical processes was widely extended.

This knowledge and the new ideas have now been put together as a basis for the initiation of discussions. Since the technological solutions of technical problems in the fields of anaerobic digestion are tending to vary according to the material treated, e.g., waste water, sewage sludge, or agricultural products, sometimes without any good reason, this book is hoped to contribute to the consolidation of knowledge in the different fields, so that learning can be accessed more easily and applications can be harmonized.

The book includes detailed descriptions of all the process steps to be followed during the production of biogas, from the preparation of the suitable substrate to the use of biogas, the end product. Each individual stage is assessed and discussed in depth, taking the different aspects like application and potential into account. Biological, chemical, and engineering processes are detailed in the same way as apparatus, automatic control, and energy or safety engineering. With the help of this book, both laymen and experts should be able to learn or refresh their knowledge, which is presented concisely, simply, and clearly, with many illustrations. The book can also be used for reference, and includes many tables and a large index. It is strongly recommended to planners and operators of biogas plants, as it gives good advice on how to maximize the potential of the plant.

Originally I collected data and information about biogas plants just out of curiosity. I wanted to know all the details in order to comprehensively teach my students at the University of Applied Sciences in Munich. For five years I surfed the internet and read many books, patents, and magazines, and also approached many companies and manufacturers of plant components, who kindly shared their

Biogas from Waste and Renewable Resources. An Introduction.
Dieter Deublein and Angelika Steinhauser

ISBN: 978-3-527-31841-4

knowledge with me. Dipl.-Ing. Angelika Steinhauser gave me invaluable assistance in the writing, but the main inspiration to publish all the know-how contained in this book was due to Dipl.-Ing. Steffen Steinhauser. We, the authors, thank him cordially for it. We also thank Dr. F. Weinreich from the publishing house WILEY-VCH Verlag GmbH & Co KGaA, who supported this idea. Last but not least, I would like to thank my wife and my son. Without their continuous motivation and very active support this book would never have been completed.

Deggendorf, January 2008 *Dieter Deublein*

Abbreviations

α	Plate inclination	°
$(\alpha_{BR})_a$	Heat transfer coefficient at the wall outside the bioreactor	$W/m^2 \cdot °C$
$(\alpha_{BR})_i$	Heat transfer coefficient at the wall inside the bioreactor	$W/m^2 \cdot °C$
$(\alpha_H)_a$	Heat transfer coefficient at the wall outside the heating pipe	$W/m^2 \cdot °C$
$(\alpha_H)_i$	Heat transfer coefficient at the wall inside the heating pipe	$W/m^2 \cdot °C$
$\Delta\vartheta_{BH}$	Average temperature difference between heating medium and substrate	°C
$\Delta\vartheta_{BR}$	Maximum temperature difference between substrate and the outside of the reactor	°C
$\Delta\vartheta_H$	Temperature difference between inlet and outlet of the heating medium to the bioreactor	°C
$\Delta\vartheta_{SU}$	Maximum temperature difference between substrate inside and outside of the reactor	°C
ΔP_{VP}	Pressure head of the preparation tank pump	bar
ΔT_E, ΔT_A	Differences in absolute temperatures	K
$\Delta G'_f$	Gibbs free energy	$kJ\ mol^{-1}$
ε	Porosity	%
ε_{FS}	Porosity of Siran	%
η_{el}	Efficiency to produce electrical energy	%
η_K	Efficiency of the compressor	%
η_{th}	Efficiency to produce heat	%
η_{VP}	Efficiency of the preparation tank pump	%
Θ	Sludge age	d
ϑ_{HA}	Temperature of the heating medium at the outlet	°C
ϑ_{HE}	Temperature of the heating medium at the inlet	°C
ϑ_S	Dewpoint temperature	°C

Biogas from Waste and Renewable Resources. An Introduction.
Dieter Deublein and Angelika Steinhauser

ISBN: 978-3-527-31841-4

ϑ	Temperature	°C
ϑ_A	Lowest ambient temperature	°C
ϑ_{BR}	Temperature of the substrate in the bioreactor	°C
λ	Air fuel ratio for stoichiometrically equivalent air fuel ratio $\lambda = 1$	–
λ_{BR}	Heat transmission coefficient of the insulation of the bioreactor	$W\ m^{-1} \cdot °C$
ρ_{MK}	Grinding ball density	$Kg\ m^{-3}$
ρ^*	Normal gas density	$Kg\ Nm^{-3}$
ρ_{FS}	Density of Siran	$g\ cm^{-3}$
ρ_G	Density of substrate	$kg\ m^{-3}$
ρ_S	Densitiy of co-ferment	$kg\ m^{-3}$
ρ_w	Density of heating medium	$kg\ m^{-3}$
$(P_{BRR})_{tot}$	Total power consumption of the agitators	kW
$(P_{SC})_{tot}$	Total power consumption of the co-ferment conveyors	kW
A	Area for cultivation of energy plants	m^2
A_{BR}	Surface of the bioreactor, where heat is lost	m^2
A_{COD}	Degree of decomposition determined by the COD value	–
A_D	Total available area	ha
A_{Dtechn}	Technically usable area	ha
A_M	Cultivation area for maize	ha
A_S	Degree of decomposition determined by oxygen demand	–
AT_4	Breathing activity	$mg\ O_2/g_{DM}$
B	Disintegration intensity	$kJ\ kg^{-1}$
B_A	Bioreactor area load	$kg_{oDM}/(m^2 \cdot h)$
B_{BR}	Average bioreactor volume load	$kg_{oDM}/(m^3 \cdot d)$
bn	billion	
BOD_5	Difference in oxygen concentration (day 1 vs. day 5)	$mgO_2\ L^{-1}$
B_R	Bioreactor volume load	$kg_{oDM}/(m^3 \cdot d)$ or $kg_{COD}/(m^3 \cdot d)$
B_{RoDMSB}	Organic sludge load	$kg/kg \cdot d$
B_{RS}	Total sludge load	$kg_{COD}/(kg_{DM} \cdot d)$
B_S	Breadth	m
c_0	Concentration of organics in the substrate	$kg_{COD}\ m^{-3}$
C_1, C_2	Constants	
COD	Chemical oxygen demand (COD value)	$mgO_2\ L^{-1}$
COD_0	COD value of untreated sample	$mgO_2\ L^{-1}$
COD_{max}	Maximum COD value	$mgO_2\ L^{-1}$
C_S	Biomass concentration in excess sludge	$kg_{COD}\ m^{-3}$
c_{SU}	Specific heat capacity of the substrate	$kJ/kg \cdot °C$

c_w	Specific heat capacity of the heating medium	kJ/kg·°C
D	Net income from fertilizer	US$ a^{-1}
D_{BR}	Diameter of bioreactor	m
D_{BRl}	Diameter of discharge pipe	m
D_{BRR}	Outer diameter of agitator	m
D_D	Decanter diameter	m
D_E	Diameter of residue storage tank	m
d_{FS}	Pore diameter of Siran	m
D_{HR}	Diameter of heating pipe	m
DIN	German industrial norm	–
D_L	Diameter of aeration pipe	m
DM	Dry matter	% or g L^{-1}
DM_{BR}	Flow rate of dry matter into the bioreactor	kg_{oDM} d^{-1}
d_{MK}	Grinding ball diameter	M
$DM_{R,e}$	Dry matter in outflow of sludge bed reactor	g L^{-1}
D_{PT}	Diameter of preparation tank	m
D_W	Diameter of windings of heating pipe	m
E	Nominal capacity of electrical power of the CHP	kW
E_{Eel}	Electrical power consumption of the plant	kW
E_{el}	Capacity of the plant to deliver electrical energy	kW
E_M	Yield of CH_4 per biomass	kmol CH_4 kg^{-1}
$E_{OILspec}$	Specific energy per volume of ignition oil	kWh L^{-1}
E_R	Theoretical yield	Mg_{DM}/ha·a
E_{Rmax}	Maximum theoretical yield	Mg_{DM}/ha·a
E_S	Solar energy	kW
E_{spec}	Specific biogas energy	kW m^{-3}
E_{th}	Capacity of the plant to deliver heat	kW
E_{tot}	Total energy	kW
f_{VBR}	Factor to increase the bioreactor volume	–
f_{VE}	Factor to increase the residue storage tank	–
f_{VPT}	Factor to increase the preparation tank	–
G	Net income from current	US$ a^{-1}
GB_{21}	Gas formation within 21 days	Nl kg_{DM}^{-1}
GVE	Animal unit	–
h_1, h_2, h_3, h_4, h_5	Specific enthalpies at different stages of the process	kJ kg^{-1}
H_{BP}	Filling height for pellet sludge	m
H_{BR}	Bioreactor height	m
H_{BS}	Height of the gas/solid separator	m
H_E	Height of the residue storage tank	m
H_{ON}, H_{UN}	Calorific value	kWh m^{-3}
H_{PT}	Height of the preparation tank	m
H_S	Height of the silo	m

IN	Inhabitant	
I_{SV}	Sludge volume index	Mg L^{-1}
K, K_1, K_2	Total investment costs	US$
KA	Plant investment costs without CHP	US$
KA_{spec}	Specific investment costs for the biogas plant per volume of the bioreactor	US$ m^{-3}
KB	Investment costs for concrete works	US$
k_{BR}	k-factor of the bioreactor wall with insulation	$W/m^2 \cdot K$
KB_{spec}	Specific price for sold current	US$ kWh^{-1}
K_{CHP}	Investment costs for the CHP	US$
k_H	k-factor of the heating pipes	$W/m^2 \cdot °C$
KK	Amortization per year for the CHP	US$ a^{-1}
KK_{spec}	Specific investment costs for CHP per capacity of electrical energy	US$ kW^{-1}
K_{OIL}	Cost for ignition oil	US$ a^{-1}
$K_{OILspec}$	Specific cost for ignition oil	US$ L^{-1}
KP	Local overhead costs	US$ a^{-1}
KP_{spec}	Specific local overhead costs	US$ h^{-1}
KR	Costs for cultivation of renewable resources	US$ a^{-1}
KR_{spec}	Specific costs for cultivation of renewable resources	US$/ha·a
KS	Costs for power consumption	US$ a^{-1}
KS_{spec}	Specific costs for power consumption	US$ kWh^{-1}
KT	Investment costs for technical equipment	US$
KV	Insurance costs	US$ a^{-1}
kW	Costs for heat losses	US$
KW_{spec}	Specific price for sold heat	US$ kWh^{-1}
KX	Maintenance costs for the concrete work	US$ a^{-1}
KY	Maitenance costs of technical equipment	US$ a^{-1}
KZ	Maintenance costs of the CHP	US$ a^{-1}
L_D	Decanter length	M
L_{HR}	Length of the heating pipe	M
L_S	Length of the silo	M
m*	Flow of gas to the compressor	m^3 h^{-1}
$\dot{M}_{BR}$	Produced flow of biogas	Mg d^{-1}
M_E	Molecular weight	Kg $kmol^{-1}$
$\dot{M}_G$, $\dot{M}_{G1}$, $\dot{M}_{G2}$	Flow rate of substrate	Mg d^{-1}
$\dot{M}_{oil}$	Flow rate of ignition oil	Mg d^{-1}
$\dot{M}_S$	Flow rate of co-ferments	Mg a^{-1}
N	Normal	
n_{BRR}	Revolutions of an agitator	rpm
Ne_{BRR}	Newton number of an agitator	
oDM	Organic dry matter	kg_{COD} or kg_{DM}
$oDM_{R,e}$	oDM in the outflow of a sludge bed reactor	g L^{-1}
O_{FSspec}	Specific surface of Siran	m^2 m^{-3}
O_{spec}	Specific surface	m^2 m^{-3}

OUR	Oxygen uptake rate	mg/(L·min)
OUR_0	Oxygen uptake rate of untreated substrate	mg/(L·min)
p_1	Biogas pressure before compressing	Bar
p_2	Biogas pressure after compressing	Bar
P_A	Power consumption of compressor	kW
P_{BRR}	Power consumption of agitator	kW
P.E.	Population equivalent	
P_{econ}	Economical potential	kWh a^{-1}
$\dot{P}_{econ}$	Specific economical potential	kWh/(ha·a)
P_K	Power consumption of the air compressor	kW
p_{K1}	Pressure before compressor	bar
p_{K2}	Pressure after compressor	bar
P_{SC}	Power consumption of a co-ferment conveyor	kW
P_{techn}	Technical potential	kWh a^{-1}
$\dot{P}_{techn}$	Specific technical potential	kWh/(ha·a)
P_{theor}	Theoretical potential	kWh a^{-1}
$\dot{P}_{theor}$	Specific theoretical potential	kWh/(ha·a)
P_{VP}	Power consumption of the pumps	kW
Q_{BR}	Heat loss of the bioreactor	kW
Q_{SU}	Required energy to heat the substrate	kW
Q_V	Total heat loss	kW
R_{CH4}	Special gas constant for CH_4	kJ/kg·°C
S	Overlapping	mm
s_{BR}	Thickness of the insulation of the bioreactor	m
T	Absolute temperature of the gas to be compressed	K
t	Residence time	d
t_B	Annual amortization for concrete works	US$ a^{-1}
t_{BR}	Residence time in the bioreactor	d
t_{BRl}	Time for discharging the reactor content	H
t_{BRR}	Time of operation of an agitator	min h^{-1}
t_E	Residence time in the residue storage tank	d
t_K	Time of amortization for the CHP	a
TLV	Treshold limit value = PEL Permissible exposure limit	
t_{min}	Minimum tolerable theoretical residence time	h
TOC	Total oxygen content in the substrate	mg L^{-1}
TOC*	Total oxygen content in the residue	% DM
t_P	Time of local work	h
T_{PT}	Residence time in the preparation tank	d
t_s	Annual operation time	H a^{-1}
t_{SC}	Running time of a co-ferment conveyor	h d^{-1}
t_T	Annual amortization for technical equipment	US$ a^{-1}
t_{TS}	Residence time in the activated sludge tank	d
v_A	Velocity of the upstream	m h^{-1}

V_{BR}	Bioreactor volume	m^3
v_{BRl}	Velocity in the discharge pipe	$m\ s^{-1}$
V_E	Volume of residue storage tank	m^3
v_F	Velocity of gas in gas pipes	$m\ s^{-1}$
v_G	Velocity of inflow	$m\ h^{-1}$
V_G*	Inflow rate	$m^3\ d^{-1}$
V_{GS}	Volume of the gas holder	m^3
v_H	Velocity of the heating medium in the pipe	$m\ s^{-1}$
V_K	Volume of compressor pressure vessel	m^3
v_L	Velocity of air in aeration pipe	$m\ s^{-1}$
V_{PT}	Volume of the preparation tank	m^3
V_S	Silo volume	m^3
v_u	Rotational velocity of the agitator system	$m\ s^{-1}$
v_W	Velocity of the substrate in heat exchanger pipes	$m\ s^{-1}$
$\dot{V}_{BR}$	Produced flow of biogas	$m^3\ d^{-1}$
$\dot{V}_E$	Feedback from the residue storage tank to the bioreactor	$m^3\ d^{-1}$
$\dot{V}_K$	Compressor throughput	$Nm^3\ h^{-1}$
$\dot{V}_L$	Volume rate of air in the aeration pipe	$Nm^3\ h^{-1}$
$\dot{V}_S$	Volumetric flow of excess sludge	$m^3\ d^{-1}$
$\dot{V}_{SC}$	Volume flow of co-ferment in the conveyor	$m^3\ h^{-1}$
$\dot{V}_{Oil}$	Volume rate of ignition oil	$m^3\ d^{-1}$
$\dot{V}_{VP}$	Flow rate of the preparation tank pump	$m^3\ h^{-1}$
$\dot{V}_w$	Flow rate of heating medium in the pipe	$m^3\ h^{-1}$
W	Net income from heat	US\$ a^{-1}
w_G	Gas velocity in empty reactor	$Nm^3/m^2 \cdot s$
W_O, $W_{O,N,}$ $W_{U,N}$	Wobbe Index, upper Wobbe index, lower Wobbe index	$kWh\ m^{-3}$
w_s	Area load	$m^3/m^2 \cdot h$
w_t	Specific work of the compressor	$kJ\ kg^{-1}$
X	Biomass concentration in the reactor	$kg_{DM}\ m^{-3}$
x_B	Fraction of the investment costs without CHP for concrete works	–
x_T	Fraction of the investment costs without CHP for technical equipment	–
y_B	Specific maintenance costs of the concrete work	US\$ a^{-1}
y_{CHP}	Specific maintenance costs for CHP	US\$ a^{-1}
y_T	Specific maintenance costs of technical equipment	US\$ a^{-1}
Z	Fraction of the liquefied methane	–
Z	Insurance rate	US\$ a^{-1}
Z_R	Interest rate	US\$ a^{-1}

Acknowledgement

The following companies, institutions and individuals have kindly provided photographs and other illustrations. Their support is gratefully acknowledged.

agriKomp GmbH (www.agrikomp.de)	Figs. 5.15, 5.23
Bekon Energy Technologies GmbH (www.bekon-energy.de)	Fig. 5.63
Bioferm GmbH (www.bioferm.de)	Fig. 6.1
Bundesverband der landwirtschaftlichen Berufsgenossenschaften	Fig. 4.3
Ceno Tec GmbH(www.ceno-tec.de)	Figs. 5.1, 6.1
Coop (www.coop.ch)	Fig. 5.1
D. Saffarini, University of Wisconsin-Milwaukee	Fig. 3.32
Filox Filtertechnik GmbH (www.filox.de)	Fig. 5.33
Gerardo P. Baron	Fig. 5.1
H. Bahl, Elektronenmikroskopisches Zentrum, University of Rostock	Fig. 3.9
Hexis AG (www.hexis.com)	Fig. 6.21
Leibniz Institute of Marine Sciences, Kiel (www.ifm-geomar.de)	Fig. 2.9
Institute of Cultural Affairs, Tokyo (www.icajapan.org)	Fig. 1.21
Ishii Iron Works (www.ishii-iiw.co.jp)	Fig. 6.1
K. O. Stetter and R. Rachel, University of Regensburg	Fig. 3.31
Klein Abwasser- und Schlammtechnik GmbH (www.klein-news.de)	Fig. 5.33
Kompogas AG (www.kompogas.ch)	Fig. 6.11
KWS Saat AG / Dr. W. Schmidt	Fig. 1.12
Landratsamt Freising	Fig. 6.2
Max-Planck-Institute for Breeding Research / Dr. W. Schuchert	Fig. 1.12
MDE Dezentrale Energiesysteme GmbH (www.mde-online.com)	Fig. 6.11
MTU-CFC GmbH (www.mtu-friedrichshafen.com)	Fig. 6.11
Pondus Verfahrenstechnik GmbH (www.pondus-verfahren.de)	Fig. 3.28
Protego (www.protego.com)	Fig. 5.7

Biogas from Waste and Renewable Resources. An Introduction.
Dieter Deublein and Angelika Steinhauser

ISBN: 978-3-527-31841-4

Reck-Technik GmbH (www.reck-agrartechnik.de)	Fig. 3.28
R. Priggen, Düsseldorf	Fig. 5.66
S. Battenberg, Technical University of Braunschweig	Figs. 3.31, 5.47
Schmack Biogas AG (www.schmack-biogas.com)	Figs. 5.1, 5.15
Scientific Engineering Centre "Biomass", Kiev (www.biomass.kiev.ua)	Fig. 1.21
Siemens AG	Fig. 6.11
South-North Institute for Sustainable Development, Beijing (www.snisd.org.cn)	Fig. 1.21
SUMA Sondermaschinen GmbH	Fig. 5.23
SunTechnics (www.suntechnics.at)	Fig. 1.21
Technical University Braunschweig / German Research Foundation	Figs. 5.11, 5.48
Thöni GmbH (www.thoeni.com)	Fig. 5.1
Turbec SpA (www.turbec.com)	Fig. 6.11
U.T.S. Umwelttechnik Süd GmbH (www.umwelt-technik-sued.de)	Fig. 5.15
University of Karlsruhe	Fig. 1.12
VORSPANN-TECHNIK GmbH & Co. KG (www.vorspanntechnik.com)	Fig. 5.39
VTA Engineering und Umwelttechnik GmbH (www.vta.cc)	Fig. 6.11
WELtec BioPower GmbH (www.weltec-biopower.de)	Fig. 5.23

Part I
General thoughts about energy supply

Human beings are the only animals with the ability to ignite and use a fire. This advantage has been important for the growth of mankind, particularly during the past few decades, when the rapid rate of innovation in industry was especially facilitated by the immense richness of oil. Today, thousands of oil platforms exist globally, which provide the oil for ca. 50 000 kWh of energy per year. Yearly, around 10 bn US$ are spent in drilling for new oilfields to secure the supply of oil and hence the base for industrial growth in future.

But, as with all fossil resources, the quantity of oil is limited and will not last for ever. A time will come for sure when all the existing accessible oil fields will have been exploited. What will then happen to mankind?

May the same happen as is observed in nature? Not only in animals but also in plants there are sudden "explosions of populations". Such growth naturally stops, however, as soon as a source of life runs dry. The organisms start suffering from deficiency symptoms and become dominated or eaten by stronger organisms.

How will human beings generate energy when all the oil resources we benefit from today are fully consumed? There is as yet no clear answer to this question. But regardless of what the answer may be, it is clear that the mankind will always want to continue building huge inventories of energy. With the declining quantity of fossil fuels it is critical today to focus on sustained economic use of existing limited resources and on identifying new technologies and renewable resources, e.g., biomass, for future energy supply.

Biogas from Waste and Renewable Resources. An Introduction.
Dieter Deublein and Angelika Steinhauser

ISBN: 978-3-527-31841-4

1
Energy supply – today and in the future[1)]

Today, globally most energy is provided by burning oil. Only a very small percentage is generated by nuclear power plants. The contribution of energy from renewable resources is almost negligible. But this will change in the future with increasing prices of oil.

In the future, countries may use different technologies, depending on their climatic and geographical location. Germany refrains from using nuclear power plants as a source of energy. This makes Germany one of the leading countries in the development of technologies for alternative and renewable energy sources.

1.1
Primary energy sources

In general, primary energy sources are classified as follows:

Fossil energy sources

- Hard coal
- Brown coal
- Petroleum
- Natural gas
- Oil shale
- Tar sand
- Gas hydrate

Renewable energy sources

- Water
- Sun
- Wind
- Geothermal heat
- Tides
- Biomass

Nuclear fuels

1) Cp. BOK 1

Biogas from Waste and Renewable Resources. An Introduction.
Dieter Deublein and Angelika Steinhauser

ISBN: 978-3-527-31841-4

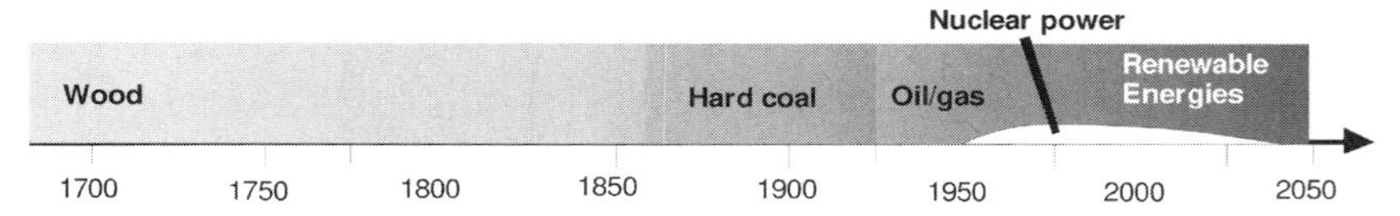

Figure 1.1 Life cycles of primary energy sources.

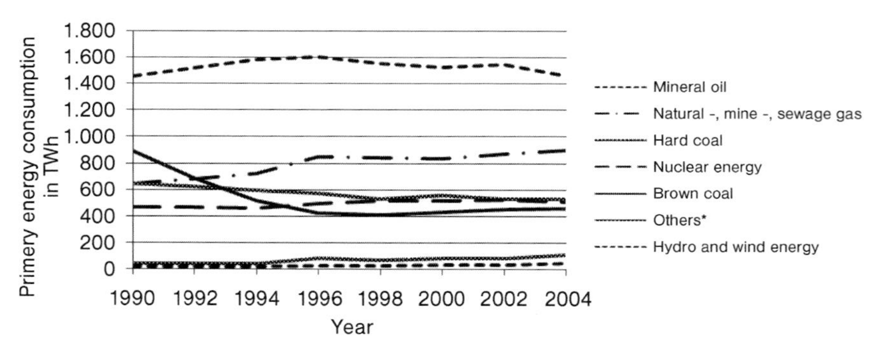

Figure 1.2 Primary energy sources related to the total consumption of primary energy resources in Germany in TWh[2] during 1990–2004 (*e.g., firewood, turf, sewage sludge, waste, and other gases).

These primary energy sources follow so called "life cycles" as shown in Figure 1.1.

Until the late 19th century, wood, the traditional biomass, was the only primary energy source used for cooking and heating. This ended when wood was replaced by hard coal, an epoch which lasted ca. 75 years. This was followed in the late 1950s by a continuously increasing use of petroleum and natural gas. Around 1950, nuclear power technology was first time industrialised, but it never became truly accepted. For some years now, this technology has remained stagnant and has not expanded because of still unresolved issues such as the storage of the radioactive waste and the risk of explosion of a reactor. This leaves "renewable energies", showing the biggest potential for securing the availability of energy in the future.

As an example: the total consumption of primary energy in Germany is ca. 4100 TWh a^{-1}, which has been provided by the use of different primary energy sources, shown in Figure 1.1. The primary energy source used during the past few years in Germany was mainly mineral oil (Figure 1.2). In the early 1990s, quite a significant part of energy in the Eastern part of Germany was also generated by processing brown coal. After the German reunification, however, the mining of brown coal was stopped because of the great environmental damage it was causing.

2) Cp. WEB 20

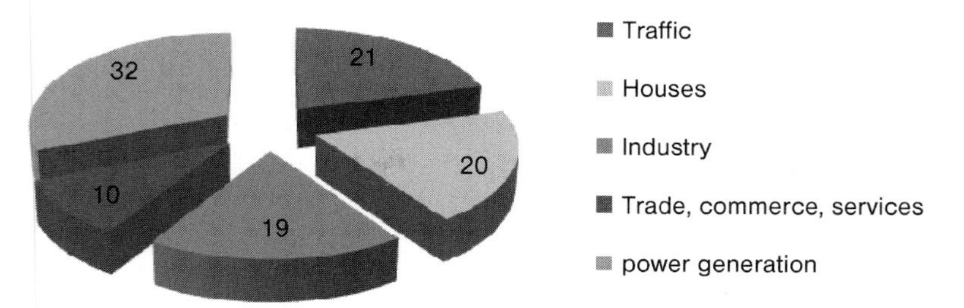

Figure 1.3 Primary energy as resource in % – segmentation in industrialized countries.

After this, the consumption of energy provided by hard coal remained almost static, while the energy from natural gas, mine gas, or sewage gas strongly increased to make up for that previously provided by brown coal. The use of renewable energy has been almost static during recent years, with a very slight though consistent upward trend.

Consumers using primary energy are shown in Figure 1.3. This chart shows that the traffic sector consumes 21% of the primary energy, which is even more than industry (19%). In fact the amount of energy supplied to industry is decreasing, and increasing amounts go to traffic. This is explained by the current trend toward a society with a high number of cars per family leading to a high demand of petrol, a secondary energy source of petroleum.

1.2
Secondary energy sources

Secondary energy sources are defined as products that have been produced by transforming primary energy carriers into higher quality products by applying processes such as refining, fermentation, mechanical treatment, or burning in power stations:

Products derived from coal

- Coke
- Briquettes

Products derived from petroleum

- Petrol
- Fuel oil
- Town gas
- Refinery gas

Products derived from renewable resources

- Biogas
- Landfill gas
- Pyrolysis gas

The secondary energy sources are converted to end-point energy.

1.3
End-point energy sources

The end-point energy is the energy used by the final consumers and provided in form of, e.g. district heating, wood pellets and electricity. In Germany, for example, the consumption of end-point energy is about 2600 TWh a^{-1}. It is important to emphasize that only electricity and not gas is defined as end-point energy since gas is the energy source that electricity is derived from.

Usually the amount of end-point energy consumed is used for calculation purposes and is taken as a base to reflect energy balances.

1.4
Effective energy

Only about 1/3 of the primary energy is effective energy which is actually used by customers in form of heating, light, processing, motion, and communication. The other 2/3 is lost when transforming the primary energy sources into effective energy. As an example, in Germany only 1400 TWh a^{-1} of energy is effectively used. About 570 TW a^{-1} of this energy is actually electricity. To cover these quantities, the electricity is produced mainly by using fossil energy sources (60%) like hard coal, petroleum, or natural gas (Figure 1.4); 30% is derived from nuclear power stations, while the amount of electricity from renewable energy sources is only about 7.25% to date.[3)]

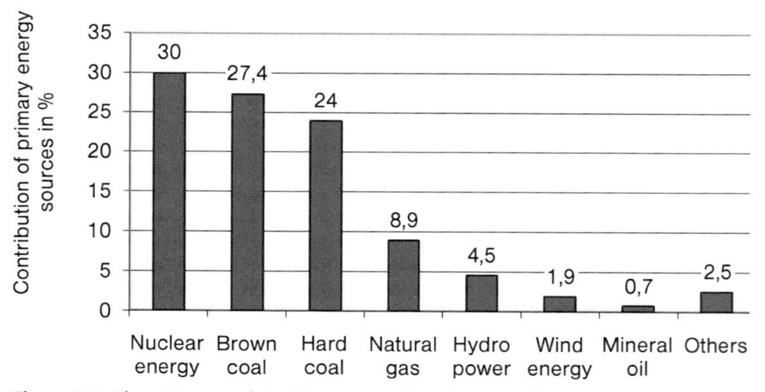

Figure 1.4 Electricity supply in Germany – Contribution of primary energy carrie's on total power supply.

3) Cp. BOK 3

2
Energy supply in the future – scenarios

Shell International[4] has published a projection for different energy sources for the years 1990 up to 2100 (Figure 2.1). Assuming the "Sustainable Growth" scenario, energy consumption will increase by 7 times (at most) during this period. Applying the "Dematerialization" scenario (= much lower consumption driven by sustained economic use), the amount of energy will increase by a factor of 3 (at least). Both scenarios can be explained and are driven by the assumptions of an increase in population from about 6 bn to around 10 bn plus a continuous fast path taken by emerging markets to accelerate their economic growth.

Further, by 2020 the technologies around renewable resources are expected to have reached the potential for full economic use. Shell foresees a fast growth for these future alternatives and has projected that by 2050 the regenerative energy resources will provide 50% of the total energy consumption worldwide. According to Shell, the main source will be solar energy and heat.

Similarly, the WEC (World Energy Council) in 1995 has put forward a scenario in which the primary energy consumption will increase 4.2-fold by 2100 (referring back to 1990), and in its "Ecological" scenario of 1995 it still talks about a 2.4-fold increase.[5]

The IPCC (International Panel on Climate Change) expects a 3 times higher energy consumption by 2100 (referring back to 1990), providing a high demand. With sustained economic use of energy, calculations suggest that almost 30% of the total global primary energy consumption in 2050 will be covered by regenerative energy sources. In 2075 the percentage will be up to 50%, and it is expected to continuously increase up to 2100. According to the IPCC report, biomass is going to play the most important role, projected to deliver 50 000 TWh in 2050, 75 000 TWh in 2075, and 89 000 TWh in 2100, in line with the calorific value derived from the combustion of more than 16 bn Mg of wood.[6]

Many other institutions have developed their own scenarios and done their own projections, as shown in Table 2.1.

The economic potential of using hydroelectric power to provide energy is already almost fully exploited. All other renewable resources, however, still have huge potential and can still be widely expanded.

4) Cp. WEB 65
5) Cp. WEB 79
6) Cp. WEB 72

Biogas from Waste and Renewable Resources. An Introduction.
Dieter Deublein and Angelika Steinhauser

ISBN: 978-3-527-31841-4

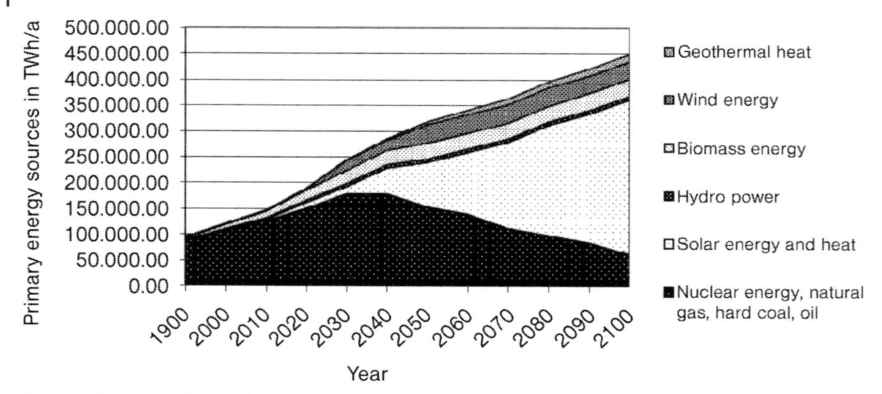

Figure 2.1 Projection of the energy supply up to year 2100 (acc. to Shell International).

Table 2.1 Perspectives for energy sources as a percentage of the total energy consumption in Germany.

	Federal Ministry for the Environment in Germany[7]				*Federal Ministry of Economy and Technology in Germany*[8]		*Greenpeace*[9),10)]	
	Today		*Projection in year 2080*		*In 2020*		*In 2010*	
Total final energy consumption	*Static at around 2600 TWh a⁻¹*				*Static*	*Declining to ca. 2115 TWh a⁻¹*	*Independent of the energy consumption*	
	Elect.	***Heat***	***Elect.***	***Heat***			***Elect.***	***Heat***
Natural gas	21.9		0		28.0	41.0	4.1	30.2
Nuclear energy	12.9		0		4.0	2.0		
Hard coal	13.2		0		22.0	11.0	2.4	5.6
Brown coal	11.6		8					
Petroleum	37.5		0		41.0	36.0		18.8
Renewable energy	1.31	1.59	24.0	68.0	4.0	10.0	18.5	20.4
Hydro-electricity	0.60		1.5				4.1	
Wind	0.40		8.5				4.8	
Photovoltaics	0.01		4.3				2.8	
Solarthermal heat		0.07		25.0			2.6	0.2
Geothermal heat		0.04	10.0	35.0			3.0	0.2
Biomass	0.30	1.48	3.0	8.0			1.2	20.0

Note: Technical final energy potential = technically usable electrical energy in the system.

7) Cp. WEB 23
8) Cp. WEB 26
9) Cp. WEB 66
10) Cp. JOU 13

Biomass is rich in carbon but is not yet a fossil material. All plants and animals in the ecological system belong to biomass. Furthermore, nutrients, excrement, and bio waste from households and industry is biomass. Turf is a material intermediate between biomass and fossil fuel.

There are several processes to transform biomass into solid, liquid, or gaseous secondary energy carriers (Figure 2.2): these include combustion, thermo-chemical transformation via carbonization, liquefaction or gasification, physico-chemical transformation by compression, extraction, transesterification, and biochemical transformation by fermentation with alcohol or aerobic and anaerobic decomposition.

Today in Germany, 65% of the heat and electricity generated with processes based on biomass are provided by combusting firewood and forest residual wood, followed by the use of industrial residual wood and matured forest. About 14% of the energy comes from the use of liquid or gaseous biological energy carriers. When considering heat only, it is even higher, as shown in Table 2.2.

Thermochemical processing or combustion are the most effective ways to maximize the generation of energy. Combustion is only efficient, however, if the water

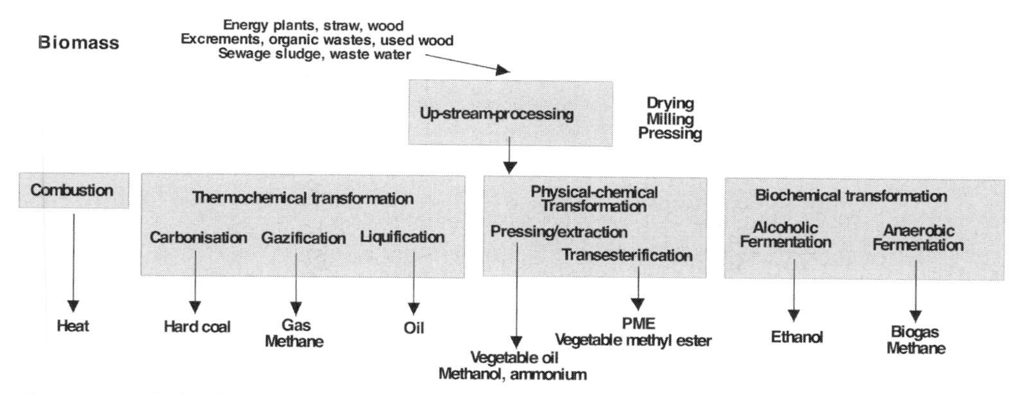

Figure 2.2 Applied technologies to transform biomass[11] into secondary energy sources.

Table 2.2 Heat generated from biomass.

Energy carrier	*Percentage*	*Generated heat*
Organic residues. By-products, waste (biogas, sewage sludge gas, landfill gas)	9.4	6.25–6.53 TWh a^{-1}
Bio fuels	1.3	0.8–0.9 TWh a^{-1}
Biogenous solid fuel (firewood, forest residual wood)	89.3	45.8 TWh a^{-1}
Industrial residual wood		11.9 TWh a^{-1}
Matured forest without recovered paper		3.3 TWh a^{-1}
Other wood-like biomass		0.3 TWh a^{-1}
Straw		0.78 TWh a^{-1}

11) Cp. JOU 16

content of the biomass is below 60% to prevent most of combustive energy from going into the evaporation of water. In the worst case, all this energy will have to be generated from the flue gas. The only chance to regain this usable energy will then be to condense the evaporated water in a condensing boiler.[12] However, this is only possible if the biomass is free from corrosive materials. From an economic point of view, the temperature of the flue gas is important. Furthermore, the composition of the combustion residue needs to be carefully evaluated for possible use.

If the biomass has very high water contents (e.g., liquid manure, freshly harvested plants), it is best to select and accept a process which provides only about 70% of the energy resulting from the combustion of dry material. As an advantage, the residues can be easily returned to nature, especially since no materials enriched with minerals and thus plant-incompatible ash are generated.

If biomass is to be used to serve as source win liquid fuel, it is best to produce ethanol and/or methanol via alcoholic fermentation. This process is more efficient than anaerobic fermentation referring to the hectare yield.

Overall, the energy balance is particularly favorable for biomass when considering the energy yield from the biomass [output] to the assigned primary energy [input], including all efficiencies as to be seen in Table 2.3.

With the output/input ratio of 28.8 MJ/MJ, biomass appears to be a very efficient source of biogas.

One of the leading countries in developing biogas plants is Germany, where hence a lot of efficiency data have been yet generated. In the following abstracts, these data are presented to show the potential of this technology and to highlight important factors that should be considered before planning a biogas plant. It is

Table 2.3 Energy balance[13],[14] for different final energy carriers.

Energy source	*Energy balance Output/ Input [MJ/MJ]*	*Remarks*
Rape oil	5.7	Energy recovery of the colza cake and green waste included
Ethanol	2.7	From wheat
Ethanol	1.6	From sugar beet
Ethanol	5.0	From sorghum Energy recovery of the bagasse included
Electricity and heat	8.5	Combustion of the whole plant From cereals
Electricity and heat	19.7	Combustion of miscanthus plants (not dried)
Electricity and heat	14.2	Combustion of energy plants
Electricity and heat	20.4	Combustion of residual straw
Electricity and heat	19.0	Combustion of forest residual wood
Biogas	28.8	From excrement (CHP cycle)

12) Cp. BRO 3

13) Cp. BRO 16

14) Cp.: BOK 62

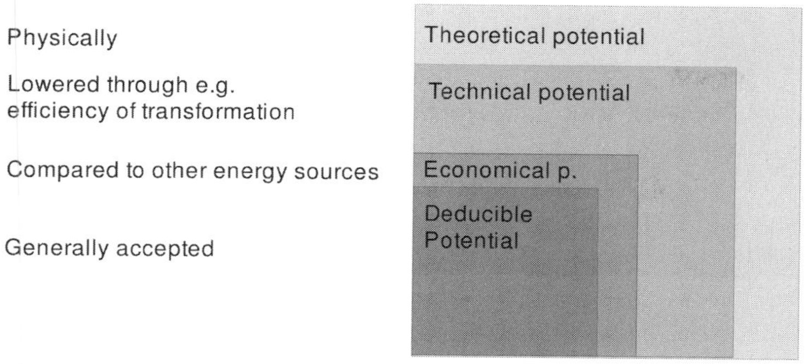

Physically | Theoretical potential
Lowered through e.g. efficiency of transformation | Technical potential
Compared to other energy sources | Economical p.
Generally accepted | Deducible Potential

Figure 2.3 Evaluation of the scope – from the theoretical to the deducible potential.

important to differentiate and carefully evaluate the theoretical, technical, economical, and realizable potential (Figure 2.3).

The theoretical potential comprises all the energy that should theoretically be physically generated within a defined time period and a defined space.

The technical potential is part of the energy of the theoretical potential. It is that specific part which can be provided within the given structural and ecological boundaries and by respecting any legal restrictions.

It may not always make sense to fully exhaust the technical potential, especially if there is no profitable return.

However, the economic potential may not be realizable without any administrative support from certain institutions.

The total yield from biomass results from the maximum area available for cultivation and the energetic yield from the biomass cultivated on this specific area.

2.1
Amount of space

The amount of space in Table 2.4 is defined as the land area plus the surface area of the water, because algae or water plants in general are biomass and may have potential in the future.

The right hand columns in the table show the amount of space that is available for cultivation of biomass and may have potential.

In theory all the amount of space A_D, including the surface of the water, can be used to produce biomass.

Technically, biomass can be cultivated on all areas except the settlement area, mining lands or badlands. This is an amount of space of $A_{Dtechn} = 0{,}88 \cdot A_D$ of the total surface of Germany.

As soon as the micro algae production is developed, then technically an even larger surface, means 95% of the entire available space, could be exploited.

Economically, the cultivation of energy plants competes with the cultivation of other agricultural products. The market will probably equilibrate itself. But overall

Table 2.4 Total available area in Germany as A_D in hectares [ha].[15]

Utilization of the amount of area today		*Total area*	*Theoretically usable*	*Technically usable*	*Economically usable*
Total		35 703 099	35 703 099	31 503 678	20 117 031
Settlements and areas used for transport and traffic		4 393 895	=	374 052	300 812
The above areas include:	Buildings and open space including residences, trade, industry	2 308 079	=	0	0
	Areas for winning substances out of the soil without mining land	73 240	=	=	0
	Area for recreation incl. parks	265 853	=	=	=
	Area for cemeteries	34 960	=	=	=
	Areas for traffic (roads, streets)	1 711 764	=	0	0
Area for agriculture including moor and heathland		19 102 791	=	=	9 551 395
Forest area		10 531 415	=	=	=
Surface of the water including sea		808 462	=	=	0
Mining land		179 578	=	0	0
Other areas		686 957	=	=	0
These include:	Badlands	266 593	=	=	0
			100%	88%	56%

= means same number as in the left column.

15) $1\,km^2 = 100\,ha = 10\,000\,a = 1\,000\,000\,m^2$

about 50% of the agricultural area is considered to be available for profitable production of biomass. Some other surfaces will never be agriculturally usable in a profitably way. So the total area for profitable agricultural use for biomass is estimated to be $A_{Dtechn} = 0.56 \cdot A_D$.

2.2 Potential yield from biomass

2.2.1 Theoretical potential[16)]

Biogas results from the microbial degradation of biomass, formed by photosynthesis by solar power E_S.

$$6CO_2 + 6H_2O + Es \rightarrow C_6H_{12}O_6 + 6O_2$$

$$\text{Carbon dioxide} + \text{Water} + \text{Solar energy} \rightarrow \text{Sugar (Glucose)} + \text{Oxygen}$$

Metabolic processes in the plants, transform the following compounds into secondary products.

Carbohydrates:	Starch, inulin, cellulose, sugar, pectin
Fat:	Fat, fatty acids, oil, phosphatides, waxes, carotene
Protein:	Protein, nucleoproteid, phosphoproteid
Others:	Vitamins, enzymes, resins, toxins, essential oils.

During the metabolism of the sugar, the plant releases energy, when necessary, to the environment, so that the possible energy yield from plants may vary greatly.

Multiplying the proportion of the main plant components (see Table 2.5) by the entire vegetation, an averaged elementary composition of plants dry matter results:

$$C_{38}H_{60}O_{26}$$

With the help of an approximate equation from Buswell (1930), the theoretical maximum yield of methane can be estimated taking the elementary composition as a base:

$$C_cH_hO_oN_nS_s + y\,H_2O \rightarrow x\,CH_4 + (c - x)CO_2 + n\,NH_3 + s\,H_2S$$

Table 2.5 Main components of plants without nitrogen N and sulfur S.

Carbohydrate	$C_6H_{12}O_6$
Fat	$C_{16}H_{32}O_2$
Protein	$C_6H_{10}O_2$

16) Cp. WEB 18

where

$$x = 0.125\,(4c + h - 2o - 3n + 2s)$$
$$y = 0.250\,(4c - h - 2o + 3n + 2s)$$

or, simplified

$$C_cH_hO_o \rightarrow (c/2 + h/8 - o/4)CH_4$$

The hectare yield of methane can hence be calculated from the hectare yield of the dry matter. This again depends on the planting, which should be as productive as possible.

The maximum theoretical possible yield is estimatedat $E_{Rmax} = 30\,MgDM/(ha \cdot a)$ when applying two harvests per year and cultivating C4 plants with an average elementary composition of $M_E = 932\,kg/kmol$. Based on the simplified equation from Buswell the yield of CH_4 is $E_M = 20\,kmol\,CH_4/kg$ biomass and the energy yield P_{theor} is calculated by the formula:

$$\overline{P}_{theor} = \frac{E_{R_{max}}}{M_E} \cdot E_M$$

to give $144.200\,kWh/(ha \cdot a)$. If one multiplies the hectare yield by the entire surface of Germany (35 703 099 hectares), the following equation

$$P_{theor} = \overline{P}_{theor} \cdot A_D$$

results in a primary energy quantity from biomass of $5.148\,TWh/a$. Theoretically the entire amount of primary energy supply in Germany could be covered by biomass alone.

Assuming that the yield of the available cultivable area on earth is proportionally the same as in Germany, an area of 7 420 Mio ha, half of the available area of 14 900 Mio ha on earth, would theoretically be enough to cover the total world primary energy consumption of $107\,000\,TWh\,a^{-1}$.

If a precondition is that the maximum yield should be guaranteed on a long-term basis, this could be facilitated by

Accurate and targeted addition of fertilizer
Water and fertilizer can be added very accurately by using hoses which are directly led to the roots. The accuracy depends on the characteristics of the local soil, but the overall yield per hectare of conventional agriculture could perhaps be doubled, particularly, when some missing nutrients are supplied with the water.

Multiple harvests per year
Yields of 25–30 Mg DM/ha.a can be obtained if the field crops shown in Table 2.6 are cultivated immediately after each other during one year.[17),18)]

17) Cp. JOU 26

18) Cp. JOU 32

Table 2.6 Crop rotation (GPS = Mixture of winter wheat and peas).

1st Planting	2nd Planting	3rd Planting
Wheat	Maize (mass-producing species)	GPS
Winter rye	Sunflower	
Winter barley	Sorghum	
Triticale,	Sudan grass	
Winter oat	Hemp	
Winter rape	Mustard	
Beets	Phacelia	
Winter peas	Radish	
Incarnat clover	Sweet pea	
Winter sweet pea	Peas	

Today the most frequently cultivated crop rotation consists of the following three crops:

1. The domestic cold-compatible C3 plants: winter rape or winter rye
2. The southern C4 plants: corn (mass-producing species),[19] as main crop during summer
3. The cold-resistant C3 plants: GPS.[20]

In order to generate energy, all the plants are harvested as soon as they finish their growth without leaving them time to fully develop. The costs of cultivation are 61–84 US$/Mg for the cultivation of winter wheat, winter barley, and triticale a crossing of wheat and rye in Germany.[21]

Overall the cultivation of energy plants has just started. Besides maize, some other C4 plants like sorghum, sugar cane, or Chinese reed seem to be efficient when used as biomass.[22] Their yield, though, still needs to be improved. Also, certain C3 plants such as grain, grasses, hemp, rape, beet, sunflower, or winter peas seem to have good potential as energy sources with a yield still to be increased, too. In future this broader range of energy plants will allow interesting new combinations and an increased level of flexibility in deciding on the crop rotation system.

2.2.1.1 **C3 plants (energy plants)**

The enzyme most important for the production of energy is RuBisCo (Rubilose 1.5-diphosphate carboxylation-oxygenase). It is the most frequently produced enzyme of all organisms and can be found in the chloroplasts of the plants in the form of proteins. Their level in the proteins amounts to 15%.

RuBisCo catalyzes photosynthesis and photorespiration. It binds oxygen as well as CO_2 and acts as oxygenase. For photorespiration to occur, the chloroplasts,

19) Cp. WEB 11
20) Cp. JOU 30
21) Cp. WEB 89
22) Cp. BOK 72

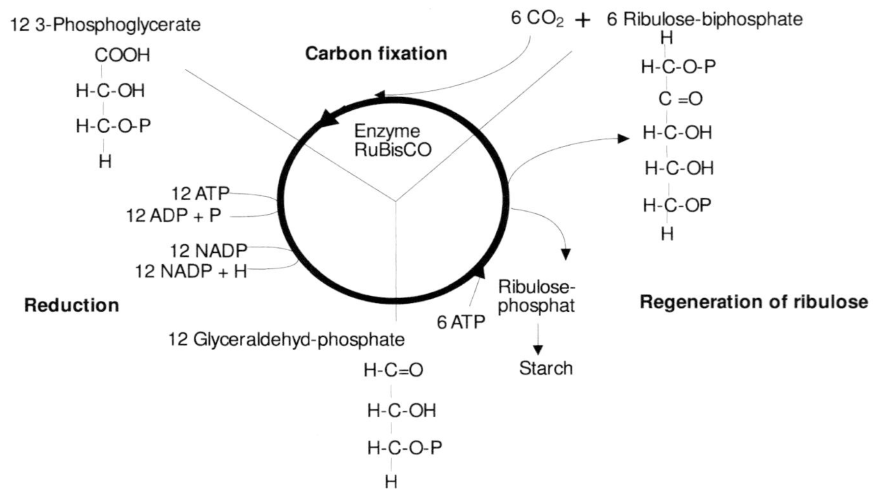

Figure 2.4 Calvin cycle.[23)]

mitochondria, and glyoxisomes, cell components around the mitochondria, need to be involved.

The ratio of photosynthesis to photorespiration is defined by the ratio of CO_2 and O_2 in the air. With a higher concentration of CO_2, the output of the photosynthesis increases.

In moderate zones, e.g., in Central Europe, photorespiration in plants plays a subordinate role. Predominantly C3 plants occur, which use the light-independent reaction, the Calvin cycle (Figure 2.4), to bind CO_2. They are called C3 plants, because the first stable product in the Calvin cycle after the CO_2 fixing 3PGS (Phosphoglycerate) has 3 C-atoms. Also the molecule which is reduced from 3PGS with NADPH+H+ to 3PGA (Phosphoglycerin aldehyde) in the following phase of the Calvin cycle contains 3 C-atoms.

The leaf structure of C3 plants is layer-like. In warm summer weather the transpiration and the evaporation at the surface of the sheets increases. In order to minimize the water loss, the plants close their pores. CO_2 cannot be absorbed by the pores any longer. Thus the photosynthesis is stopped and the biomass yield is limited.

In addition, the biomass yield depends on the soil as well as the entire climatic conditions: in some regions of the world the yield can be up to five times higher than in Germany. It is not possible, however, to obtain the theoretically projected yields just by cultivating C3 plants (Table 2.7).

Other typical representatives of C3 plants are onions, wheat, bean, tobacco.

Most C3 plants are well adapted to the moderate climatic zones but not to arid, saline areas with hot and dry air. Under such climatic conditions the ratio of photosynthesis to photorespiration increases from 2:1 and negatively impacts the yield.

23) Cp. WEB 31

Table 2.7 Yield per hectare of C3 plants.

Plant	*Yield [Mg DM (fruit + haulm)/ ha.a]*[24]	*Water content [%] related to the total mass*	*Advice for plantation*
Trees (stored)	1–2	15–20	Cut every 150 years
Fast-growing wood (poplar, willow)	15	30–60	Cut every 6 years
Eucalyptus	15–40	High	–
Rape (whole plant)	4.2–6.9	12–34	Crop rotation every 4 years
Sunflower (mature plant)	2.5	15	Crop rotation every 5 years
Hemp	3–4	65–75	Crop rotation yearly
Sugar beet	7.2–18.2	74–82	Crop rotation every 4 years
Potato	5.8–12.5	75–80	Crop rotation every 4–5 years
Jerusalem artichoke	12–27	72–81	Crop rotation yearly
Straw and grain	4–15	14–16	Crop rotation yearly
Bastyard Grass[25]	13.7	65–80	For 5 cuts per year
Meadow	7.7	65–80	For 5 cuts per year

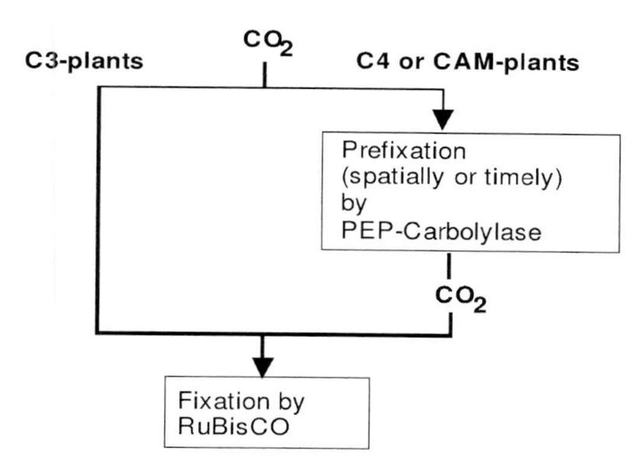

Figure 2.5 Different ways for the CO_2-Fixation.

2.2.1.2 C4 plants and CAM plants

There is a large group of 1700 variants of C4 plants and/or CAM plants which are all well adapted to hot and dry climates and do grow in arid, saline areas. This is possible since the CO_2 fixing occurs in C4 plants spatially separated from where the Calvin cycle occurs. In CAM plants the CO_2 fixing happens at a different time of the day from that when the Calvin cycle occurs (Figure 2.5). Such plants can utilize even the smallest CO_2 concentrations.

The separation of the CO_2 fixing occurs with the help of the enzyme PEP carboxylase (PEP = phosphoenolpyruvate), which possesses a substantially higher affinity

24) $1\,km^2 = 100\,ha = 10\,000\,a = 1\,000\,000\,m^2$

25) Cp. WEB 90

to CO_2 than RuBisCo. The first product of the photosynthesis which is stable is oxalacetate (see Figure 2.7), a C4 product. This characterizes the so-called C4 plant.

Compared to C3 plants, the leaves of C4 plants are anatomically different. The spatial separation of the CO_2 fixation takes place in cells at a distance from each other, the bundle sheath cells and the mesophyll cells, both containing chloroplasts but different types: the mesophyll cells contain normal chloroplasts while the budle sheath consist of chloroplasts with grana. The vascular bundles to transport the cell liquid are covered by a layer of thick bundle sheath cells which are surrounded by mesophyll cells.

An intensive mass transfer is continuously happening between the bundle sheath cells and the mesophyll cells. This starts with the formation of oxalacetate (Figure 2.6), a result of the enzymatic reaction of PEP-carboxylase binding CO_2 to PEP = (phosphoenolpyruvate). Oxalacetate is then enzymatically transformed into malate and transferred to the chloroplasts of the bundle sheath cells. In the bundle sheath cells it degrades into pyruvate and CO_2 while forming NADPH+H+ as a by-product. CO_2 is introduced into the Calvin cycle while pyruvate is transported back into the mesophyll cells.

CAM plants actually belong to the group of C4 plants. The name "CAM plants" is derived from the Crassulaceae Acid Metabolism (acid metabolism of the Crassulaceae), since the metabolism was first observed in the plant species "Crassulaceae".

Because of the high water loss, these plants open their stomata only at night to take up CO_2 which is stored in form of malate. During the day, CO_2 is released and transformed in the Calvin cycle forming ATP as a by-product.

Like C3 plants, the CAM plants have layer-like structured leaves.

Some species of CAM plants are cactuses, pineapple, agave, Kalanchoe, Opuntia, Bryophyllum, and the domestic Sedum spec. or Kalanchoe (Crassulaceae).

C4 and/or CAM plants show the following advantages, compared to C3 plants:

- C4 and/or CAM plants can generate biomass twice as fast if conditions are favorable (see Table 2.8).
- The upper leaves of C4 and/or CAM plants are perpendicularly directed to the sun, so that the low-hanging leaves still get sufficient light even under unfavorable light conditions.

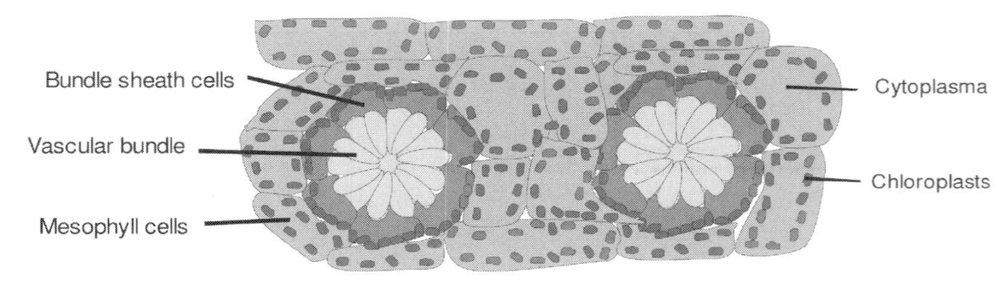

Figure 2.6 Anatomy of the leaves of a C4 plant.[26]

26) Cp. BOK 4

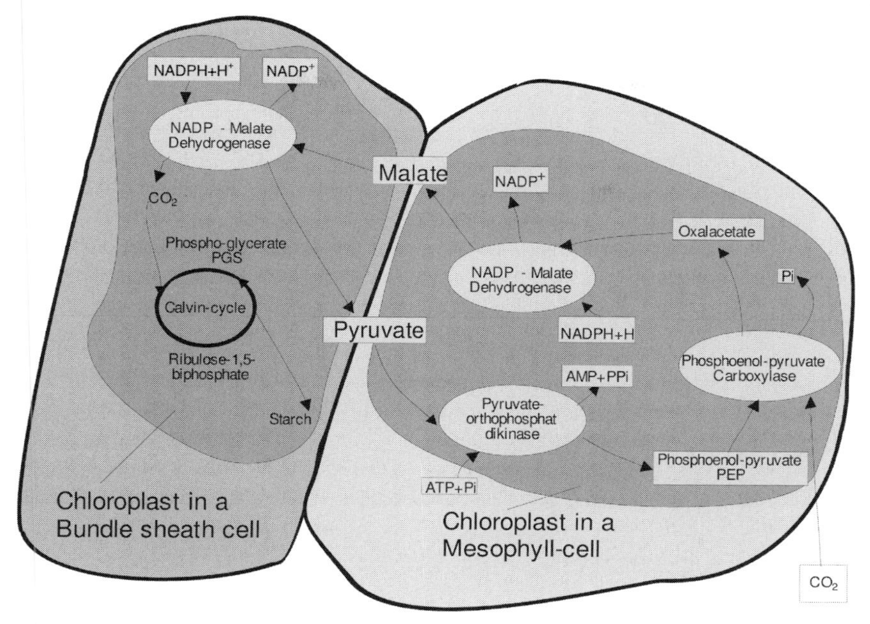

Figure 2.7 Mass transfer in C4 plants (C4-dicarbon acid path).[27]

Table 2.8 Yields of C4 plants (CAM plants are less productive).

Plants	*Max. yield (approximate)*	*Water content (depends on harvest time)*	*Advice for plantation*
	[Mg_{DM}/ha.a] (fruit+haulm)	*[%]*	*[Years]*
Miscanthus	25–30	15–45	Harvesting 20–25 from the 3rd year on
Sorghum spec.	17	17–60	Harvesting 20–25 from the 3rd year on
Sorghum spec.	5–32	70–80	One year plant
Maize	30[28]	15	One year plant

- C4 and/or CAM plants need only half the water.
- C4 and/or CAM plants adapt to dry and warm locations.
- C4 and/or CAM plants do not need pesticides but only some fertilizers in the first year.

27) Cp. BOK 4

28) Cp. JOU 18

Figure 2.8 Sorghum (above right),
Micro-algae chlorella in glas tuber (below).
Energy maize abt. 5 m heigh (above left).

- C4 and/or CAM plants once planted grow again after biomass has been harvested.

2.2.1.3 **Micro-algae**

By cultivating micro-algae, even the surface of water as well as the area of rooftops can be exploited in a profitable way.

A yield of 15–17 Mg biomass per year seems theoretically be achievable by planting micro-algae and cultivating them in well-lit bioreactors.[29]

29) Cp. WEB 91

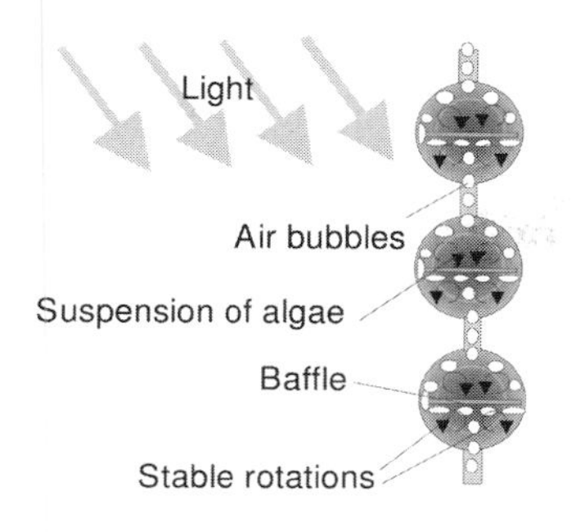

Figure 2.9 Reactor for micro-algae growth.

Most of the micro-algae naturally grow much better when the light is somewhat diffused rather than in direct clear light. Sun may even limit the growth. In order to control the light in the latest bioreactors, the micro-algae are cultivated in airlift reactors in which a circular flow is caused by changing the direction of the gas bubbles (Figure 2.9). The light reflects at the outer wall of the reactor.

The circular flow is set in such a way that the algae are located mainly in the outer area of the incidence of light. The algae absorb just enough light to keep the Calvin cycle alive for a maximum yield of biomass.

In reactors erected in the sea, the sea water can actually be used to help maintain a moderate temperature inside the reactor. The micro-algae may also serve to clean the water, especially in cases where the reactor is located close to a river mouth and the water is led through the reactor.

Micro-algae can not only be used to produce biogas but also to provide lipids, fatty acids, vitamins, e.g., vitamin E, beta-carotene, or even pigments like phycocyanin or carotenoids. Antioxidants like tocopherols or omega fatty acids may also be extracted, which are very interesting from a pharmaceutical point of view.

In 2000 the first farm for micro-algae was inaugurated close to Wolfsburg, in the middle of Germany. Within a fully closed system of bio-reactors (about 6000 m^3 in total) chlorella algae are converted into about 150–200 Mg animal food produced annually.[30]

2.3
Technical potential[31]

Technically it should soon be feasible to achieve a yield of $E_R = 1/2\ E_{Rmax} = 15$ Mg/ha.a of biomass.[32] This final effective outcome may be lower than the theoretical potential, since certain losses have to be taken into consideration because

30) Cp. WEB 92
31) Cp. BOK 5
32) Cp. WEB 69

- Quite often the biomass that is used to generate the energy is just a leftover after having been consumed as food. Some other parts of biomass had been used to construct houses. . . . Overall most of the quantity of biomass effectively used has served other purposes before being taken for energy supply, so that part of the energy has already been wasted.
- Technically the transformation from primary energy to effective energy goes along with quite immense losses of around 20–70%.
- Energy plants need to be cultivated in a sustainable way to ensure the continuous energy supply over time. It is important to ensure that the soil is not getting leached.

The real technical potential hence after the equations

$$\overline{P}_{techn} = \frac{E_{R\max}}{2 \cdot M_E} \cdot E_M$$

$$P_{techn} = \overline{P}_{techn} \cdot 0{,}88 \cdot A_D$$

results in 72.100 $kWh/(ha \cdot a)$ or 2.265 TWh/a when multiplied by the area of land that is technically available.

The technically realistic yield of energy provided by biomass should provide about 50% of the total energy consumption in Europe.

Humans themselves would be part of a closed CO_2 cycle (Figure 2.10). Excrement and/or waste are directed into a separator to separate solids and water. The water flows into a constructed wetland and is purified there to drinking water. The concentrated solid is converted into energy by being processed in an anaerobic reactor with a generator attached to it. The fermented residue is composted and used as a fertilizer for food plants. The constructed wetland may be run with water hyacinth and/or common duckweed which can be returned to the cycle.

Water hyacinths are fast-growing plants which should be cut quite frequently. In that way they are well suited to be utilized as an effective renewable source to provide biomass for energy supply.

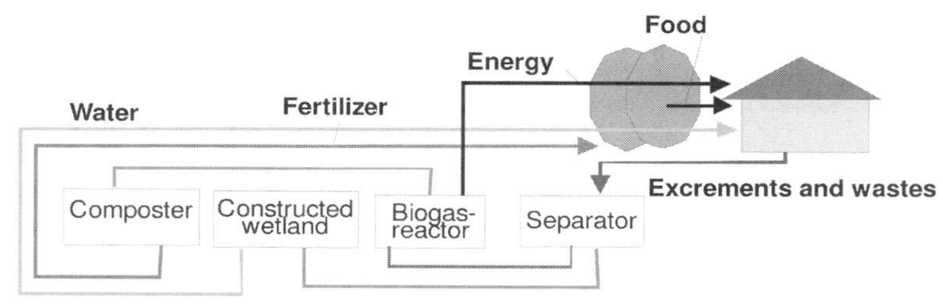

Figure 2.10 Closed CO_2 cycle.

2.4
Economic potential

Prices for crude oil and energy are rising globally. This trend anticipates that any technical feasibility will be profitable sooner or later.

The economic potential hence equals the technical potential

$$\overline{P}_{econ} = \frac{E_{R\max}}{2 \cdot M_E} \cdot E_M$$

giving 72.100 $kWh/(ha \cdot a)$. When multiplied by the area which is economically available

$$P_{econ} = \overline{P}_{econ} \cdot 0.56 \cdot A_D,$$

this results in 1.441 TWh/a, about 35% of the total primary energy supply of Germany.

2.5
Realizable potential

There is a huge gap between the technical and profitable potential and the realizable potential. A lot of what is technically feasible is rejected for various reasons, mainly special interests, e.g., landscape protection or job safety. A lot can be explained rationally but a lot is just based on emotion.

Today, almost 20 Mio ha of the agricultural area is cultivated only to produce food without considering the possibility of using it for energy supply. Just about 5% of the agricultural area (about 1.2 Mio ha) is disused. About 30% of this specific area is planted with energy-affording plants. In the next few years it may well be possible that the agricultural area used to produce biomass for energy supply will increase to about 2–2.6 Mio ha, even if we bear in mind that the forest area certainly cannot be simply transformed into an area of cultivable land for energy-affording plants. Such areas are expected only to deliver about 5 Mg/ha.a of dry biomass material.

In the same way, parks will remain on a long-term basis and may only provide a very small, almost negligible amount of biomass.

In the future, additional yield of biomass can only be achieved by exploiting those areas that are agriculturally used today. From a technology point of view, however, this is the only area that can be used to cultivate biomass or to provide output for the production of bio-diesel fuel. Today, about 70% of the non-food rape is consumed by the bio-diesel fuel industry. So just about 0.6 Mio ha are left and realizable for the cultivation of energy plants. The target of 4 Mio ha or about 20% of the total agicultural area in Germany available seems unrealistic and overestimated.[33)]

33) Cp. BOK 8

Further realizable potential may be provided by waste and sewage, which are already partly exploited in biogas plants. The fermentation of waste materials has to be seen in competition with being fed to animals, combusted, or composted (Table 2.9).[34]

The potential to provide biogas is the most important data point and is specific to the kind of organic material used. For example, in Germany the materials shown in Figure 2.11 will be available by taking 2/3 of the entire volume of excrement (liquid manure) of the German agricultural livestock into consideration:

Table 2.9 Possibilities to exploit bio waste (– = not suited; 0 = partially suited; + = well suited).

	Feeding	*Combustion*	*Composting*	*Fermentation*
Liquid manure	–	–	0	+
Sewage sludge	–	0	0	0
Bio waste	–	–	0	+
Grass from lawns	0	–	+	+
Sewage from industry, biologically contaminated	+	–	0	+
Waste grease	–	–	–	+
Waste from slaughterhouse	–	–	0	+
Wood	–	+	+	–
Excrement	–	–	+	+
Straw	0	0	+	0

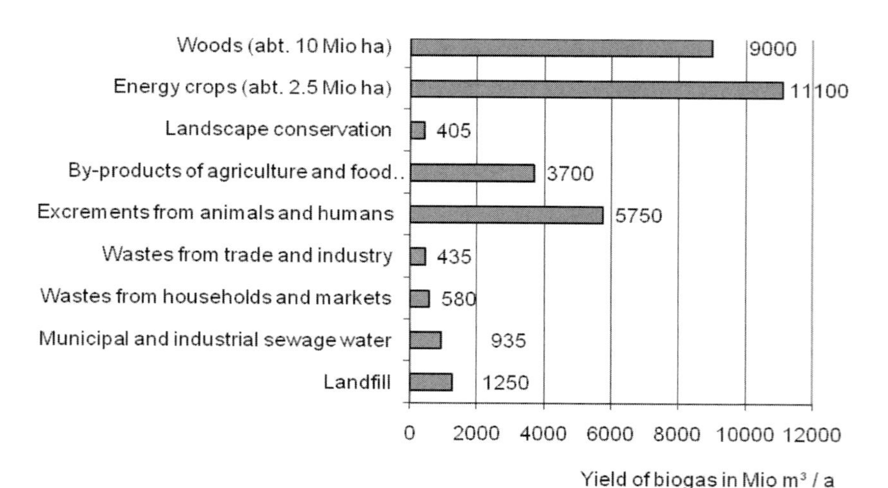

Figure 2.11 Yield of biogas from different sources[35] in Mio m^3/a.

34) Cp. BOK 6

35) Cp. WEB 15

From these sources, the energy potential as to be seen in Table 2.10 can be derived:

190 Mio Mg of excrement may deliver the same yield as 500 000 ha of land to cultivate energy plants. The power generated by two power nuclear plants may be provided by fully exploiting agriculture and forestry.[36]

Another source of energy is fermented plants of created wetlands, which deliver a much lower but still appreciable amount of biomass, or the wastewater from the paper industry.

Even human urine may be exploited (Figure 2.12). Around 40 Mio Mg a^{-1} of urine (500 L/person.a)[37] could be made available by investing in changing the complete system of sewage disposal; so-called "gray water" from private households, e.g., from washing dishes, laundry, or bathing, should be separated from brown water (containing excrement) and urine. Even the toilet flushing would need to be omitted to avoid too much dilution. A potential solution may look as shown in Figure 1.16.

The importance of input from landfills will decrease in the next few years. Legal regulations and restrictions have become much stricter, which may finally render this source unprofitable.

Table 2.10 Potential of energy from biogas from different sources.

Sources for biogas production	***Energy potential [TWh a^{-1}]***
Landfill	6
Communal and industrial sewage water	
Organic wastes from households and markets	18
Organic wastes from industry	
Excrement (190 Mio Mg a^{-1})	
Byproducts of agriculture and food production	47
Material from landscape conservation	
Plantations of energy plants (area ca. 2.5 Mio ha) (15 Mg/ha.a)	141
Wood (10 Mio ha forest area) (5 Mg/ha.a)	187
Urine	4
Nutrition in sewage water	5
Total	408

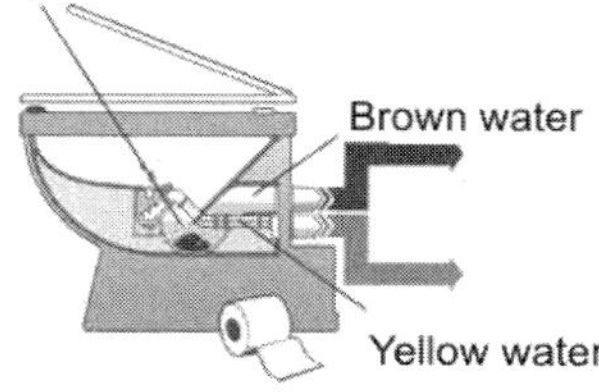

Figure 2.12 Separation of human excrements.[38]

36) Cp. BOK 7
37) Cp. WEB 70
38) Cp. WEB 60

Overall, the biggest output of energy from renewable resources, however, will be provided by using bio waste from the food industry. Pomace, for example, from wine making, consists of grape pods, stones, and stems, which can be used for energy recovery. Today pomace serves as a base for the production of alcohol and/or as animal feed, as does the waste from breweries, sugar refineries, and fruit processing plants. All this bio waste can be a source of profit by fermentation. Even the wastewater from dairies or waste from slaughterhouses will be fermented in the future. The potential is huge. Annually 0.9 Mio sheep and horses, 3.8 Mio cows, 0.4 Mio calves, and 43 Mio pigs are slaughtered in Germany.

The total yield of realizable biogas sources in Germany should be around 408 TWh a^{-1}, which is about 10% of today's primary energy supply and about 48.5% of today's primary energy consumption of natural gases (natural gas, mine gas, sewage gas) – about 840 TWh a^{-1}.

When all biogas is used to generate electrical power, the potential yield of power from the biogas amounts to about 143 TWh a^{-1}, assuming an efficiency of 35% for the power generators. Biogas may hence contribute to 10–12% of the total power supply.[39]

In some literature, lower yields of biogas of around maximum 74 TWh a^{-1}[40] only are estimated.

In Western Europe, France and Germany are leading just looking at the potential yield of energy resulting from cultivating energy-affording plants and exploiting agricultural by-products. France has the highest potential with 178 TWh a^{-1} (Figure 3.1).

This means that so called "passive houses", with a very low energy requirement for heating (primary energy consumption of 120 kWh/m^2.a), can be heated by using biogas, e.g., by generating power by a fuel cell. The required amount of energy will be supplyable by animals. One animal unit (abbreviation used: GVE) 500 kg in weight produces 550 m^3 biogas per year or, depending on the energy content of the biogas, ca. 3500 kWh_{th}. One cow can hence supply a small apartment of about 30–60 m^2 living area with enough heat. Or a passive house of about 400–800 m^2, needing ca. 40 000 $kWh_{th}\,a^{-1}$, can be heated with energy plants growing on one ha of field.

39) Cp. BRO 6

40) Cp. BOK 63

3
History and status to date in Europe

Very old sources indicate that using wastewater and so-called renewable resources for the energy supply is not new, but was already known before the birth of Christ.

Even around 3000 BC the Sumerians practiced the anaerobic cleansing of waste.

The Roman scholar Pliny described around 50 BC some glimmering lights appearing underneath the surface of swamps.

In 1776 Alessandro Volta personally collected gas from the Lake Como to examine it. His findings showed that the formation of the gas depends on a fermentation process and that the gas may form an explosive mixture with air.

The English physicist Faraday also performed some experiments with marsh gas and identified hydrocarbon as part of the it. A little later, around the year 1800, Dalton, Henry, and Davy first described the chemical structure of methane. The final chemical structure of methane (CH_4), however, was first elucilated by Avogadro in 1821.

In the second half of the 19th century, more systematic and scientific in-depth research was started in France to better understand the process of anaerobic fermentation. The objective was simply to suppress the bad odor released by wastewater pools. During their investigations, researchers detected some of the microorganisms which today are known to be essential for the fermentation process. It was Béchamp who identified in 1868 that a mixed population of microorganisms is required to convert ethanol into methane, since several end products were formed during the fermentation process, depending on the substrate.

In 1876, Herter reported that acetate, found in wastewater, stoichiometrically forms methane and carbon dioxid in equal amounts. Louis Pasteur tried in 1884 to produce biogas from horse dung collected from Paris roads. Together with his student Gavon he managed to produce 100 L methane from 1 m^3 dung fermented at 35 °C. Pasteur claimed that this production rate should be sufficient to cover the energy requirements for the street lighting of Paris. The application of energy from renewable resources started from that time on.

Biogas from Waste and Renewable Resources. An Introduction.
Dieter Deublein and Angelika Steinhauser

ISBN: 978-3-527-31841-4

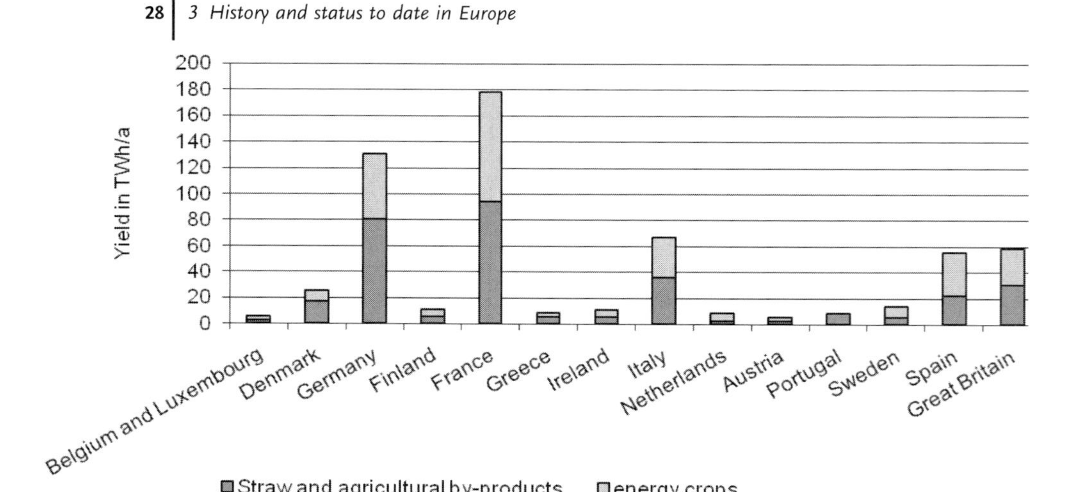

Figure 3.1 Biogas production across European countries.

3.1 First attempts at using biogas

While Pasteur produced energy from horse dung, in 1897 the street lamps of Exeter started running on gas from waste water. This development suggested that more and more biogas could be produced by anaerobic purification plants for wastewater. Most of the biogas, however, was still wasted to the atmosphere.

In 1904 Travis tried to implement a two-step process which combined the purification of waste water with the production of methane.

In 1906 Sohngen accumulated acetate in a two-step process. He found that methane was formed from three basic materials: formate plus hydrogen plus carbon dioxide.

In 1906 the technician Imhoff started constructing anaerobic waste water treatment units in the Ruhr, Germany. He installed so-called "Imhoff tank" (Figure 3.2) with separate spaces for sedimentation and digestion. The residence time of the bio waste was 60 days.

In Germany methane gas was first sold to the public gas works in the year 1923.[41)] In the next years this practise became more and more common in Europe. A further development was the installation of a CHP near the biogas production and to produce the current necessary for the waste water treatment plant and to heat houses with the excess heat from the CHP.

Until the 2nd world war, the use of biogas was progressing very fast and much effort went into developing more efficient systems, e.g., floating-bell gasholders, efficient mixers, and heating systems to increase the yield of digestion. In Europe, highly technical spherical digestors agitated with intermittent vertical screw

41) Cp. JOU 15

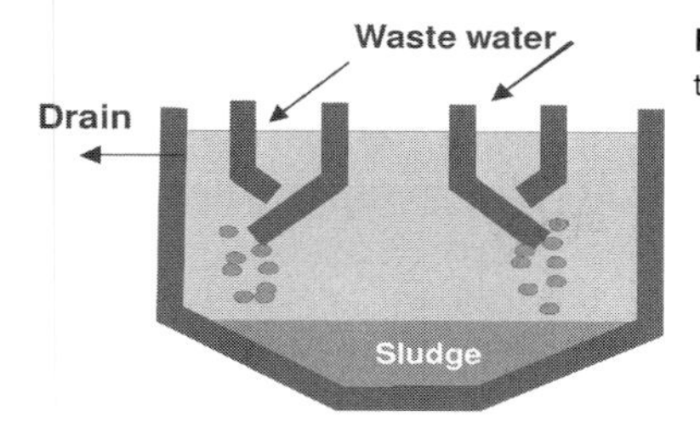

Figure 3.2 Imhoff tank – a sedimentation tank for the mechanical sewage treatment.

conveyors and a haul-off in the cover was preferred. In the United States, simple cylindrical vessels were used with flat bottoms, continuously circulating mixing systems, and collecting pipes at the top.

Around 1930 it was first tried to remove water, carbon dioxide, and sulfide from the biogas, to compress it in gas bottles, and to use it as fuel for automobiles. In order to maximize the efficiency of such a procedure, so-called co-fermenters, i.e. solid organic waste, e.g., food, cereals, and silage were added. Different combinations were tried, but only in 1949 (Stuttgart) the addition of fat after fat separation enabled the yield of biogas to be increased.

In Halle, experiments on digestion were performed by adding waste liquorice, rumen, lignin and/or cereals. Lignin was the least efficient material, providing 19 L gas per kg dry matter with a dwell period of only 20 days. Rumen provided 158 $L\,kg^{-1}$, liquorice even 365 $L\,kg^{-1}$ but with a dwell period of 45 days. Around 1950 Poebel conducted some extensive research on co-fermentation in the Netherlands by including organic waste of households in his experiments.

Around the same time (1930–1940) the idea came up to use agricultural waste to produce biogas. Buswell's target was to provide the whole amount of gas consumed by Urbana, a small city in Illinois. He examined many different natural materials. In parallel, Ducellier and Isman started building simple biogas machines in Algeria to supply small farmhouses with energy. This idea was brought to France, and many people installed their own small and technically very simple biogas plants.

Around 1945, only Germany started using agricultural products to produce biogas. Imhoff again was leading. In 1947 he claimed that the excrement of one cow delivered 100 times more biogas than the sewage sludge of one single urban inhabitant. He projected how much biogas the excrement of cows, horses, pigs, and potato haulms would supply. The first small biogas plant with a horizontal cylindrical vessel for fermentation was developed in Darmstadt, and in 1950 the first larger biogas plant was inaugurated in Celle. In total, about 50 plants were installed during the following years in Germany.

While expanding the number of biogas plants, globally researchers deepened their knowledge about the chemical and microbial processes contributing to fermentation. Doing very fundamental biochemical research in 1950, Barker detected the methane-forming bacteria *Methanosarcina* and *Formicicum methanobacterium*. Very important also was the finding from Bryant et al. in 1967 showing that methane-forming microbial cultures consisted of a minimum of two kinds of bacteria. One type was said to be responsible for converting ethanol to acetate and hydrogen, and the other for forming methane via chemical reaction of carbon dioxide and the free hydrogen. Today it is known that four specific and different kinds of bacteria must work in synergy to produce biogas.

Around 1955 the importance of biogas was significantly reduced, as biogas was not profitable any longer due to an excess of oil. The price of fuel oil was very low, ca. 0.10 L^{-1}. At the same time, more mineral fertilizer was used in mass. Almost all the biogas plants were shut down except two: that in Reusch/Hohenstein (1959) and the Schmidt-Eggersgluess plants close to the monastery of Benediktbeuren, built around 1955.

This plant, consisting of two digesters, one storage tank, a gasholder, and the turbine house, was originally constructed for 112 animal units (GVE) and a gas production of 86400 $m^3 a^{-1}$. In the last few years, however, it was only used for about 55 GVE. The plant cost was 72000 US$ but it has cost around 12000 US$ for maintainance annually during the past 25 years. The ratio of straw chaff to the excrement and urine was 1:2. The dung was flushed into a dump, mixed with anaerobic sludge, and pumped daily into the digestors. The principle of the "change container" procedure is that while the material digested in the first fully filled fermenter for ca. 20 days, the second was filled. If the second container was full, the content of the first digestor was transferred into the storage tanks. In that way the first digestor was refilled while material was digesting in the second container.

Temperatures of 38–39 °C were considered as optimal for the digestion. The resulting biogas was used in the monastery kitchen for cooking. Any surplus was connected to a 70-HP MAN diesel engine. In the end, however, the plant was shut down in 1979 when cattle breeding was abandoned.

3.2
Second attempts at using biogas

In 1970 the demand for biogas increased, driven by the oil crisis. The number of facilities went up to 15 in Bavaria and up to 10 in Baden-Wuerttemberg.

Later, in the 1990s, biogas technology was stimulated for two reasons:

- The profitability of using power derived of biogas
- The recycling management and Waste Avoidance and Management Act which was implemented in 1994 and resulted in higher costs for disposal of solid waste.

The agricultural sector observed the trend and accepted it very conditionally, since the biogas facilities did not work in a profitable way, mainly because of the high costs in constructing the facilities. Only after the farmers had learned to work themselves and to pool their experience were the facilities run economically.

Table 3.1 Number and total amount of digested material of large fermentation and co-fermentation biogas plants of bio waste (>2500 Mg a^{-1}) in Europe in 1997.[42]

	Number of biogas plants	*Mg of digested waste per year*
Austria	10	90 000
Belgium	2	47 000
Denmark	22	1 396 000
Finland	1	15 000
France	1	85 000
Germany	39	1 081 700
Italy	6	772 000
Netherlands	4	122 000
Poland	1	50 000
Spain	1	113 500
Sweden	9	341 000
Switzerland	10	76 500
England	1	40 000
Ukraine	1	12 000
Total	108	4 241 700

In 1954, Ross, in Richmond, USA, reported about the process of digesting communal waste with sludge. Apparently, a closed facility was running in Chicago, USA, to digest the waste.

At the end of the 1990s, numerous plants were built and implemented for the mechanical-biological treatment of garbage. The technology was based on anaerobic with some aerobic composting. The aerobic process proved to be advantageous, since it enabled enough energy to be provided to run the plant itself.

Not only Germany but other European countries applied the same technology for the disposal of waste (Table 3.1). For example, in Denmark several large biological gas facilities were built for processing of liquid manure together with residues from the food industry.

About 44 anaerobic fermentation plants with a capacity of about 1.2 Mio Mg bio waste in total existed in Germany in April 1999. Of these plants, 31 were running by the wet-fermentation procedure (18 single-stage, 13 multi-stage procedures); the other 13 facilities worked according to the dry-fermentation process (9 single-step, 4 multi-level procedures). At the same time, around 550 aerobic bio waste composting plants were functioning, with an overall capacity of approximately 7.2 Mio Mg bio waste.[43]

42) Cp. BOK 22

43) Cp. WEB 46

3.3
Third attempts at applying biogas

In 2000, the law of "Renewable Energies", which stated the rules for the subsidization of the power supplied by biogas facilities, became effective. Over the past few years, the number of biogas facilities has continuously been rising, especially after implementing even higher subsidies. About 1500 biogas facilities were in use in Germany, most of them in Bavaria.

Electrical power was supplied from biogas into the network out of the sources shown in Table 3.2.

3.4
Status to date and perspective in Europe

In the year 2005 in Austria a biogas plant were constructed to feed $10\,m^3h^{-1}$ crude biogas (giving to $6\,m^3h^{-1}$ clean biogas) into the natural gas network, equivalent to $400\,MWh\,a^{-1}$.[44] The gas is produced from the excrement of ca. 9000 laying hens, 1500 poultry, and 50 pigs. In Sweden, communal vehicle fleets and even a train are running on biogas.[45]

In Germany, the number of biogas plants has increased during the past few years following a governmental promotion handed out for the installation of plants. In fact the number was tripled from 850 plants connected to the electricity network in 1999 to 2700 plants to date in 2006 (Figure 3.3). In the agricultural sector alone, more than 600 plants were put on stream in 2006, contributing to a total power output of 665 MW and a total energy generation of 3.2 TWh provided by all biogas plants. It is planned to construct 43 000 biogas plants in Germany until the year 2020.

Almost all waste water is already fed into central sewage water treatment plants area-wide with facilities to produce sewage gas. Several small plants with a volume of waste water of less than $8\,m^3$ per day still exist. The objective is, however, to integrate these, too, into the central system as soon as the appropriate pipework is installed.

Table 3.2 Electric power supply from biomass in Germany in the year 2000.

	Number	*Installed electric power [MW]*	*1000 MWh a^{-1}*
Sewage gas	217	85	61
Landfill gas	268	227	612
Biogas	1040	407	127
Total	1525	407	800

44) Cp. WEB 93

45) Cp. WEB 94

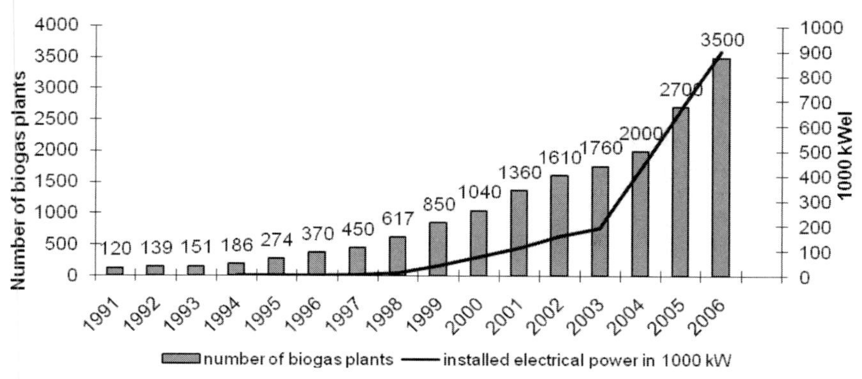

Figure 3.3 Expansion of biogas production in Germany.[46)]

Overall, the agricultural sector is seen to be a rich source of biomass. Projections suggest that the agricultural waste alone will enable more than 220 000 additional individual plants and communal facilities to be run, provided that an investment of 25–40 bn US$ is allocated. This will provide farmers the opportunity to become more independent from the food trade and get additional incomes working as "energy farmers".

For example, in Hungary in December 2005 a biogas plant with a capacity of 2.5 MW was inaugurated. The plant is fed with liquid manure from several cattle farms and wastes from poultry farming.

46) Cp. WEB 95

4
History and status to date in other countries

In the rich industralized countries, biomass represents on average about 3% of the total amount of primary energy carriers. In the emerging markets it accounts for 38%. In some particularly poor countries it reaches even more than 90%.

In the United States, the percentage of biomass related to the total consumption of primary energy is about 4%, in Finland it is 2%, in Sweden 15%, and in Austria 13–15%.[47),48)] In contrast, Nepal, a developing country, has 145 000 biogas plants for a population of about 20 Mio with ca. 9 Mio cows and ca. 7 Mio other useful animals. It is hence the country with the highest number of biogas plants per inhabitant. The number is expected to increase by another 83 500 plants, financed through the world bank. In Vietnam, ca. 18 000 biogas plants were built by the year 2005 and another 150 000 are planned to be constructed by 2010.[49)] Similar projections are available for India and China. The financing could be managed through the sales of CERs (certified emission reductions),[50)] because methane mitigation saves carbon emissions and can be traded as carbon credits.

In the years around 10 BC, biogas was first used in Assyria for heating baths. Little information is available about later years. But, as early as 1859, a hospital for leprosy patients in Mumbai, India, inaugurated their purification plant for waste water to provide biogas for lighting and to assure power supply in case of emergencies. At the end of the 19th century the first biogas plants were constructed in southern China. Here, Guorui developed a digester of 8 m^3 capacity in the year 1920 and founded the Sanzou Guorui Biogas Lamp Company. He moved to Shanghai in the year 1932 and named his new enterprise Chinese Guorui Biogas Company, with many subsidaries in the south of China. Also, in the region of Wuchang, the building of biogas plants was started in order to solve the problems of disposal of liquid manure and improve hygiene.[51)] In India Jashu Bhai J Patel developed a floating cup bioreactor in the year 1956, known as "Gobar Gas plant" 1962 this construction were acknowledged by the "Khadi and Village Industries Commission (KVIC) of India" and distributed worldwide. Only in the 80th this construction were replaced by the Chinese "dome" bioreactor.

47) Cp. WEB 89
48) Cp. BOK 2
49) Cp. WEB 96
50) Cp. WEB 97
51) Cp. WEB 98

Biogas from Waste and Renewable Resources. An Introduction.
Dieter Deublein and Angelika Steinhauser

ISBN: 978-3-527-31841-4

In Africa, 64 Mio ha of forest was cut between the years 1990 and 2005 for fuel. One biogas plant with a 10 m^3 bioreactor could lead to the reforesting of 0.26–4 ha of land.

4.1 History and status to date in China

In China, the consumption of primary energy is about 11500 TWh per year today and is expected to double[52] within the next 20 years, given the strong development of the country, especially in the more rural areas, where 70% of the population live and 40% of primary energy is required. The use of biomass as the primary energy carrier is considered to be very important as to be seen in Figure 4.1. Already, in the year 2001, waste water from different industries, e.g., the alcohol, sugar, food, pharmaceuticals, paper industries, and even slaughterhouses was fermented in biogas plants. 600 plants with reactor volumes of 1.5 m^3 were running, with a total capacity of 150 Mio m^3 for waste disposal, generating about 1 bn m^3 of biogas.

The yield of biogas in China in the future is estimated to be much higher. Some projections give the total annual realizable potential of biogas as 145 billion m^3,[53] providing about 950 TWh. Other calculations[54] foresee 250 TWh (Table 4.1) only from agricultural residuals as a main source. In this assumption, industrial sewage water, communal wastes, and energy plants were not considered.

Given the long history and experience in the field of biogas technology in China, the growth figures only reflect a continued effort in driving any progress in this sector, which started a very long time ago.

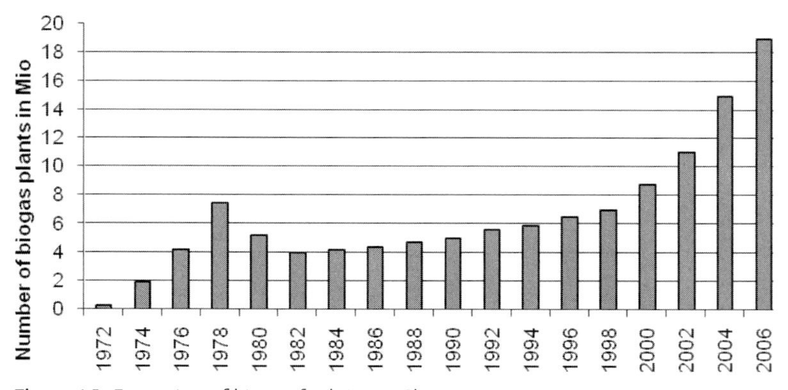

Figure 4.1 Expansion of biogas facilities in China in Mio.

52) Cp. BOK 64
53) Cp. WEB 99
54) Cp. DIS 7

Table 4.1 Potential for biogas production in China.

	TWh
Excrement	20
Residuals from agriculture	200
Residuals from forestry	30
Total	250

4.1.1
Period from 1970 to 1983

In the years around 1970, the first wave of installations of small self-made biogas plants started in China. These so-called "power plants at home" were attached to private rural houses. The costs for such constructions were quite significant. On average it took 17 working days for construction and a family had to invest around 5% of their annual income. But within only a few years the investment was amortized by savings on fuel costs and some additional income from selling the fermented residues as fertilizer.

About 6 Mio biogas plants were set up in China, promoted by the Chinese government to provide energy, for environmental protection, and to give an improvement in hygiene. The "China dome" bioreactor became a standard construction (see Part V) and an example for other developing countries. Such a typical plant consisted of a concreted pit of a few cubic meters volume.[55] The raw materials – feces, wastes from the pig fattening, and plant residuals – were introduced via a gastight inlet into the interior of the reactor. The gas resulting from fermentation collected in the storage space above the substrate. Whenever needed, a slight overpressure was applied to direct some of the gas via hoses to the kitchen. The biogas was usually not used for power supply. This practice was quite common, and about 7 Mio households, about 6% of all the households in the country, were applying this principle in 1978.

The plants were usually integrated in productive agricultural units, i.e. cooperatives (Figure 4.2). Only a few of these had their own power plant supplied by biogas.

In a cooperative, about 90 families lived together. They cultivated sugar cane and bananas, carried on freshwater fishing, and bred silkworm. Any waste was collected and brought to a central biogas plant of 200 m^3 volume, where it was transformed into fuel. The fuel was used for cooking, to heat the living rooms, and to drive electrical generators. At the same time, the process in the biogas plant killed the germs in the feces, leaving a hygienic residue for use as fertilizer. The people suffered less from parasitic infections and the nutritional value of the soils was improved, yielding large crops. The cooperative society was able to survive on

55) Cp. BOK 11

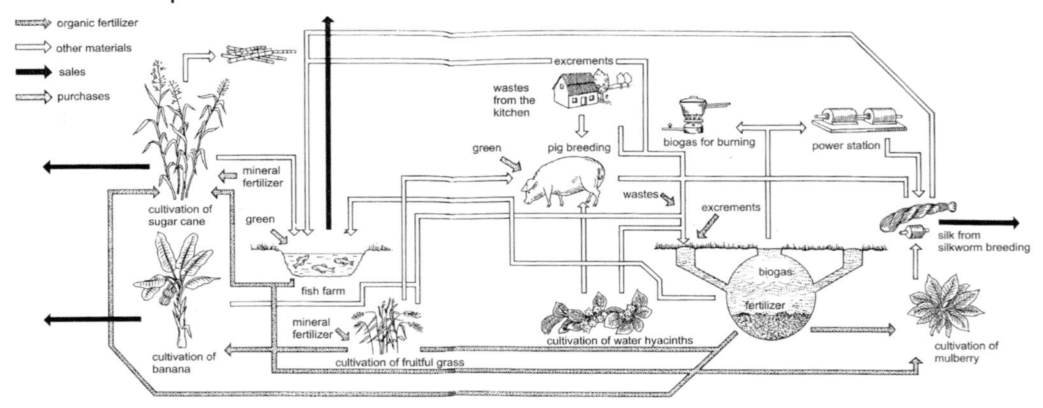

Figure 4.2 Autarc system in developing countries.

its own and only imported a few small amounts of grass for feeding, some chemical fertilizer, and some liquid manure from the neighborhood.

From a political and economic point of view, however, there was no pressure to continue making progress in this field. The importance and interest declined at the end of the period (around 1983), and more and more biogas plants were shut down.

4.1.2 Period from 1984 to 1991

During this period, quite a few new plants were installed, while about the same number of old plants were shut down. The biogas technology, however, gained again more importance since universities started to be engaged in the recovery of biogas and acquired new insights and knowledge.

4.1.3 Period from 1992 to 1998

Building on the new results of the latest research, more biogas plants were set up again, starting in the year 1992. This vogue was supported by the following three political slogans and campaigns:

4.1.3.1 "A pit with three rebuildings"

The campaign named "A pit with three rebuildings", encouraged people to build a pit serving as a bioreactor and to rebuild three rooms: the sty, the toilet facilities, and the kitchen.

The sty and the toilet room were rebuilt to have direct drainage into the 8–10 m^3 bioreactor. The kitchen was set up with a biogas cooker directly connected with suitable pipework to the bioreactor. The toilet water for rinsing was withdrawn from the top of the reactor using a scoop.

4.1.3.2 "4 in 1"

Especially in the north of China, the campaign named "4 in 1" was strongly accepted as it was based on a concept that was developed in a small city, Pulandian,[56] in the province Liaoning. This concept took into account the continental climate in the north of China, where there are huge temperature differences between summer and winter. The significant temperature drop and the cold climate during winter only allowed the common biogas plant to be run for about 5 months.

In order to operate the plant during the whole year, the following four different sections were built:

- Bioreactor with a volume of about 8 m^3
- Greenhouse with about 300–600 m^2 of space
- Sty with about 20 m^2 of space
- Toilet.

The yield of biogas was about 0.15 to 0.25 m^3 per cubic meter of reactor volume. A volume of 8 m^3 for the bioreactor was sufficient to produce enough gas for a family of 4 members. The fermented residue from the reactor was used as a fertilizer in the greenhouse, while the biogas itself was used for cooking and to heat and light the greenhouse. Overall, the investment for such a plant and concept was usually recovered after 1–2 years.

4.1.3.3 "Pig-biogas-fruits"

The campaign named "Pig-biogas-fruits" was strong in the south of China. This plant, which the government promoted,was similar to the "4 in 1" concept, but, because of the milder temperatures, it was not mandatory to have all the different sections consolidated together. Further, the greenhouse was not required, and the fertilizer was used for fruit trees.

4.1.4 Period from the year 1999 onwards

In the year 1999, the following two projects were established to fight against the worsening environmental crisis: "Energy and environment" and "Home-bio and wellbeing". Similarly to the actions taken in Germany the programs included financial aid to motivate people in the rural areas of China to build biogas plants. The concept is working and the number of plants is rapidly increasing.

In the year 2003, "China's 2003–2010 National Rural Biogas Construction Plan" was announced. Objectives were set aimed at increasing the number of biogas plants in China to 20 Mio by 2005, giving 10% of all farmers' households the use of their own biogas plant, and to 50 Mio by the year 2010. Each small biogas plant earns an award of 150 US$ from the government. China plans to supply 15% of its total energy consumption from renewable resources by the year 2020, which means that 200 Mio biogas plants have to be built. An investment of 187 bn US$ is foreseen.

56) Cp. WEB 100

Near the city of Meili in the province of Zhejiang, biogas is produced from the excrement of 28000 pigs, 10000 ducks, 1000000 chickens, and 100000 hens. In Mianzhu in the province Sichuan a biogas plant closely connected to an ethanol production plant produces some MW electricity from biogas.

Nanyang in the province Henan is one of the leading biogas cities in the world because of its location in the center of a rank soil area. Here, there is an abundance of corn, and 1.75 Mio Mg cereals of second quality can be used for the production of biogas.[57)]

4.2
History and status to date in India

India's[58)] consumption of energy today is at 6500 TWh a^{-1}, but this is expected to double in the near future. Today, about 2.5 Mio biogas plants are running, with an average size of 3–10 m^3 of digester volume. Depending on the substrate, the plants generate 3–10 m^3 biogas per day, enough to supply an average farmer family with energy for cooking, heating, and lighting.

The national advisory board for energy in India has published a report forecasting the required quantity and the manner of supply of energy in the future. The board estimates that India has enough resources to sustain 16–22 Mio small biogas plants with 2 m^2 reactor volumes, each to supply sufficient energy for a farmer family with 4 cows. The estimates suggest that the plants will together provide an energy yield that corresponds to 13.4 Mio Mg of kerosene oil. The amount of fertilizer is projected at 4.4 Mio Mg.

In plants in India, the substrate, cattle dung, and biogenous waste, are manually mixed with water in a ratio of 10% dry matter to 90% water. The mix is filled into the digester by simply pushing. The reactor is neither heated nor isolated, enabling the fermentation process to take place at temperatures in the the region of 14 °C during winter and 25 °C during summer. In the reactor itself, the substrate is mixed by a simple mixer which is operated manually. After a dwell time of the substrate in the reactor of around 100 days the fermented residue is removed with buckets or scoops. Pumping systems are not used.

In general, such a small biogas plant costs around 5000 Indian rupees (about 120 US$) per cubic meter of digester. The plants are constructed with the help of local artisans who receive a daily wage of 50 rupees (1.20 US$).

The construction of more and more biogas plants has revealed several beneficial side effects, such as a significant reduction in the exhaustive cultivation of forests. Unexpected successes were noted in the medical sector also. Since respiratory systems and eyes were no longer exposed to aggressive wood smoke from fires, the number of cases of acute asthma and eye diseases was significantly reduced.

Overall, the use of biogas for energy supply provides economic but also ecological and hygienic advantages.

57) Cp. WEB 98

58) Cp. WEB 13

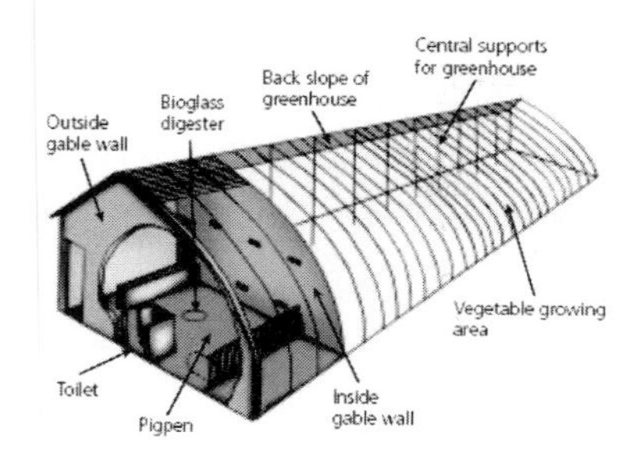

Figure 4.3 Biogas plants.
China: cross section of a "four in one" plant in China (1st row, left), "Four in one" plant in China (1st row, right)
Green house in China, part of a "Four in one" plant (2nd row, left), pig pen with loophole to digester and digester cover in the floor (2nd row, right)
Plastic biogas reactor in South Africa (3rd row, left), Moderm biogas plant in Kerala, India, (Suntechnics GmbH) producing 6.75 kW to lighten street-lamps (3rd row, right), Feeding of a biogas plant in Peru (4th row, left), Bioreactors in the Ukraine (Elenovka/Dnipropetrovsk).

4.3
Status to date in Latin America

In countries in Latin America, e.g., Argentina, Peru, Brazil, Chile, and Mexico, the implementation of biogas plants is just starting. The simple constructions are similar to those in Asia, with a reactor volume of 2–10 m^3 Instruction manuals can be found on the Internet.[59)]

4.4
Status to date in the CIS states

In the CIS states, energy was available in abundance over many years. This changed only very recently. Following attempts to adapt the price of oil and gas to world prices, energy prices have already risen considerably regionally and will raise drastically (by a factor of 5) within the next few years.[60)] People can no longer afford to heat with fossil sources of energy, and politicians are moving only very slowly toward regenerative sources of energy.

In Russia, the company "AO Stroijtechnika Tulskij Sawod" in Tula has been offering two standardized small plants for the production of biogas from domestic waste since the year 1992.[61)] One plant, with a bioreactor volume of 2.2 m^3, processes 200 $kg\,d^{-1}$ substrate. This can be the excrement from 2–6 cows, 20–60 small domestic animals and pigs, or 200–600 head of poultry, and can also include plant material, straw, corn, sunflowers, and leftovers. The other plant consists of two bioreactors of 5 m^3 volume each, in which fermentation takes place at a temperature of 55 °C, and a biogasholder with a volume of 12 m^3. Thus, up to 80 KWh of electrical power is produced per day. The plants can be transported by truck to the place of installation. Optionally provided can be a generator, a heating system, or an infrared emitter. Because of the cold winters in Russia, the plants are provided with particularly thick thermal insulation.

Up to now, more than 70 plants have been installed in Russia, more than 30 in Kazakhstan, and 1 plant in the Ukraine.

In the Ukraine, 162 000 m^3 volume of bioreactors were formerly installed in sewage treatment plants, but because of their poor condition, these have had to be shut down. However, there are new plans to produce ca. 5000 m^3 biogas equal to 28.2 TWh from animal husbandry and poultry farming (47%), sewage sludge (6%) and landfills (47%).[62)] In the Ukraine, an additional 2–3% of the agricultural land can easily be used for the production of biogas. About 3000 biogas plants would have to be built for this purpose. Thus, the Ukraine could become independent of imports of natural gas.

A first plant was erected in 2003 at a pig-fattening station in the village Eleniwka in the province of Dnipropetrovsk. It supplies 3300 m^3 biogas daily. The invest-

59) Cp. WEB 101
60) Cp. WEB 102
61) Cp. WEB 103
62) Cp. WEB 104

ment costs of 413 300 US$ were financed by a Netherlands investor. The power is sold. It is estimated that the break-even point will be reached within ca. 8 years.

Presently, some similar plants are under construction or planned in the Ukraine. The Bortnichi sewage treatment plant is planned to provide the capital Kiev with 9 MW, ca. 10% of the total electrical power consumption. An Austrian company will perhaps fund the investment costs of 10 Mio US$. Break even is foreseen within 2.5 years.

5
General aspects of the recovery of biomass in the future

In developed countries it is difficult for an agriculturist to decide how to best use his land.

If the object is to use the land in the best ecological way possible, a slow downcycling of the biomass will be recommended in the first place. The agriculturist should decide to do silviculture. For example, wood can be used to build houses. After the house has been deconstructed, the quality of the wood is still good enough to serve as material for wardrobes or rail tracks and then to make boxes or art works. Only after such a long cycle should the wood be combusted.

If the object is to use the land in the best economic way possible, the financial aid of governmental institutions will be particularly considered. Depending on the institutional programs, it may well be most profitable to cultivate energy plants to produce biodiesel or ethanol. But maybe it will be most efficient even to have the agricultural area lying idle, as the government may have assigned higher financial benefits to it as if it was used to cultivate, e.g., food plants.

If the agricultural area is to be used to maximize the yield of renewable energy, a combination of different technologies will be most efficient, e.g., wind turbines installed to generate electricity, with photovoltaic cells underneath and the cultivation of grass or energy-affording plants at an even lower level (Figure 5.1). The

Figure 5.1 Wind-driven generator, photovoltaic and energy crops.

Biogas from Waste and Renewable Resources. An Introduction.
Dieter Deublein and Angelika Steinhauser

ISBN: 978-3-527-31841-4

cultivation of two different energy-affording plants in sequence is most profitable. If liquid manure is available, the biomass is recommended to be fermented in a biogas plant; otherwise it is to be combusted.

If the yield of fuel is to be maximized, it is recommended that starch-based plants rather than fat-containing plants should be cultivated to minimize the loss of energy during transformation processes.

Biomass in general will achieve much greater importance as a primary carrier for energy supply in the near future. This will lead to significant changes in the personal habits of people and in the agricultural cultivation methods applied today. Humankind is almost forced to face those changes given the fact that resources of fossil energy carriers are running short. It may be possible in the remote future to meet the energy demand by using biomass only.

It is highly critical that the developed markets should adjust to the required changes as quickly as possible. The emerging markets, in contrast, should not imitate the developed markets, but must take different approaches immediately to recover and secure the supply of energy.

Part II
Substrate and biogas

The formation of methane is a biological process that occurs naturally when organic material (biomass) decomposes in a humid atmosphere in the absence of air but in the presence of a group of natural microorganisms which are metabolically active, i.e. methane bacteria. In nature, methane is formed as marsh gas (or swamp gas), in the digestive tract of ruminants, in plants for wet composting, and in flooded rice fields.

Biomass which is suitable to be fermented is named "substrate".

Biogas from Waste and Renewable Resources. An Introduction.
Dieter Deublein and Angelika Steinhauser

ISBN: 978-3-527-31841-4

1
Biogas

Biogas consists mainly of methane and carbon dioxide, but also contains several impurities. It has specific properties which are listed in Table 1.1.

Biogas with a methane content higher than 45% is flammable.

1.1
Biogas compared to other methane-containing gases

In general, methane has the following characteristics (Table 1.2):

The most well-known methane-containing gas is natural gas. There are several variants of natural gas of different origins. They differ in their chemical composition, their ratios of the chemical elements and the fuel-line pressure.[1)]

Of the natural gas in Germany, 51% is sourced from Western Europe (Denmark, Netherlands, Norway, United Kingdom) and 31% is imported from Russia. Depending on the country of origin, 5 slightly different qualities can be distinguished – GUS-Gas, North Sea gas, Compound gas (all three are so-called H-gases), Holland gas, and Osthannover gas (both are L-gases) – which differ in their Wobbe Indices, calorific value, and methane number.

There are four different levels in the supply chain network, which is grouped into local, regional, nationwide, and international areas for distribution (Table 1.3):

Industrial end customers source gas from pipes with up to 4 bar pressure. Even with varying gas consumption, the private end customers should be supplied with gas of at least 22 mbar pressure at the point at home when the purchased quantity is at maximum. Biogas in the reactor is pressure-free, but a gas finishing plant is followed by a unit in which the pressure can be arbitrarily set and adjusted to the respective needs.

1) Cp. BOK 63

Biogas from Waste and Renewable Resources. An Introduction.
Dieter Deublein and Angelika Steinhauser

ISBN: 978-3-527-31841-4

Table 1.1 General features of biogas.

Composition	***55–70% methane (CH_4) 30–45% carbon dioxide (CO_2) Traces of other gases***
Energy content	6.0–6.5 kWh m^{-3}
Fuel equivalent	0.60–0.65 L oil/m^3 biogas
Explosion limits	6–12% biogas in air
Ignition temperature	650–750 °C (with the above-mentioned methane content)
Critical pressure	75–89 bar
Critical temperature	−82.5 °C
Normal density	1.2 kg m^{-3}
Smell	Bad eggs (the smell of desulfurized biogas is hardly noticeable)
Molar Mass	16.043 kg kmol^{-1}

Table 1.2 Features of methane.

	Temperature		***Pressure [bar]***	***Density [kg L^{-1}]***
Critical point	−82.59 °C	(190.56 K)	45.98	0.162
Boiling point at 1.013 bar	−161.52 °C	(111.63 K)	–	0.4226
Triple point	−182.47 °C	(90.68 K)	0.117	–

Table 1.3 Grid-type networks for the natural gas supply.

Pressure stage	***Pressure [bar gauge]***	***Pipe diameter [mm]***	***Flow rate [M s^{-1}]***
Low pressure	<0.03	50–600	0.5–3.5
”	0.03–0.1	50–600	1–10
Medium pressure	0.1–1.0	100–400	7–18
High pressure	1–16	300–600	<20
”	40–120	400–1600	<20

Depending on its quality, natural gas is divided into two groups that are characterized by the Wobbe Indices:[2)]

Group: L-Gas

Group: H-Gas

Table 1.4 shows a comparison of specifications of natural gases with biogases (agricultural biogas, biogas from sewage plants, landfill gas). Because of its vari-

2) The Wobbe-Index W is a characteristic value to describe the quality of methane gases. In correlation with the calorific value, gross and net Wobbe Indices are distinguished, $W_{O,N}$ and $W_{U,N}$ given in kWh/m^3. The relationship between the calorific values and the Wobbe Indices is given in the equations including the relative density ρ* (air = 1)

$$W_{O,N} = \frac{H_{O,N}}{\sqrt{\rho}} \qquad W_{U,N} = \frac{H_{U,N}}{\sqrt{\rho}}$$

Table 1.4 Methane gas – comparative qualities.

Gas composites/ features	*Formula*	*Units*	*Natural gases*			*Biogases*[3]		
			Group H (GUS)	*Group H (North Sea)*	*Group L (Holland)*	*Sewage gas*[4]	*Agricultural gas*	*Landfill gas*
Methane	CH_4	% by vol.	98.31	86.54	83.35	65–75	45–75	45–55
Ethane	C_2H_6	% by vol.	0.50	8.02	3.71			
Propane	C_3H_8	% by vol.	0.19	2.06	0.70			
Butane	C_4H_{10}	% by vol.	0.08	0.21–0.60	0.22			
Pentane	C_5H_{12}	% by vol.	0.02	0.10	0.06	<300 mg/Nm³(mandatory limit in Germany)[5]		
Hexane	C_6H_{14}	% by vol.	0.01	0.05	0.05			
Heptanes	C_7H_{16}	% by vol.	<0.01	<0.01	<0.01			
Octane	C_8H_{18}	% by vol.	<0.01	<0.01	<0.01			
Benzene	C_6H_6	% by vol.	<0.01	<0.01	<0.01			
Carbon dioxide	CO_2	% by vol.	0.08	1.53	1.27	20–35	25–55	25–30
Carbon monoxide	CO	% by vol.	0.00	0.00	0.00	<0.2	<0.2	<0.2
Nitrogen	N_2	% by vol.	0.81	1.10	10.64	3.4	0.01–5.00	10–25
Oxygen[6]	O_2	% by vol.	0.05/3.00	0.05/3.00	0.05/3.00	0.5	0.01–2.00	1–5
Hydrogen	H_2	% by vol.	0.00	0.00	0.00	traces	0.5	0.00
Hydrogen sulfide	H_2S	mg/Nm³	5.00	5.00	5.00	<8000	10–30.000	<8000
Mercaptan sulfur	S	mg/Nm³	6.00	6.00	6.00	0	<0.1–30	n.a.
Total sulfur	S	mg/Nm³	30.00	30.00	30.00	n.a.	n.a.	n.a.
Ammonium	NH_3	mg/Nm³	0.00	0.00	0.00	traces	0.01–2.50	traces
Siloxanes		mg/Nm³	0.00	0.00	0.00	<0.1–5.0	traces	<0.1–5.0
Benzene, Toluene, Xylene		mg/Nm³	0.00	0.00	0.00	<0.1–5.0	0.00	<0.1–5.0
CFC		mg/Nm³	0.00	0.00	0.00	0	20–1000	n.a.
Oil		mg/Nm³	0.00	0.00	0.00	traces	traces	0.0
Gross calorific value[7]	$H_{o,n}$	kWh/Nm³	11.7	11.99	10.26	6.6–8.2	5.5–8.2	5.0–6.1
Net calorific value	$H_{u,n}$	kWh/Nm³	9.98	10.85	9.27	6.0–7.5	5.0–7.5	4.5–5.5
Normal density	ρ^*	kg/Nm³	0.73	0.84	0.83	1.16	1.16	1.27
Rel. density relatedto air	d	–	0.57	0.65	0.64	0.9	0.9	1.1
Wobbe Index	$W_{o,n}$	kWh/Nm³	13.27–14.72	13.5–14.1	10.5–13.0	7.3	n.a.	n.a.
Methane number	MZ	–	ca. 94	ca. 70	ca. 88	134	124–150	136
Rel. Humidity		%	60	60	60	100	100	<100
Dewpoint	ϑ_s	°C	$ts<t_{average, bottom}$	$ts<t_{average, bottom}$	$ts<t_{middle, bottom}$	35	35	0–25
Temperature	ϑ	°C	12	12	12	35–(60)	35–(60)	0–25

3) Cp. WEB 105
4) Cp. LAW 10
5) Cp. LAW 3
6) in dry natural gas networks / in wet natural gas networks
7) Cp. JOU 14

able composition, landfill gas is not allowed to be mixed with natural gas. Therefore the table also includes a typical composition of landfill gas.

1.2 Detailed overview of biogas components

Table 1.5[8] gives an overview of the typical gas components and their impacts on the gas quality.

The gas components are specific to the plant and substrate and should be checked regularly on a long-term basis, as shown by the evaluations of 10 selected biogas plants (60–110 animal units) in Bavaria that use different substrates (Figure 1.1). In plants Nos. 1–5, about 6.5–62.5% of fat removal tank flotated material were added, depending on the entire dry weight of the substrate. In plant No. 1 about 62.5% of fat removal tank flotated material were fermented with a variety of organic materials (e.g., wool, biogenous waste, aromas). For co-fermentation purposes, small quantities of around 2–6.5% of primarily potato, pulp, and vegetable waste was added to liquid manure (refer to plants Nos. 6–7). For plants Nos. 8–10, only liquid manure and/or muck was required.

Table 1.5 Typical components and impurities in biogas.

Component	*Content*	*Effect*
CO_2	25–50% by vol.	– Lowers the calorific value – Increases the methane number and the anti-knock properties of engines – Causes corrosion (low concentrated carbon acid). if the gas is wet – Damages alkali fuel cells
H_2S	0–0.5% by vol.	– Corrosive effect in equipment and piping systems (stress corrosion); many manufacturers of engines therefore set an upper limit of 0.05 by vol.%; – SO_2 emissions after burners or H_2S emissions with imperfect combustion – upper limit 0.1 by vol.% – Spoils catalysts
NH_3	0–0.05% by vol.	– NO_x emissions after burners damage fuel cells – Increases the anti-knock properties of engines
Water vapour	1–5% by vol.	– Causes corrosion of equipment and piping systems – Condensates damage instruments and plants – Risk of freezing of piping systems and nozzles
Dust	>5 µm	– Blocks nozzles and fuel cells
N_2	0–5% by vol.	– Lowers the calorific value – Increases the anti-knock properties of engines
Siloxanes	0–50 mg m^{-3}	- Act like an abrasive and damages engines

8) Cp. BOK 34

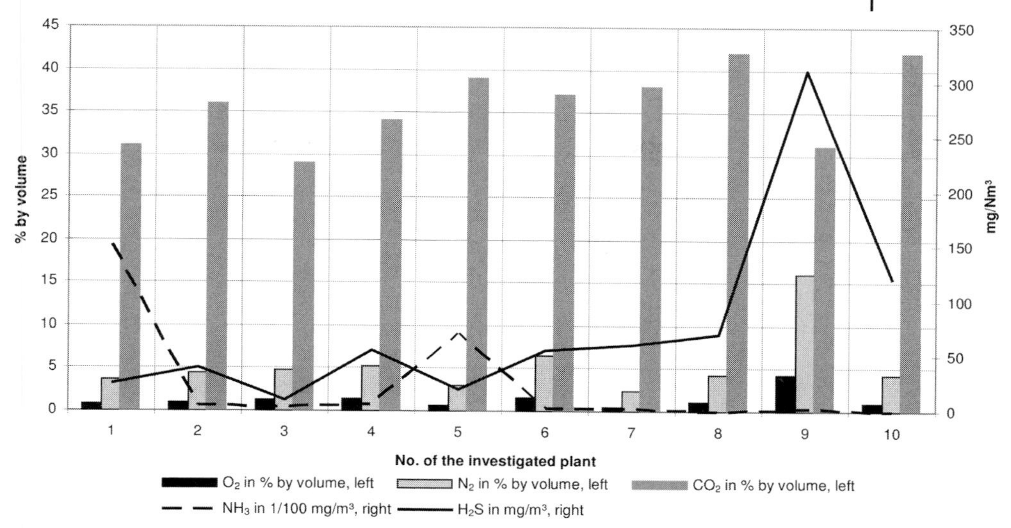

Figure 1.1 Main composition of biogas produced with biogas plants and added substrates (No. 1–5: Flotate from grease removal tank (6.5–62.5%), No. 6–7: Liquid manure with co-ferments (potatoes, pulp, small amounts of vegetables), No. 8–10: Liquid manure and/or excrement.

A combination of biomass and co-substrate helps in decreasing the content of CO_2 produced during the fermentation process. With FAF as co-ferment, the content of CO_2 is about 35% – lower than that obtained by just fermenting liquid manure (about 40%). If maize and muck are used as co-ferments the CO_2 content is around 45%.

1.2.1 Methane and carbon dioxide

The composition of the gas (referring mainly to the ratio of carbon dioxide to methane) can be only partially controlled. It depends on the following factors:

- The addition of long-chain hydrocarbon compounds, e.g., materials which are rich in fat, can help to improve the quality of the gas, provided that the quantities are reasonable and not to large to avoid acidity.
  As shown in Figure 1.2, the methane content increases the higher the number of C-atoms in the substrate.
- Generally, the anaerobic decomposition of biomass improves with longer time of exposure. Toward the end of the residence time the content of methane increases disproportionately, especially as soon as the CO_2-releasing process of hydrolysis starts to become deactivated.
- The fermentation process takes place much faster and more evenly if the material in the bioreactor is well and

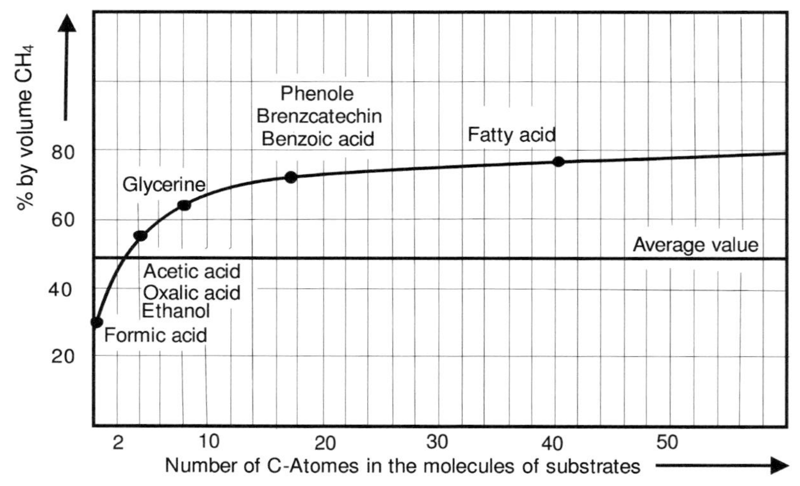

Figure 1.2 Correlation between methane content in biogas and carbon chain length.[9]

homogeneously activated. The time of exposure can be shorter.

- If the substrate is well enclosed in lignin structures, the type of disintegration of the substrate becomes important. The structure should be disrupted or defibrated rather than cut.
- A higher content of liquid in the bioreactor results in a higher concentration of CO_2 dissolved in water, reducing the level of CO_2 in the gas phase.
- The higher the temperature during the fermentation process the lower is the concentration of CO_2 dissolved in water.
- A higher pressure leads to a higher concentration of CO_2 dissolved in water. It may influence the quality of the gas in a positive way if the material from the bottom of the reactor is removed because CO_2 is discharged.
- A breakdown in power supply is to be avoided for complete and sufficient hydrolysis of material. It is important that the material for discharge be completely decomposed.
- The substrate has to be well prepared to expedite and intensify the decomposition.

1.2.2 Nitrogen and oxygen

In general, biogas contains nitrogen and oxygen in a ratio of 4:1. Both these components of air are introduced when the ventilation is switched on to remove the sulfide. Small quantities of air can infiltrate if the gas pipes are not fully tight.

9) Cp. BOK 28

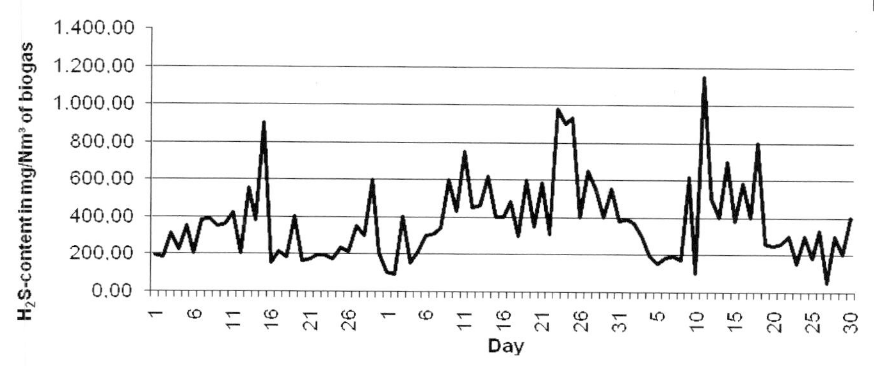

Figure 1.3 Daily variation of the H_2S-content in biogas.

1.2.3 Carbon monoxide

The amount of carbon monoxide is below the detection limit of 0.2% by vol.

1.2.4 Ammonia

The concentration of ammonia is usually very low (<0.1 mg m^{-3} biogas). It only may exceed 1 mg m^{-3} up to max. 1.5 mg m^{-3} (Plant 1 in the graph above) when a high amount of co-substrates (62.5%) is used in the plant.

This happens especially when excreta from poultry or special types of waste are fermented. Concentrations up to 150 mg m^{-3} have been reported, which may impair the burn behavior and the life of the engines.[10]

1.2.5 Hydrogen sulfide

The content of hydrogen sulfide in the exhaust gas depends on the process and the type of waste. Without a desulfurizing step, the concentration of H_2S would often exceed 0.2% by volume.[11] If a viscous substrate is fermented, the content of H_2S is lower than when using a liquid substrate. It is also lower with bio waste compared to leftovers, liquid manure, and agricultural co-substrates.

The hydrogen sulfide content in the biogas is subject to considerable unexplainable variation in the course of the day (Figure 1.3).

It is one objective to keep the hydrogen sulfide content at the lowest level possible, since plant components downstream are damaged by H_2S. Therefore the biogas is usually desulfurized while still in the bioreactor.

But it can be reduced by applying a gross pre-desulfurization in the reactor. This intermediate step can help to reduce its content to below 70 mg m^{-3} when using

10) Cp. BOK 65

11) Cp. BOK 37

co-substrates or to below 310 mg m^{-3} in plants using pure liquid manure for fermentation. But even with desulfurization the H_2S-content in the biogas is relatively high.

1.2.6
Chlorine, fluorine, mercaptans

The concentration of chlorine, fluorine, and mercaptans in biogas is usually below the detection limit of 0.1 mg m^{-3}. Halogen hydrocarbons are present in sewage and/or dump gases.

1.2.7
BTX, PAK, etc.

The concentrations of benzene, toluene, ethylbenzene, xylene, and cumene (BTX) are always below the detection limit of 1 mg m^{-3}, with a few exceptions in the case of toluene. A higher toluene value (7.4 mg m^{-3}) may result if special wastes are co-fermented.

The concentrations of PAH (polycyclic aromatic hydrocarbons) may exceed the detection limit of 0.01 µg m^{-3} sometimes, but always only slightly, up to maximum of 0.03 µg/m^3.

1.2.8
Siloxanes

Various compounds of silicon, the "siloxanes", form a separate group of materials present in biogas. Siloxanes can particularly be found in cosmetics, detergents, printing inks, and building materials, and hence in household refuse and in waste water which is directed into the waste water plant. During processing treatment plant in the digestion tower, quite high concentrations of traces of siloxane are carried over into the sewage gas. Likewise, the content of siloxane in the gas can be very high in co-fermenting plants, which contain a large amount of sewage sludge. Concentrations as high as 18 mg m^{-3} of methylcyclosiloxane have been measured. Such values exceed the recommended limit of 0.2 mg m^{-3} [12] for polysiloxane in biogas for heating and power plants.

At high temperatures, siloxanes and oxygen chemically form SiO_2, which remains on the surfaces of machine parts. Usually it results in flow reduction and friction, but in heating and power plants it may cause abrasion of the pistons.[13] How much such phenomena affect the power generated in fuel cells has not yet been investigated.[14]

12) Cp. WEB 10
13) Cp. JOU 21
14) Cp. WEB 25

2
Substrates

In general, all types of biomass can be used as substrates as long as they contain carbohydrates, proteins, fats, cellulose, and hemicellulose as main components. It is important that the following points are taken into consideration when selecting the biomass:

- The content of organic substance should be appropriate for the selected fermentation process.
- The nutritional value of the organic substance, hence the potential for gas formation, should be as high as possible.
- The substrate should be free of pathogens and other organisms which would need to be made innocuous prior to the fermentation process.
- The content of harmful substances and trash should be low to allow the fermentation process to take place smoothly.
- The composition of the biogas should be appropriate for further application.
- The composition of the fermentation residue should be such that it can be used, e.g., as fertilizer.

Lignin, the main constituent of wood, and most synthetic organic polymers (plastics) simply decompose slowly.

Table 2.1 shows the maximum gas yields per kg dry matter of different substrates. Some substrates legally require a proper sanitization before and after the fermentation process.

In the following sections some of the substrates are described more in detailed.

2.1
Liquid manure and co-substrates

Most agricultural biogas plants are used to ferment liquid manure (Table 2.2.),[15] nowadays quite often combined with co-substrates to increase the biogas yield.

15) Cp. BOK 33

Biogas from Waste and Renewable Resources. An Introduction.
Dieter Deublein and Angelika Steinhauser

ISBN: 978-3-527-31841-4

Table 2.1 Co-ferments, their hazardousness (U = harmless, H = to be hygienized, S = trash-containing, SCH = contaminent-containing, complexity of the pretreatment (I = no complexity, II = little complexity, III = high complexity, and advice for the production. (– = not suitable).

Substrate for biogas production[16],[17],[18],[19],[20]	*DM [%] oDM in DM [%]*	*Biogas yield [m³ kg⁻¹ oTS] Retention time [d]*	*Production advice*
	Residuals from beverage production		
Spent grain, fresh or ensilaged	20–26	0.5–1.1	U, I
	75–95	–	
Spent grain, dry	90	0.6	U, I
	95	–	
Yeast, boiled	10	0.72	U, I
	92	–	
Marc	40–50	0.6–0.7	U, I
	80–95	–	
Spent diatomite (beer)	30	0.4–0.5	U, I
	6.3	–	
Spent hops (dried)	97	0.8–0.9	U, II
	90	–	
Spent apples	22–45	0.56–0.68	U, II
	85–97	–	
Apple mash	2–3	0.5	U, II
	95	3–10	Inhibition through volatile fatty acids
Spent fruits	25–45	0.4–0.7	U, I
	90–95	–	–
	Animal waste		
Slaughterhouse waste	–	0.3–0.7	H, I
	–	–	
Meat and bone meal	8–25	0.8–1.2	H, I
	90	–	
Fat from the separator used in gelatine production	25	–	H, I
	92	–	
Animal fat	–	1.00	H, I
	–	33	
Homogenized and sterilized animal fat	–	1.14	H, I
	–	62	
Blood liquid	18	680	H, I
	96	–	
Blood meal	90	0.65–0.9	H, I
	80	34–62	

16) Cp. BOK 60
17) Cp. WEB 33
18) Cp. WEB 86
19) Cp. JOU 3
20) Cp. WEB 106

Table 2.1 *Continued*

Substrate for biogas production[16),17),18),19),20)]	*DM [%]* *oDM in DM [%]*	*Biogas yield [m³ kg⁻¹ oTS]* *Retention time [d]*	*Production advice*
Stomach content of pigs	12–15	0.3–0.4	H, I
	80–84	62	Abrasion through sand
Rumen content (untreated)	12–16	0.3–0.6	H, I
	85–88	62	Floating sludge formation
Rumen content (pressed)	20–45	1.0–1.1	H, S, II
	90	62	Floating sludge formation
Greens, gras,[21)] cereals, vegetable wastes			
Vegetable wastes	5–20	0.4	U, S, II
	76–90	8–20	
Leaves	–	0.6	U, S, II
	82	8–20	Inhibition through trash
Greens (fresh)	12–42	0.4–0.8	U, S II
	90–97	–	
Grass cuttings from lawns	37	0.7–0.8	U, S, II
	93	10	Earth content
Grass ensilage	21–40	0.6–0.7	U, S, II
	76–90	–	
Hay	86	0.5	U, S, II
	90–93	–	
Meadow grass, clover	15–20	0.6–0.7	U, S, II
	89–93	–	
Market wastes	8–20	0.4–0.6	U, S, III
	75–90	30	pH value decrease
Leaves of sugar beet/fodder beet ensilaged	15–18	0.4–0.8	U, S, II
	78–80	–	
Sugar beet/fodder beet	12–23	0.7	U, S, II
	80–95	–	
Fodder beet mash[22)]	22–26	0.9[23,24)]	U, S, II
	95	–	
Potato haulm	25	0.8–1.0	U, S, II
	79	–	
Clover	20	0.6–0.8	U, S, II
	80	–	
Maize ensilaged	20–40	0.6–0.7	U, S, II
	94–97	–	
Colza	12–14	0.7	U, S, II
	84–86	–	
Sunflower	35	–	U, S, II
	88	–	
Sorghum	24–26	–	U, S, II
	93	–	

21) Cp. BOK 25
22) 15 000 m³ gas / hectare cultivated area
23) Cp. WEB 82
24) Cp. PAT 2

Table 2.1 *Continued*

Substrate for biogas production[16),17),18),19),20)]	***DM [%] oDM in DM [%]***	***Biogas yield [m^3 kg^{-1} oTS] Retention time [d]***	***Production advice***
Diverse kinds of cereals	85–90	0.4–0.9	U, S, II
	85–89	–	
Straw from cereals	86	0.2–0.5	U, II
	89–94	–	
Maize straw	86	0.4–1.0	U, II
	72	–	
Rice straw	25–50	0.55–0.62	U, II
	70–95	–	
	Wastes from the food and fodder industry		
Potato mash, potato pulp, potato peelings	6–18	0.3–0.9	U, S, II
	85–96	3–10	Inhibition through volatile fatty acids
Potato pulp dried, potato shred, potato flakes	88	0.6–0.7	U, S, II
	94 –96	–	
Cereal mash	6–8	0.9	U, S, II
	83–90	3–10	Inhibition through volatile fatty acids
Mash from fermentations	2–5	0.5–85	S, II
	90–95	35–60	
Mash from distillations	2–8	0.42	U, I
	65–85	14	
Mash from fruits	2–3	0.3–0.7	S, II
	95	–	
Oilseed residuals (pressed)	92	0.9–1.0	S, II
	97	–	
Colza/flax extraction shred	88–89	0.4–0.9	S, II
	92–93	–	
Colza/flax cake	90–91	0.7	S, II
	93–94	–	
Castor oil shred	90	–	S, II
	81	–	
cocoa husks	95	–	S, III
	91	–	
Pomace	63	–	S, II
	53	–	
Molasses	77–90	0.3–0.7	S, II
	85–95	–	
Molasse of lactose	30	0.7	S, II
	74	–	
Waste from tinned food industry	–	–	U, II
	–	–	
Wheat flour	88	0.7	U, I
	96	–	

Table 2.1 *Continued*

Substrate for biogas production[16),17),18),19),20)]	*DM [%]* *oDM in DM [%]*	*Biogas yield [m^3 kg^{-1} oTS]* *Retention time [d]*	*Production advice*
Wheat bran, wheat powder bran	87–88	0.5–0.6	U, I
	93–95		
Malt germ	92	0.6	U, I
	93	–	
Wastes from households and gastronomy			
Bio waste	40–75	0.3–1.0	H, S, SCH, III
	30–70	27	Extensive pretreatment
Leftovers (canteen kitchen)	9–37	0.4–1.0	H, S, SCH, III
	75–98		
Dry bread	65–90	0.8–1.2	H, I
	96–98		
Sewage sludge (households)	–	0.20–0.75	H, S, I
	–	17	
Leftovers, overstored food	14–18	0.2–0.5	H, SCH, III
	81–97	10–40	Trash (bones, packings)
Fat removal tank flotate, fat residues from gastronomy	2–70	0.6–1.6	H, II
	75–98	–	Trash possible (bones, packings)
Mixed fat	99.9	1.2	H, I
	99.9	–	
Sewage sludge (industry)	–	0.30	H, I
	–	20	
Flotation sludge	5–24	0.7–1.2	H, II
	90–98	12	
Wastes from pharmaceutical and other industries			
Vegetable extraction residues	–	0.2–0.75	H, S, I
	–	–	
Egg waste	25	0.97–0.98	H, I
	92	40–45	Sediments of the eggshell Ammonia from proteins
Blood plasma	30–40	0.66–1.36	H, I
	95–98	43–63	Ammonia from proteins
Waste from paper and carton production	–	0.2–0.3	U, III
	–	–	
Pulp	13	0.65–0.75	U, III
	90		
Biological oils and lubricants	–	>0.5	U, III
	–	–	
Crude glycerine (RME production)	>98	1.0–1.1	U, I
	90–93	–	
Biologically degradable packaging/ plastics (e.g. Polyhydroxybutyric acid)	–	0.64	U, I
	–		

Table 2.1 *Continued*

Substrate for biogas production[16),17),18),19),20)]	*DM [%] oDM in DM [%]*	*Biogas yield [m^3 kg^{-1} oTS] Retention time [d]*	*Production advice*
	Productive livestock	husbandry	
Liquid manure from cattle	6–11	0.1–0.8	U, I
	68–85	–	
Excreta from cattle (fresh)	25–30	0.6–0.8	U, I
	80	–	
Liquid manure from pigs	3–10	0.3–0.8	U, I
	77–85	–	
Excreta from pigs	20–25	0.27–0.45	U, I
	75–80	–	
Excreta from chicken	10–29	0.3–08	U, I
	67–77	–	
Excreta from sheep (fresh)	18–25	0.3–0.4	U, I
	80–85	–	
Excreta from horses (fresh)	28	0.4–0.6	U, I
	25	–	
Low-fat milk	8	0.7	
	92	–	
Whey	4–6	0.5–0.9	U, I
	80–92	3–10	pH value decrease
Whey without sugar	95	0.7	
	76	–	
Starch sludge	–	–	U, I
	–	–	

The liquid manure from all animal species may contain foreign matter. Some of these substances can be processed in the biogas plant, e.g.:

- Litter
- Residues of fodder

Others are unwanted foreign matter because they impair the fermentation of the liquid manure; e.g.:

- Sand from mineral materials present in feed of pigs and poultry
- Sawdust from scattering
- Soil from roughage
- Soil which is carried from meadows
- Skin and tail hair, bristles, and feathers
- Cords, wires, plastics and stones and others.

Table 2.2 Yield of liquid manure per GVE (livestock unit).

	GVE[25]	*Liquid manure [m³] per animal*			*DM content [%]*	*Gas production [m³/GVE/day]*
		Per day	*Per month*	*Per year*		
Cattle						
Feeder cattle, cow	1	0.05	1.5	18.0	7–17	0.56–1.5
Dairy cow, stock bull, trek ox	1.2	0.055	1.65	19.8	"	"
Feeder bull	0.7	0.023	0.69	8.3	"	"
Young cattle (1–2 years)	0.6	0.025	0.75	9.0	"	"
Calf breeding (up to 1 year)	0.2	0.008	0.24	2.9	"	"
Feeder calf	0.3	0.004	0.12	1.4	"	"
Pigs						
Feeder pigs	0.12	0.0045	0.14	1.62	2.5–13	0.60–1.25
Sow	0.34	0.0045	0.14	1.62	"	"
Young pig up to 12 kg	0.01	0.0005	0.015	0.18	"	"
Young pig 12–20 kg	0.02	0.001	0.03	0.36	"	"
Young pig >20 kg	0.06	0.003	0.09	1.08	"	"
Young pig (45–60 kg), feeder pig, young sow (up to 90 kg)	0.16	0.0045	0.14	1.62	"	"
sow + 19 young pigs / a (over 90 kg)	0.46	0.014	0.42	6	"	"
Sheep						
Up to 1 year	0.05	0.003	0.09	1.08	n.a.	n.a.
Over 1 year	0.1	0.006	0.18	2.16	n.a.	n.a.
Horses						
Up to 3 years, small horses	0.7	0.023	0.69	8.3	n.a.	n.a.
Horses over 3 years	1.1	0.033	0.99	11.9	n.a.	n.a.
Poultry						
Young feeder poultry, young hens (up to 1200 g)	0.0023	0.0001	0.006	0.07	20–34	3.5–4.0
Young feeder poultry, young hens(up to 800 g)	0.0016	0.0001	0.006	0.07	"	"
Laying hen(up to 1600 g)	0.0030	0.0002	0.006	0.07	"	"

25) Cp. BOK 66

The presence of foreign matter leads to an increased complexity in the operating expenditure of the plant. For example, during the process of fermentation of liquid manure from pigs and cattles the formation of scum caused by feed residues and straw and/or muck is expected. Likewise, the addition of rumen content and cut grass can contribute to its formation.

Pig liquid manure rather causes aggregates at the bottom as the feed contains a certain proportion of sand and consists of undigested parts of corn or grain. Likewise, excreta from hens lead to a similar phenomenon due to the high content of lime and sand.

In general, organic acids, antibiotics, chemotherapeutic agents, and disinfectants found in liquid manure can impair and even disrupt the fermentation process in biogas plants (see Part III). In the liquid manure of pigs the high content of copper and zinc derived from additives in the feed can be the limiting factor.

The degree to which the organic substance in the biomass is decomposed in the bioreactor depends on the origin of the liquid manure (Table 2.3). The organic content in liquid manure derived from cattle is only 30% decomposed because of the high content of raw fibers in the feed, while about 50% of pig liquid manure and more than 65% of chicken liquid manure is broken down. The more decomposable the organic substance, the higher is the content of ammonia in the liquid manure compared to the untreated material. The amount of ammonia in hen liquid manure represents about 85% of the total original nitrogen content.

Table 2.3 Usual parameters for biomass fermentation.

Substrate	*Degradation of organic substances [%]*	*Degradation of organic acids [%]*	*Percentage of ammonium nitrogenfrom total nitrogen [%]*	*pH value*
Liquid manure from pigs	54	83	70	7.7
Liquid manure from pigs + waste fat	56	88	65	7.8
Liquid manure from pigs	40	76	72	7.9
Liquid manure from pigs, separated	–	–	73	7.9
Liquid manure from dairy cattle, separated	24	68	50	7.9
Liquid manure from dairy cattle	37	–	58	7.8
Liquid manure from cattle	30	–	47	–
Liquid manure from cattle, separated	–	–	63	8.3
Liquid manure from bulls	52	–	74	8.0
Liquid manure from poultry	67	–	85	8.2

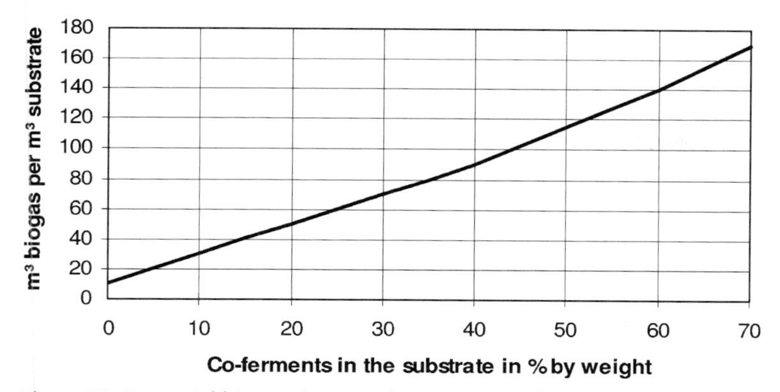

Figure 2.1 Biogas yield in correlation to the percentage of co-ferments in agricultural plants.

Table 2.4 Characteristic values of substrates with different co-ferments.

Input	*Percentage of substrate [%]*	*DM [%]*	*Total N [% of DM]*	*GVE/ha*
Liquid manure from pigs	30	7.7	7.7	1.8
Fat removal tank flotate	30	20	1.1	0.7
Leftovers	20	25	4.0	2.3
Flotated fat	20	9.0	5.1	1.3

By co-substrates added to the liquid manure the content of organic substrate is increased, hence the yield of biogas. From an economic point of view it is only profitable, however, when the materials are sourced from a location within a distance of 15–20 km.

In general, the content of dry matter (liquid manure and co-substrates) in the substrate should be below 2–12% to ensure the functionality of standard pumps and a proper mixing in the bioreactor, what is important for an efficient transformation process.

But the addition of co-substrates poses a higher hygienic risk (see Part IV). If the residue from the fermentation process is to be used as fertilizer for agricultural areas, the co-substrates should meet the national laws and should not pose any hazard from exposure, e.g., they must be free of pathogens.

Some co-substrates, like residues from separated fat, leftovers, and flotated material from a fat removal tank, contain nitrogen-rich nutrients, (Table 2.4). When distributing this to fields, the upper limit for nitrogen can be exceeded, given e.g., by the German law for water protection. The limit is set at 210 kg ha^{-1}, which corresponds to the quantity of nitrogen in excreta of 3.5 GVE ha^{-1}. In this case, the residue from fermentation may not be used on the agricultural areas from which the residue came, but must be transported to, e.g., cattle-less agricultural enterprises in the neighbourhood, what often involves additional costs.

Besides liquid manure and muck, domestic waste water with excreta is accumulated in agricultural enterprises. In Germany[26] it is permitted to ferment it in a biogas plant if the total amount of waste water is less than $8\,m^3a^{-1}$.

For the farmer, however, it is no more profitable, since the domestic waste water has to be analyzed for pathogens, and additionally parts of the biogas plant have to be adapted.

2.2
Bio waste from collections of residual waste and trade waste similar to domestic waste

Residual waste is the term used for waste generated by households. As shown by Table 2.5, the composition of the residual waste depends on the location of the household. Waste from shops or trade can also be considered as residual waste because of its very similar composition.

In the 1990s, Germany, Austria, and Switzerland introduced a new system for waste management, the bio waste container. This made it possible to collect the organic bio waste suitable for composting or fermentation separately from the inorganic waste.

Only about 30–45% of garbage is bio waste, which means about 50–100 kg a^{-1} bio waste per inhabitant in Germany. But about 50% of all the organic materials (not only related to bio waste) can actually be fermented based on their known characteristics and technical knowledge to date (Table 2.6).[27]

In the central congested areas of cities (the location of multi-storey housing), the bio waste is poor in structure and quite pasty. This waste includes leftovers, spoiled food, market waste, and different industrial wastes (e.g., mash, waste liquor, waste from the food industry and the industry of luxury articles).

In the outskirts of a town or in rural settlements the bio waste is fairly rich in structure and fibrous, and hence is well suitable for composting.

Different types of bio waste accumulate over the year depending on the season (Table 2.7).

The monthly amount of bio waste, depending on seasonality, is as shown in Figure 2.2.

2.3
Landfill for residual waste

Until the 1990s, residual waste (i.e. household waste) was discarded on landfills, by default. Biological components of the waste were degraded quite slowly and the fermentation process took about 20–40 years. The landfill gas produced during the process was gathered by using horizontal drainages and gas pits for disposal. About 12–300 m^3 of landfill gas was produced in total per Mg of residual waste, but it contained quite a high level of toxic and corrosive organic components, so that damage to combined heating and power units (CHPs) often resulted.

26) Cp. BOK 23

27) Cp. BOK 31

Table 2.5 Components of waste in 5 actual examples and 3 standardized examples.[28]

	Actual examples					*Standardized examples*		
	Waste from an area with industry and houses	*Waste from a rural area*	*Waste from a small city*	*Waste from shops*	*Waste from industry similar to waste from houses*	*Rural standard waste*	*City standard waste*	*Standard waste*
Org. waste >8 mm	35.8	n. a.	25.6	26.3	26.5	26.9	25.9	26.4
Org. waste <8 mm	3.2	n. a.	5.6	6.1	2.4	7.3	5.7	6.5
Total org. matter	39.0	34.2	31.8	32.4	28.9	34.2	31.6	32.9
Paper/carton/wood	11.8	17.0	14.0	25.0	5.9	14.2	17.3	15.7
Glass	2.0	4.8	6.7	7.1	1.0	5.4	7.3	6.3
Metals	4.4	2.2	5.5	3.7	0.5	3.7	5.4	4.6
Foils	4.8	0.8	2.8	3.5	n.a.	2.6	3.7	3.2
Other plastics	8.6	4.0	7.2	6.3	4.9	6.4	6.9	6.6
Texties / Shoes	5.7	2.6	4.6	1.9	n.a.	4.2	3.8	4.0
Sanitary products	6.5	6.7	8.3	0.9	n.a.	8.5	6.0	7.3
Inorg. waste <8 mm	8.7	4.0	4.6	4.7	n.a.	5.2	4.6	4.9
Minerals, earth	6.0	5.4	3.7	2.2	11.8	3.8	3.2	3.5
Problematic waste	n.a.	1.3	1.1	1.0	n.a.	1.2	1.0	1.1
Other fractions	2.5	17.0	9.7	11.3	47.0	10.6	9.2	9.9
Total	100.0	100.0	100.0	100.0	100.0	100.0	100.0	100.0

28) Cp. BOK 24

Table 2.6 Comparison of processes for the disposal of bio waste.[29]

Parameter	*Fermentation*	*Burning*	*Land filling*	*Composting*
Biological degradability	Not all org. matter can be fermented	All org. matter	All org. matter but not all will rot down	Not all org. matter compostable
Energy yield	Material with high water content can be treated. Energy production more than 700 kWh Mg^{-1}	Energy production depends on the water content since water has to be evaporated	Low energy production	Energy-consuming
Compliance with legal requirements	Ignition loss often >5%[30] Residue odourless	Ignition loss <5% Gas purification necessary (dioxin, furans)	odor, toxic exhausts (NH_3, CH_4)	Odor, toxic exhausts (NH_3, CH_4)
Residues resulting from the processing	Little amount of residue Shorter time for rotting Residue to be used as fertilizer	Ash as residue	Residue needs large landfill volume	Long time for rotting Residue to be used as fertilizer
Impurities in the residues	Reduction of germs Elimination of infectious agents (Salmonella)	Destruction of germs	Germs unaffected	Savings on fertilizer and energy (about 90 kWh Mg^{-1})
Exhaust	Little greenhouse effect due to CO_2	Greenhouse effect due to CO_2	Stronger greenhouse effect due to CH_4	Little greenhouse effect
Waste water	Proportional to the amount of waste	Water evaporates to the environment	Has to be cleaned	Water remains in the residue
Investment costs	Medium	High	Low	Low

29) Cp. WEB 12
30) According to German laws the ignition loss of waste for landfill has to be <5% – this value can only be attained by combustion.

Table 2.7 Bio waste over the year.

Type of bio waste	*Jan*	*Feb*	*Mar*	*Apr*	*May*	*Jun*	*Jul*	*Aug*	*Sep*	*Oct*	*Nov*	*Dec*
Bio waste from households												
Green waste from gardens and vegetable wastes												
Grass cuttings from lawn, meadows, and waysides												
Grass cuttings from roadsides[a]												
Branches cut from trees and woods												
Branches from woods at roadsides[a]												
Leaves												
Plant waste												
Green for covering												
Bio waste from cemeteries												
Christmas trees												
Creeks and lakes												
Green waste from water												
Waste from cutting of wood												
Bio waste from agriculture												
Liquid manure, excreta												
Crack												
Straw												
Foul hay												
Beet tops												
Bio waste from industry												
Residues from pharmaceuticals production												
Residues from oilseed processing												
Residues from food production												
Spent fruits												

a contaminated with heavy metals

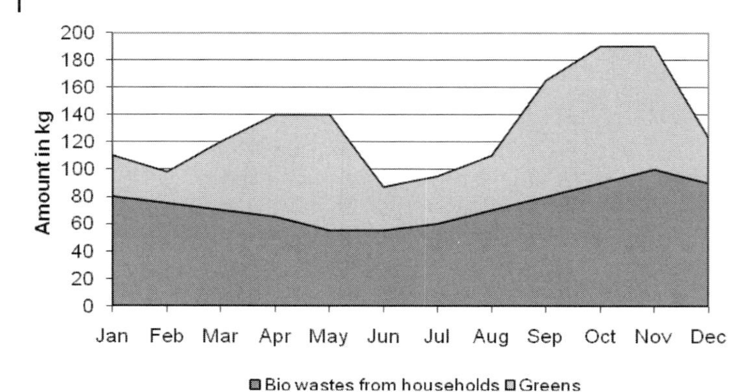

Figure 2.2 Bio waste over the year in a small village with gardens.

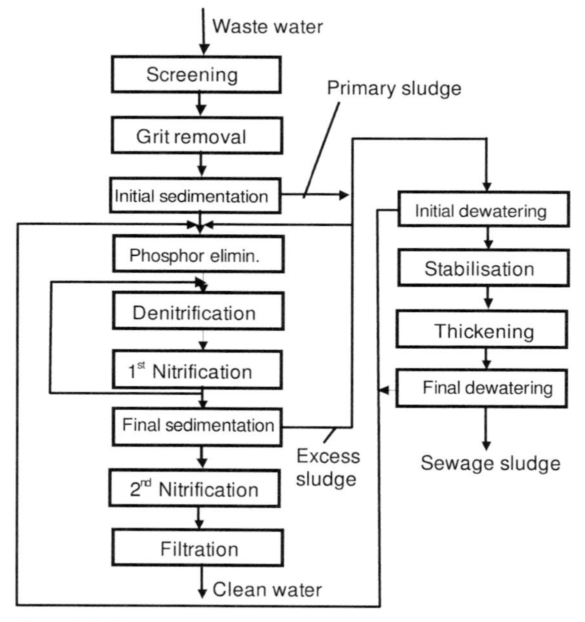

Figure 2.3 Sewage treatment.

2.4
Sewage sludge and co-substrate

The anaerobic degradation of sewage sludge is called digestion, stabilization, or sewage sludge fermentation. Figure 2.3 shows a sewage treatment plant.

The first step in the whole process is the removal of large impurities such as wood, clothing, etc., using a rake. The waste water then passes through the sand trap where it is pre-purified. Fast-sinking particles are separated, and this is

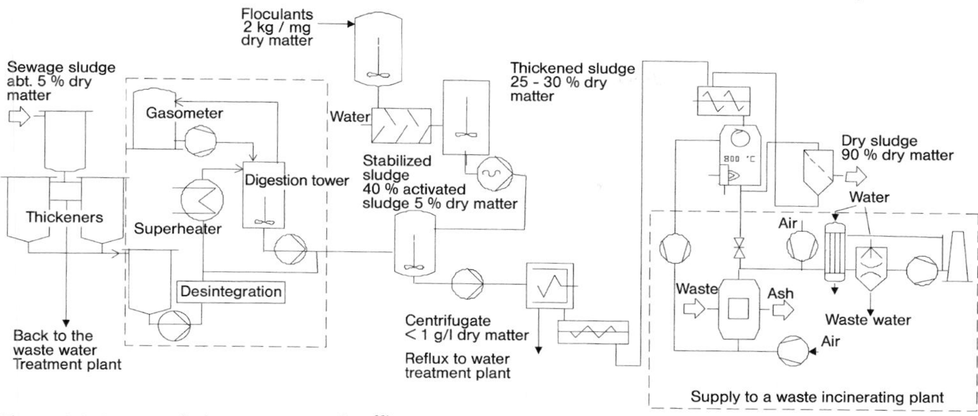

Figure 2.4 Sewage sludge treatment plant[31)]

followed by the segregation of slow-sinking organic particles. Phosphorus compounds, which are quite commonly found in domestic waste water, are separated by precipitation to metal phosphates. In two other basins, nitrogen-fixing and nitrogen-removing bacteria decompose the nitrogen. In the final clarification basin, these bacteria are then separated. The last impurities are removed by a second nitrogen-removing process and a fine filtration at the end of the process.

The sewage sludge from the pre-purifier (primary sludge) and from the final clarification basin (excess sludge) is segregated with pumps, then dehydrated after sedimentation and stabilized by forming biogas (also called sewage gas or fermentation gas). With a second dehydration step and a mechanical coagulation, it is concentrated up to 30% of dry matter. The material composition of the sewage gas depends on both the origin and composition of the waste water and on the mode of operation of the sewage plant. In detail, the sewage sludge treatment for optimal disposal follows the process flowchart shown in Figure 2.4.

First surplus water is decanted in thickeners. Then the sewage sludge is pumped to the digestion tower to

- decrease smell-forming components
- decrease organic sludge solids
- improve the degree of dehydration
- reduce the pathogens
- increase the safety for disposal
- produce sewage gas with anaerobic stabilization.

The fermentation can be improved if the sewage sludge is disintegrated first.

The resulting anaerobic sludge is quite often concentrated to thick sludge by adding flocculating agents in filters or separators. Subsequently, the thick sludge with approximately 30% solids content is thermally dried, increasing the solids

31) Cp. BOK 48

Table 2.8 Biogas yield depending on the applied process engineering according to German regulations.[32] IN includes EGW.

Residence time t_{TS} in the activated sludge tank		Load in the supply to the activated sludge tank		Expected sewage sludge gas amount	
t_{TS}	Process technology	$BOD_5/(IN.d)$	Pretreatment	Variation	Average value
[d]		[g]		[l/(IN.d)]	[l/(IN.d)]
8	Nitrification in summer; sometimes part denitrification[33]	35	enlarged first sedimentation	16.5–25	20.7
15	Max. possible nitrification, denitrification over the year	35	enlarged first sedimentation	14.5–22	18.3
15	Max. possible nitrification; denitrification over the year	48	short first sedimentation; sludge removal	10.5–15.9	13.2
15	Max. possible nitrification; denitrification over the year	60	without first sedimentation	6.2–9.4	7.8
25	Aerobic stabilization	60	without first sedimentation	3.5–5.3	4.4

content to 90%. It is important in the process to quickly pass through the phase in which sludge particles tend to become sticky. Therefore, some dry sludge is added before the dryer to quickly concentrate the material in the high-efficiency mixer to give a solids content of 60–70%.

The dry sludge is used agriculturally as fertilizer. Nowadays it is also quite often burned, e.g., in an incineration plant together with residual waste.

The resulting quantity of sewage gas per inhabitant (IN) is called the population equivalent (EGW). The EGW includes gas from industry and trade as well as domestic gas. Usually the EGW is 1.55 L/IN/d (sewage sludge). With a solid DM content of 50 g/L in the sewage sludge, approximately 78 g DM is produced daily per inhabitant, leading to the quantities of sewage sludge listed in Table 2.8. These quantities are dependent on the mud age and the load in the inlet to the aeration basin. About 18.3 L biogas on average can be generated per inhabitant daily,

32) Cp. LAW 17

33) Cp. LAW 1

Table 2.9 Operation parameters of different sewage sludge digestion tower plants (typical figures).[34)]

Sewage sludge from different sewage treatment plants	*Time for digestion*	*Temperature for digestion*	*Ratio primary/excess sludge for digestion*	*Residence time of the sludge in the activated sludge tank*
	d	°C		D
1	17	38	n.a.	16–17
2	45	37	0.4:1	24
3	20	31	2.9:1	25–26
4	13	38	1.4:1	9
5	28	38–40	3:1	n.a.

provided that the process technology of a plant with a potential of more than 50 000 IN is taken as a base.

Flotated materials from fat removal tanks from poultry and pig slaughter can principally be used as co-substrates[35)] for sewage sludge. The co-substrates should first be added in a slowly increasing manner, then continuously, to allow a quick equilibration of the content of organic dry matter (oDM). The load B_R of the reactor, i.e. the daily supplied quantity of organic material related to the total reactor volume, should not exceed $B_R = 3\,\mathrm{kg_{oDM}/m^3.d}$. The flotates are almost fully decomposed. The pH value remains constant thereby at pH = 7. The gas productivity in the digester can almost be doubled from $V_{FR} = 0.8\ \mathrm{m^3/d\,gas/m^3}$ (only sewage sludge) to $V_{FR} = 1.5\,\mathrm{m^3/d\,gas/m^3}$ if flotates are added. It may be even possible to increase it further.

The plants work in their optimum range if the load of microorganisms is very high (Table 2.9). Such a status is obtained by feedback or immobilization of the microorganisms, e.g., through

- Immobilization of the biomass on carriers fixed in the reactor
- Immobilization of the biomass on free-floating objects in the reactor, which are removed and added back in.
- Segregation of the biomass and feedback.

Immobilized microorganisms proved to be particularly more resistant in the digestion tower than suspended microorganisms. They form connected systems, called flakes, which are large enough to remain on the bottom of the digester without being washed out with the clean water. From time to time they should be removed, however, and disposed of.

34) Cp. BOK 69

35) Cp. WEB 29

2.5
Industrial waste water

A slow but highly loaded stream of waste water allows the production of biogas directly without any concentration to sewage sludge. But it needs to be decided whether this is economic. The organic matter measured as COD value serves as a basis to calculate the yield. According to the chemical equation

$$CH_4 + 2O_2 \rightarrow CO_2 + 2H_2O$$

the consumption of $1\,kg_{COD}$ of oxygen relates to about $0.35\,Nm^3$ methane.[36] This theoretical energy potential of the waste water is not the actual yield, however, since it cannot be fully exhausted. Organic substances which are not yet fully decomposed or not decomposable at all are washed out. Further, part of the organic matter is used for aerobic breathing to ensure the desired quality of the process.

2.6
Waste grease or fat[37]

Waste fat is another source for energy (Table 2.10). Approximately 150 000–280 000 $Mg\,a^{-1}$ of dripping and chip fat can be collected in Germany alone. This quantity of waste fat could be increased by 0.65 kg per inhabitant and year by including the waste oil and waste fat from private households. Taking again Germany as an example, around 330 000 $Mg\,a^{-1}$ of waste fat of excellent quality could be considered for the production of biogas. These fats should not be used as co-substrates, however, but fermented in a separate plant to avoid high costs for maintenance and cleaning.

2.7
Cultivation of algae

Cultures of micro-algae are suitable as substrates for fermentation gas production. Test results with micro-algae suggest higher yields than those from energy-affording plants (C4 plants or energy forests).[38],[39] *Chlorella vulgaris* proved to be a particularly good choice.[40]

Micro-algae provide for their need for carbon for photosynthesis through atmospheric CO_2 and their need for energy from sunlight. They are easy to rear in simple basins. The basins can be set up extensively and can even be placed on

36) 16 kg CH_4 + 64 kg O_2 $\rightarrow$ 44 kg CO_2 + 36 kg H_2O, and 1 kmol CH_4 corresponds to $22.4\,Nm^3$. Therefore $22.4/64\,Nm^3$ CH_4 relates to the consumption of 1 kg O_2

37) Cp. DIS 10

38) Cp. WEB 3

39) Cp. WEB 3

40) Cp. DIS 6

Table 2.10 Specification of waste fats (ffa = free fatty acid).

	FFA [%]			Phosphorus [mg/kg]			Sulfate ash [%]		
	min.	average	max.	min.	average	max.	min.	average	max.
Gastronomy	0.36	1.16	2.59	1.10	93.29	848.70	0.006	0.34	2.630
Butchery	0.07	1.71	3.30	1.72	23.92	56.05	0.008	0.57	2.489
Hotel/canteen	0.16	0.42	0.76	1.95	5.06	10.26	0.003	0.01	0.023
Recycling yard	0.08	0.65	0.89	2.24	10.55	17.83	0.004	0.12	0.510
	Iodine no. [g_{Iod}/100g]			Unsaponifiables [%]			Chlorides [mg/kg]		
	min.	average	max.	min.	average	max.	min.	average	max.
Gastronomy	45	68	115	0.64	0.89	1.15	4.5	3'619.0	1'5494.0
Butchery	49	63	84	0.61	0.78	1.17	40.5	5'001.0	1'2631.0
Hotel/canteen	77	92	108	1.21	1.26	1.33	<0.1	36.0	76.3
Recycling yard	30	75	132	0.92	1.06	1.31	123.2	262.0	548.4

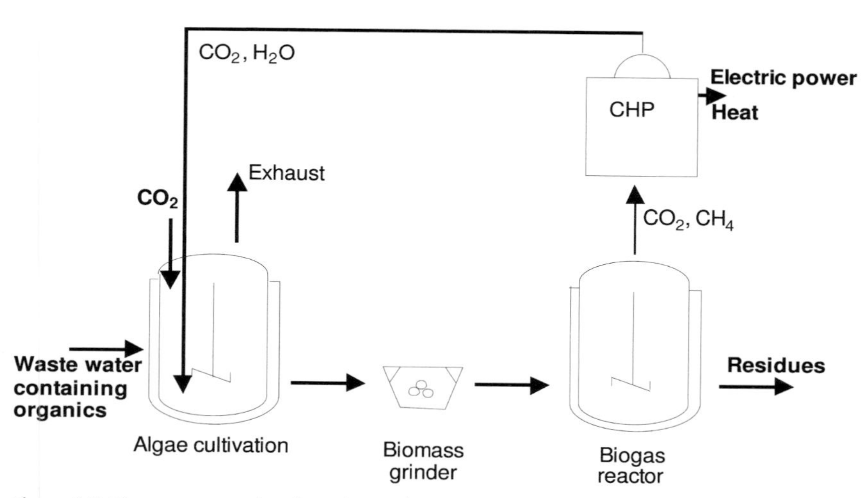

Figure 2.5 Biogas reactor plant based on cultiuation micro-algae.

unused large surfaces, e.g., on roofs, house walls, or also in the sea. Their overall height is only limited by the requirement to have sufficient light. Therefore, the hectare yield can be very high.

An algae reactor plant consists of rearing basins, a biogas reactor, and a CHP (combined heat and power unit) (Figure 2.5). The rearing basins are fed with nutrient-rich waste water, so that the number of micro-algae increases quickly. A part of the micro-algae is continuously removed and pumped to the biogas reactor to produce CH_4 and CO_2 during the methane process. In the CHP the CH_4 is transformed into power. The CO_2 partly already exists in the biogas and partly is generated when burning CH_4. It is fed back into the rearing basins, where it is used again for the growth of new micro-algae. In this way, a small CO_2 cycle is set up which allows reduction of the emission of CO_2 to the environment.

2.8 Plankton

Plankton[41] could also be used to produce methane in large quantities economically. The living conditions in the Arab sea would have to be imitated, which is a good challenge for process engineers. There, high methane concentrations are to be found in the suface water (0 to 200 m depth of water). This methane is formed by bacteria which live in oxygen-free niches outside the plankton and/or by intestinal bacteria inside the plankton animal.

2.9 Sediments in the sea

Methane does not necessarily have to be produced in technical plants. It is to be found in the form of gas hydrates (Figure 2.6) in sediments in the sea.[42],[43],[44]

Gas hydrates are sediments into whose crystal structure gas molecules (e.g., methane, carbon dioxide, hydrogen sulfide) are compactly incorporated. One m^3 gas hydrate can contain up to 164 m^3 gas. Gas hydrates are also named flammable ice.

Gas hydrates develop from dead plants and animal remains which sink to the bottom of the sea and are decomposed on the way and/or at the bottom of the sea by microorganisms. During this decomposition process, so much oxygen is

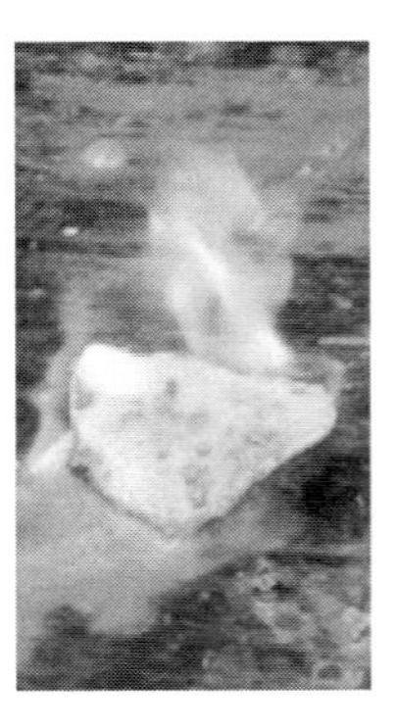

Figure 2.6 Gas hydrate.[45]

41) Cp. WEB 8
42) Cp. JOU 22
43) Cp. WEB 2
44) Cp. WEB 43
45) Cp. WEB 87

extracted from the water that other microorganisms must withdraw their oxygen demand out of sulfate occurring there. The products of their metabolism are hydrogen sulfide and carbon dioxide, which then is converted into methane by methanogenic microorganisms and deposited as gas hydrat.[46]

Gas hydrates are stable at 20 bar and 0 °C. Under atmospheric pressure and temperature they evaporate rapidly. Gas hydrates are therefore to be found only at a depth of water of 500 m and below. At much lower depths, however, higher water temperatures prevail because of the terrestrial heat, so that gas hydrates are not to be found there.

Gas hydrates occur particularly at submarine slopes. The gas hydrate layer is often widespread parallel to the earth's surface, and even below land masses. It is estimated that the total quantity of gas hydrates is approximately twice as large as the quantity of oil and coal together.

Gas hydrates have a great importance for geology, because they solidify the sediments. For meteorology, they are important because they bind enormous quantities of methane and carbon dioxide. Should the sea levels change for any reason, then these carbon compounds will be set free, with immeasurable effects on the climate and the coastal landscapes.

2.10 Wood, straw[47]

Lignocellulose-containing biomasses, as for example wood and straw, are not fermentable in a biogas plant without special pretreatment. They must be disintegrated thermally and/or chemically.

For comparison, the decomposition of three different difficultly-degradable lignocelluloses containing biomasses is shown in Figure 2.7. Maple wood, cereal straw, foliage, and/or leaves have a lignin content of 10–15%. All three biomasses were degraded with the same volume load of $B_R = 10-11\,\mathrm{kg_{COD}/m^3/d}$[48] under thermophilic conditions with a methane content of 60–65% in the biogas.

Today, straw is often burned in fields without any energy recovery. In the future, straw could make a substantial contribution to the power supply if it is fermented in a biogas plant.

The energy-efficient fermentation, particularly of rice straw, seems to be an obvious idea for developing countries. It would be economic and would contribute to environmental protection. Little harmful or unpleasant material would be released. Organic pollutants cannot yet be demonstrated. Table 2.11 indicates the content of heavy metals. Fermentation is to be preferred to burning, because burning straw releases harmful gases, so that the exhaust gas from a straw-burning plant would have to be cleaned extensively.

Silage from straw can be stored without problems, e.g., in a silo with plastic foil cover. After conclusion of the silage (4–6 weeks), it can be used directly in the

46) Cp. WEB 8
47) Cp. JOU 10
48) $\mathrm{kg_{COD}}$ stands for the amount of organic material

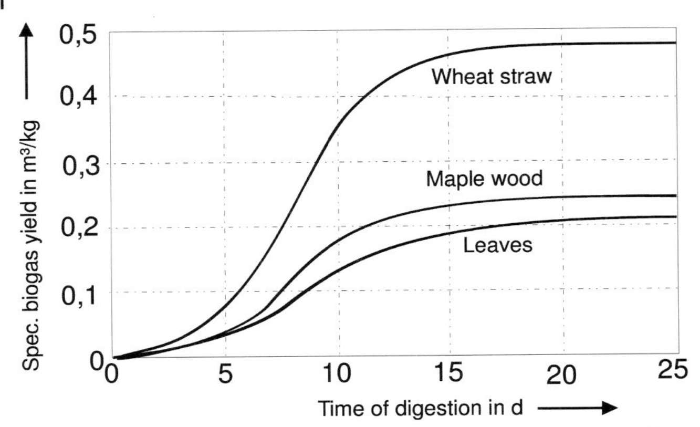

Figure 2.7 Biogas yield in relation to the type of cellulose biomass.[49]

Table 2.11 Features of rice straw.

DM	*oDM*	*Nutrients*			*CH₄ in the gas*	*Heavy metals*					
[%]	*[%]*	*[mg/kg DM]*			*[mg/kg DM]*	*[mg/kg DM]*					
		N	*NH_4*	*P_2O_5*		*Cd*	*Cr*	*Cu*	*Ni*	*Pb*	*Zn*
25–50	70–95	3.5–6.9	6.9–19.8	0.4–0.8	54–55	0.2	1.4	8.1–9.5	2.1	3.9	38–53

biogas plant. The fermentation of silage from straw is possible without any additives. It is, however, preferential to ferment it with liquid manure as co-substrate, since the fermentation runs off more stably and the degradability and/or the methane yield can be increased.

49) Cp. JOU 10

3
Evaluation of substrates for biogas production[50),51)]

The practically attainable methane yield depends on many factors like composition, grain size, and proportions of the assigned substrates, on the microbial degradability of the biomass, the content of dry matter and organic dry matter (oDM), and the relationship of the nutrients to each other. Also, the parameters of the technology of fermentation are of importance, e.g., the number of stages, the temperature, the residence time of the substrate in the bioreactor, the kind and frequency of the mixing of the substrate, and the quantity and frequency of the substrate addition.

These parameters must be analyzed in a laboratory test (Endiometer-test) and in a pilot plant as well before the construction of a production plant (Figure 3.1). In a first simple fermenting test, the basical degradability of a substrate, the graph of the degradation, and the biogas yield have to be determined. Sometimes the maximum recommendable volume load and the changes of the concentrations of certain materials have to be measured. These are important if the large-scale installation is to be continuously operated. In a test, possible and practical substrate mixtures can be determined. Therefore a test station with an agitated bioreactor with a capacity of $V_R = 4-8$ L is used in the laboratory.

A test can be in principle continuous or batchwise. First, the biomass samples must be homogenized, cut up, and diluted. They are then loaded into the laboratory reactor and the temperature and the pH value are adjusted.

Before a large-scale plant is constructed, the results from the laboratory test should be confirmed in a pilot plant with reactors of size more than 50 L. A pilot plant for the preliminary test of the fermentation (see Figure 3.2) consists of, e.g., a hydrolyzer, a methane reactor, and a storage tank. All three vessels should be individually equipped with arrangements for maintaining moderate temperatures, and with filling and cleaning devices. The vessels must be provided with the necessary measuring instruments.

For the tests in the laboratory or in the pilot plant, the following measurements are recommended. The appropriate measuring devices can be fixed to the reactor or installed separately.

- Temperature
- p_H value and redox potential

50) Cp. WEB 77

51) Cp. WEB 78

Biogas from Waste and Renewable Resources. An Introduction.
Dieter Deublein and Angelika Steinhauser

ISBN: 978-3-527-31841-4

Figure 3.1 Test station to assess fermentation progress.

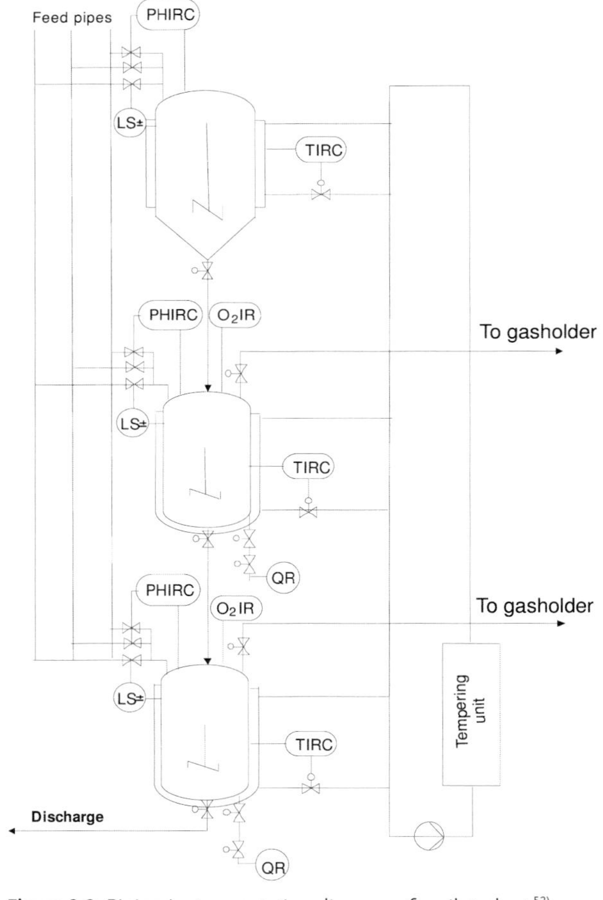

Figure 3.2 Piping instrumentation diagram of a pilot plant.[52]

52) Cp. DIS 1

Table 3.1 Energy release by degradation of biomass (C_{org} = organically bound carbon C_1, C_2 = short chain carbonic acids (C_1-, C_2-molecules) EG = released energy[53] as a % of the total energy content; ΔE_o = difference between the standard redox potentials in mV for pH = 7).

Type of degradation	*Biochemical reactions*		*EG%*	*$\Delta E_0 mV$*
Aerobic	Glucose ($C_6H_{12}O_6$) + $6O_2$	→ $6CO_2 + 6H_2O$	100	1230
Anaerobic				
Denitrification	Glucose ($C_6H_{12}O_6$) + 12NO	→ $6CO_2 + 6\ N_2 + 6H_2O$	94	950
Desulfurication	Glucose ($C_6H_{12}O_6$) + 6SO	→ $6CO_2 + 6\ H_2S$	15	250
Alcohol fermentation	Glucose ($C_6H_{12}O_6$) + C_{org}	→ Alcohols,reducing acids + C_{org}+H_2O	7	190
Methane production	Acetate C_1, C_2 + CO_2	→ $CO_2 + CH_4 + H_2O$	5	190–260

- dry matter, water content
- Content of organic dry matter (ignition loss)
- Degradability as total content of organic acids/acetic acid equivalent and inhibitors
- Salt content
- Total content of nitrogen (N), phosphorus (P), potassium (K), magnesium (Mg), sulfur (S), p_H-decreasing components
- Availability of plant nutrients such as nitrate nitrogen (NO_3), ammonium nitrogen (NH_4), phosphate (P_2O_5), potassium oxide (K_2O), and magnesium (Mg)
- Granulation (maximum grain size), gross density
- Heavy metals like lead (Pb), cadmium (Cd), chromium (Cr), copper (Cu), nickel (Ni), zinc (Zn), mercury (Hg)
- Content of short-chain fatty acids, principally acetic acid, propionic acid, butyric acid, and *iso*-butyric acid
- C/N ratio

The measurements give, among other things, information on the biogas yield, the nutrients which can be expected, the extent of decomposition of the biomass which can be provided during the fermentation, the fertilization value of the residue, and also the preferable type, dimensions, and mode of operation of the production plant.

53) Also named Gibbs Free Energy.

4
Benefits of a biogas plant

Like natural gas, biogas has a wide variety of uses, but, as it is derived from biomass, it is a renewable energy source. There are many other benefits to be derived from the process of converting substrates in a biogas plant.

- The economic pressure on conventional agricultural products always continues to rise. Many farmers are forced to give up their occupation, since their land no longer brings sufficient yield. However, the production of biogas is subsidized in many countries, giving the farmer an additional income. For the farmer, biogas production does not mean major reorientation, because microorganisms for methanation require similar care to that needed for livestock in the stable.
- With the present tendency for farms to become large-scale enterprises and with the widespread abandonment of agricultural areas, the cultural landscape is changing. Biogas production from corn or grass could contribute to the maintenance of the structure of the landscape with small farmyards.
- Biomasses that are not needed are often left to natural deterioration, but energy can be generated from these biomasses (Table 4.1).
- With aerobic degradation, the low-energy compounds CO_2 and H_2O are formed at least, i.e., much energy is lost to air – about twenty times as much as with an anaerobic process. In the case of the anaerobic degradation metabolism products of high-energy (e.g., alcohols, organic acids, and, in the long run, methane) result, which serve other organisms as nutrients (alcohols, organic acids) or are energetically used (biogas).
- Reduction of landfill area and the protection of the groundwater: the quantity of organic waste materials can be reduced down to 4% sludge when the residue is squeezed off and the waste water from the biogas plant is recycled into the waste water treatment plant (Figure 4.1).

Biogas from Waste and Renewable Resources. An Introduction.
Dieter Deublein and Angelika Steinhauser

ISBN: 978-3-527-31841-4

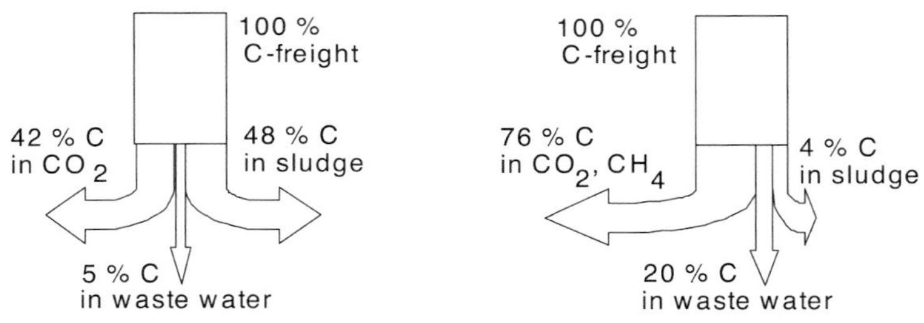

Figure 4.1 Biomass degradation with aerobic (left) and anaerobic (right) processing.[54]

- Substantial reduction of the disposal costs of organic wastes, even including meaningful re-use (e.g., as fertilizers), because the quantity of biomass decreases so significantly.
- If plants are used as co-substrates for biogas production and the residues are recycled to agriculture, no mineral fertilizer need be bought. A cycle of nutrients is reached. Nitrate leaching is reduced. Plant compatibility and plant health are improved.
- When storing the residue from the fermentation plant much less odor – methane and nitrous oxide emissions (N_2O) – are released with unfermented substrates: odor-active substances and the organic acids are reduced, giving a volume contraction of $c=0.5\,g\,L^{-1}$. The ammonium content, which is high compared with untreated liquid manure, and the higher pH value can lead, however, to higher ammonia emissions.
- CO_2-neutral production of energy (especially electrical power and heat) is achieved. The climatic protection goal agreed upon in the Kyoto minutes is thereby effectively supported.

CO_2 equivalent: Methane is a climatically active trace gas in the terrestrial atmosphere and contributes to the greenhouse effect.[55],[56] The content of methane in the atmosphere has doubled since the last ice age to $1.7\,mL\,m^{-3}$ at present. This value has remained constant in recent years. Methane contributes ca. 20% to the anthropogenic greenhouse effect. Among sources of methane caused by humans, more than 50% is from cattle farming and up to 30% is from the cultivation of rice. The mean decay time of CH_4 in the atmosphere (8 years) is much shorter than that of CO_2 (50–200 years). Efforts to decrease methane formation could therefore be effective much earlier. In order to be able to compare the effect of different greenhouse gases, each one is allocated a factor which represents a measure of its greenhouse effect in comparison to the "guidance gas" CO_2 (Table 4.1). The CO_2 equivalent of greenhouse gases can be calculated by multiplication

54) Cp. BOK 20
55) Cp. BOK 76
56) Cp. BOK 67

of the relative greenhouse effect potential with the mass of the respective gas. It indicates the quantity of CO_2 that would produce the same greenhouse effect in 100 years; i.e., CH_4 is a worse greenhouse gas than CO_2 by a factor of 21.

- Reduction of the acidification of soils, because acids are degraded.

SO_2 equivalent: Similarly to the CO_2-equivalent, the potential for acidification of gases is determined. The SO_2 equivalent of these gases indicates what amount of SO_2 develops the same acidifying effect (Table 4.2).

- Enhancement of the fertilizer quality: since during the fermentation, carbon is degraded, the C/N ratio narrows itself, e.g., from liquid manure. Thus, the nitrogen becomes better available and its effect better calculable. Fermented, dehydrated inorganic sewage sludge possesses a material composition favorable for plants ($N:P_2O_5:K_2O=9:3:1$, C/N ratio=15 : 1). By the use of residues, up to 34% higher yields can be obtained, and the time of maturation can be shortened (Table 4.3). Because of its favorable composition, degraded liquid manure is applicable not only for the fertilization of grassland, but also as a fertilizer for sensitive cultures. But the fertilization of vegetables (field vegetables, soft fruits, spice herbs, and tea) with liquid manure is not allowed. The application of the fermented liquid manure usually takes place by hauling hoses. This careful treatment is necessary in order to avoid a possible loss of ammonia by degassing.

Table 4.1 Relative greenhouse effect[57] of greenhouse gases.

CO_2	1
CH_4	21
N_2O	310
SF_4	23 900
PFC	6 500–9 200
HFC	140–11 700

Table 4.2 SO_2-equivalent.[58]

SO_2	1
NO_x	0.696
HF	1.601
HCl	0.878
H_2S	0.983
NH_3	3.762

57) Cp. BOK 22

58) Cp. BOK 13

Table 4.3 Degradability of liquid manure in %.[59]

Substrate	*Degradation of organic substances*	*Degradation of organic acids*	*N-content (NH_4-N) from N_{ges}*	*PH value*
Liquid manure from pigs	54	83	70	7.7
Liquid manure from pigs and waste fats	56	88	65	7.8
Liquid manure from dairy cattle	24	68	50	7.9
Liquid manure from cattle (bulls)	30/52	n. a.	47/74	8
Liquid manure from poultry	67	n. a.	85	8.2

- Possible avoidance of a sewer connection and fees for the disposal of the waste water, particularly with rural dwellings.
- Substantial reduction and/or complete inactivation of pathogenic germs in the finished biomass. In the generally usual mesophilic temperature range, hygienization is of subordinate importance. Fermented dehydrated sewage sludge is germ-free (number of coliforms reduced to 0.2%), since the bacteria in the digestion tower over-grow and displace all other microorganisms, bacteria, and protozoa.
- Elimination of weed seeds: The longer weed seeds remain in the liquid manure, the more they are eliminated. Therefore the growth of weeds and the spread of undesired plants in the fields is minimized.
- Liquid manure is more highly liquid after fermentation in a biogas plant: thus it possesses better distribution features and penetrates better into the soil. The latter decreases loss of N-containing gases.

59) Cp. WEB 107

Part III
Formation of biogas

Biogas from Waste and Renewable Resources. An Introduction.
Dieter Deublein and Angelika Steinhauser

ISBN: 978-3-527-31841-4

1 Biochemical reaction

The formation of methane[1] from biomass follows in general the equation:

$$C_cH_hO_oN_nS_s + y\,H_2O \rightarrow x\,CH_4 + n\,NH_3 + s\,H_2S + (c-x)CO_2$$

$$x = 1/8\cdot(4c + h - 2o - 3n - 2s)$$

$$y = 1/4\cdot(4c - h - 2o + 3n + 3s)$$

The products include, for example, the following:

Carbohydrates:	$C_6H_{12}O_6 \rightarrow 3CO_2 + 3CH_4$
Fats:	$C_{12}H_{24}O_6 + 3H_2O \rightarrow 4.5CO_2 + 7.5CH_4$
Proteins:	$C_{13}H_{25}O_7N_3S + 6H_2O \rightarrow 6.5CO_2 + 6.5CH_4 + 3NH_3 + H_2S$

Because the sulfur remains in the residue and part of the CO_2 binds to NH_3, the result in general is a biogas composition of

$$CH_4 : CO_2 = 71\% : 29\%$$

The ratio of CO_2 to CH_4 is determined by the reduction ratio of the organic raw material. During the fermentation of glucose ($C_6H_{12}O_6$), CH_4 and CO_2, for example, develop in the ratio 1 : 1, since only if this is so is the balance of the redox values fulfilled: glucose has a reduction ratio of +24,[2] CH_4 of +8, CO_2 of 0.

The energy balance can be calculated as follows:

Organic material, which is built up by photosynthesis

$$CO_2 + H_2O \rightarrow CH_2O + O_2$$

contains the energy.

1) Cp. BOK 4

2) Chemically: $C_6H_{12}O_6$; C- molar: $C_1H_2O_1$; molecular weight C-molar: $12 + 2 + 16 = 30$; degree of reduction: $4 + 2 - 2 = 4$

Biogas from Waste and Renewable Resources. An Introduction.
Dieter Deublein and Angelika Steinhauser

ISBN: 978-3-527-31841-4

$$\text{Carbon dioxide} + \text{water} + \text{solar engergy} \rightarrow \text{carbohydrates} + \text{oxygen}$$
$$(-394\,\text{kJ}) + (-237\,\text{kJ}) + (\text{free energy}\Delta G_f/\text{mol}) \rightarrow (-153\,\text{kJ}) + 0\,\text{kJ}$$
$$\Delta G_f = 478\,\text{kJ}\,\text{mol}^{-1} \text{ with } p_H = 7$$

The degradation of organic material to biogas

$$CH_2O \rightarrow 0.5CH_4 + 0.5CO_2$$

results in the following release of energy:

$$\text{Carbohydrate} \rightarrow 0.5 \cdot \text{methane} + 0.5 \cdot \text{carbon dioxide}$$
$$(-153\,\text{kJ}) \rightarrow 0.5 \cdot (-51\,\text{kJ}) + 0.5 \cdot (-394\,\text{kJ})$$
$$\Delta G_f = -70\,\text{kJ}\,\text{mol}^{-1}$$

Combusting methane CO_2 and H_2O are developed, which can serve for the photosynthesis

$$0.5CH_4 + O_2 \rightarrow 0.5CO_2 + H_2O$$

and the energy circle is closed.

$$0.5 \cdot \text{methane} + \text{oxygen} \rightarrow 0.5 \cdot \text{carbon dioxide} + \text{water}$$
$$0.5 \cdot (-51\,\text{kJ}) + 0\,\text{kJ} \rightarrow 0.5 \cdot (-394\,\text{kJ}) + (-237\,\text{kJ})$$
$$\Delta G_f = -408\,\text{kJ}\,\text{mol}^{-1}$$

Overall not one of the substances is thus enriched or lost.

478 kJ kmol^{-1} of carbon can be won as free energy by the degradation of 1 kmol glucose to CO_2 and H_2O, while burning 0.5 kmol methane results in 408 kJ kmol^{-1}. The energy content of glucose is thus 85% retained in the methane produced.

As the energy balances show, very little heat is released with the anaerobic bioreaction. Therefore bioreactors must be heated and well insulated.

The energy, which is set free when burning biomass, corresponds theoretically overall to the energy set free in the biogas production plus the energy set free burning methane. This sum is equal to the energy which was needed for photosynthesis. However, the heat energy which is produced during the biogas production, is not completely used, and the conversion to biogas is not complete. The volume of biogas which can practically be won from substrates results out of:

- Fraction of material with a high energy content in the organic mass
- Content of organic dry material (oDM) in the entire dry biomass (DM)

- DM content of the substrate
- Methane content of the biogas
- Actual grade of decomposition of the respective biogas plant.

So, for example, for an average plant substance with 25% DM and 90% oDM, we can assume 123 m^3 biogas yield per Mg fresh substrate with a methane content of 60% in a biogas plant which is able to degrade 70% of the organic matter. For a substrate with the same consistency but abt. 50% fat in the organic matter, 200 m^3 biogas per Mg substrate can be achieved with the same methane content, provided the same biogas plant is used.

2
Biology

Methane fermentation is a complex process, which can be divided up into four phases of degradation, named hydrolysis, acidogenesis, acetogenesis, and methanation, according to the main process of decomposition in this phase (Figure 2.1). The individual phases are carried out by different groups of microorganisms, which partly stand in syntrophic[3] interrelation and place different requirements on the environment.

In principle, the methane formation follows an exponential equation. The course of the biogas production (the biogas yield: $\dot{V}_{BR}[m^3 d^{-1}]$) can be theoretically described by the following equation, where C_1 and C_2 are constants.

$$\dot{V}_{BR} = C_1 \cdot (1 - e^{-C_2 \cdot t_{BR}})$$

2.1
Bioreactions

The first and second as well as the third and fourth phase are linked closely with each other. Therefore, one can accomplish the process well in two stages. In both stages the rates of degradation must be equal in size.

If the first stage runs too fast, the CO_2 portion in the biogas increases, the acid concentration rises and the pH value drops below 7.0. Acidic fermentation is then also carried out in the second stage. If the second stage runs too fast, methane production is reduced. There are still many bacteria of the first stage in the substrate. The bacteria of the second stage must be inoculated.

With biologically difficultly degradable products, the hydrolytic stage limits the rate of degradation. In the second stage, the acetogenesis possibly limits the rate of decomposition.

3) Syntrophic: some species of microorganisms acting together degrade certain compounds of substrates which they cannot degrade on their own, e.g., *Nitrosomonas* and *Nitrobacter*

Biogas from Waste and Renewable Resources. An Introduction.
Dieter Deublein and Angelika Steinhauser

ISBN: 978-3-527-31841-4

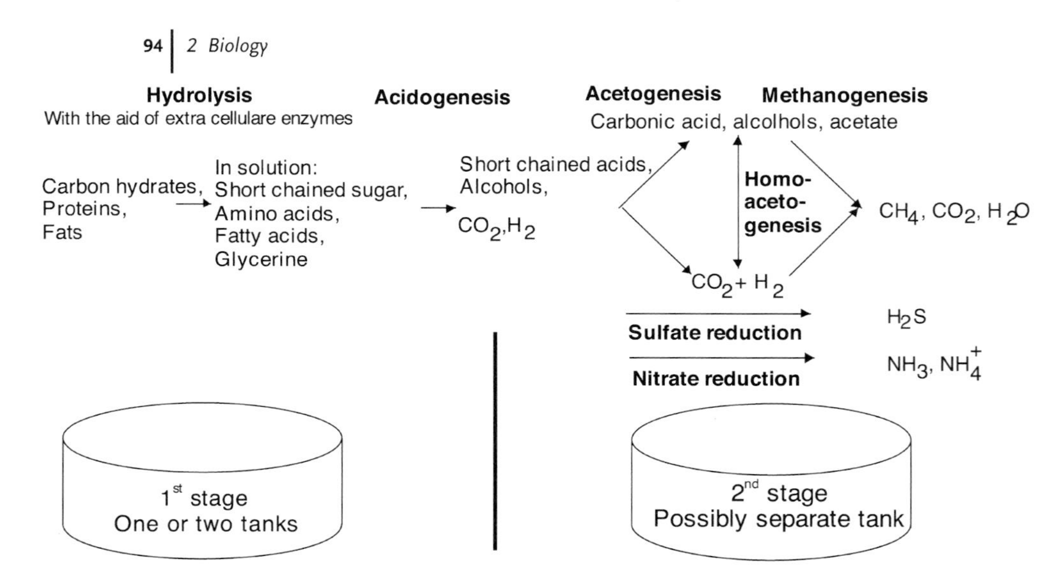

Figure 2.1 Biochemistry of the methane gas production.

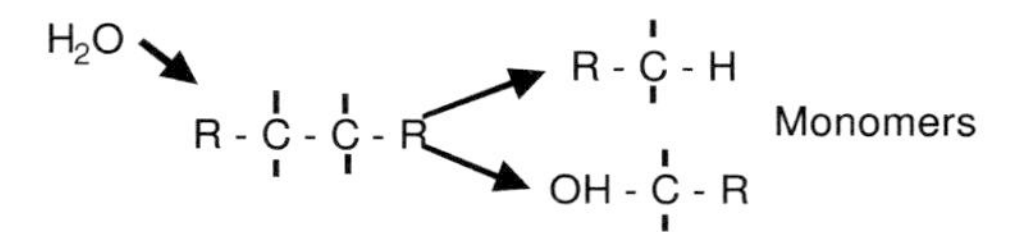

Figure 2.2 Formation of monomers (R-Rest).

2.1.1 Hydrolysis

In the first phase (the hydrolysis), undissolved compounds, like cellulose, proteins, and fats are cracked into monomers (water-soluble fragments) by exoenzymes (hydrolase) of facultative and obligatorily anaerobic bacteria. Actually, the covalent bonds are split in a chemical reaction with water (Figure 2.2).

The hydrolysis of carbohydrates takes place within a few hours, the hydrolysis of proteins and lipids within few days. Lignocellulose and lignin are degraded only slowly and incompletely.

The facultative anaerobic microorganisms take the oxygen dissolved in the water and thus cause the low redox potential necessary for obligatorily anaerobic microorganisms.

2.1.2 Acidogenic phase

The monomers formed in the hydrolytic phase are taken up by different facultative and obligatorily anaerobic bacteria and are degraded in the second, the acidogenic phase, to short-chain organic acids, C1–C5 molecules (e.g., butyric acid, propionic acid, acetate, acetic acid), alcohols, hydrogen, and carbon dioxide. The concentra-

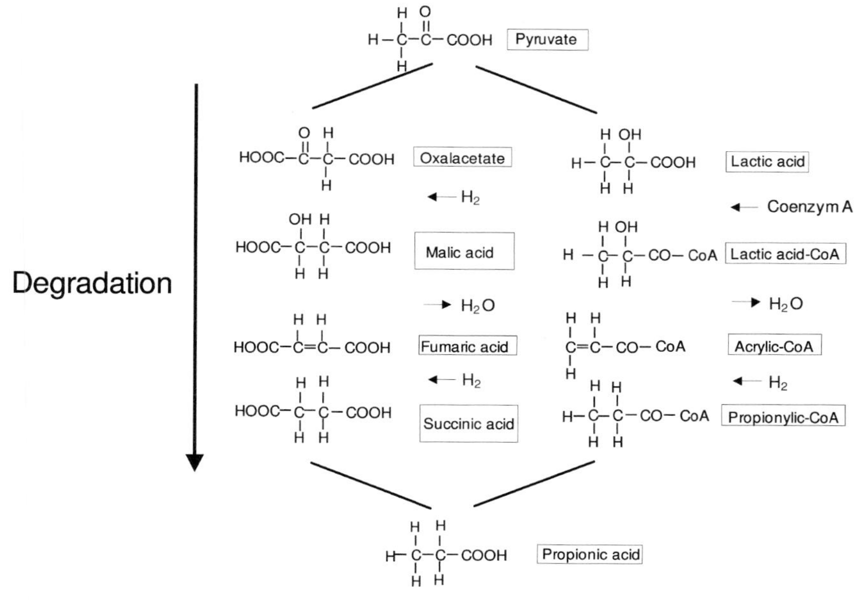

Figure 2.3 Degradation of pyruvate.

tion of the intermediately formed hydrogen ions affects the kind of the products of fermentation. The higher the partial pressure of the hydrogen, the fewer reduced compounds, like acetate, are formed.

The pathways of degradation are as follows:

a. Carbohydrates:

Formation of propionic acid by propioni bacterium via the succinate pathway and the acrylic pathway (Figure 2.3)

Formation of butyric acid (butyric acid pathway) above all by clostridium

Acetic acid → 2-hydroxy butyrate → *trans*-2-butenic acid → butyric acid → butanol (Figure 2.4)

b. Fatty acids:

These are degraded e.g. from acetobacter by β-oxidation. Therefore the fatty acid is bound on Coenzyme A and then oxidizes stepwise, as with each step two C atoms are separated, which are set free as acetate.

c. Amino acids:

These are degraded by the Stickland reaction by *Clostridium botulinum* taking two amino acids at the same time – one as hydrogen donor, the other as acceptor – in coupling to acetate, ammonia, and CO_2. During splitting of cysteine, hydrogen sulfide is released.

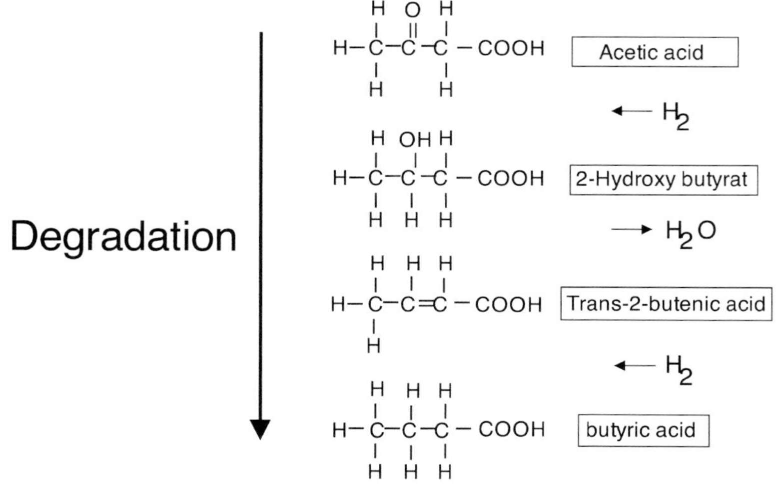

Figure 2.4 Degradation of acetic acid via the butyric acid pathway.

Table 2.1 Acetogenic degradation.

Substrate	***Reaction***
Propionic acid	$CH_3(CH_2)COOH + 2H_2O \rightarrow CH_3COOH + CO_2 + 3H_2$
Butyric acid	$CH_3(CH_2)_2COO^- + 2H_2O \rightarrow 2CH_3COO^- + H^+ + 2H_2$
Valeric acid	$CH_3(CH_2)_3COOH + 2H_2O \rightarrow CH_3COO^- + CH_3CH_2COOH + H^+ + 2H_2$
Isovaleric acid	$(CH_3)_2CHCH_2COO^- + HCO_3^- + H_2O \rightarrow 3CH_3COO^- + H_2 + H^+$
Capronic acid	$CH_3(CH_2)_4COOH + 4H_2O \rightarrow 3CH_3COO^- + H^+ + 5H_2$
Carbondioxid/hydrogen	$2CO_2 + 4\ H_2 \rightarrow CH_3COO^- + H^+ + 2H_2O$
Glycerine	$C_3H_8O_3 + H_2O \rightarrow CH_3COOH + 3H_2 + CO_2$
Lactic acid	$CH_3CHOHCOO^- + 2H_2O \rightarrow CH_3COO^- + HCO_3^- + H^+ + 2H_2$
Ethanol	$CH_3(CH_2)OH + H_2O \rightarrow CH_3\ COOH + 2H_2$

2.1.3
Acetogenic phase

The products from the acidogenic phase serve as substrate for other bacteria, those of the acetogenic phase.

The acetogenic reactions (Table 2.1) are endergonic. With the degradation of propionic acid are needed $\Delta G_f' = +76.11\,\text{kJ mol}^{-1}$, and with the degradation of ethanol $\Delta G_f' = +9.6\,\text{kJ mol}^{-1}$.[4)]

In the acetogenic phase, homoacetogenic microorganisms constantly reduce exergonic H_2 and CO_2 to acetic acid.

$$2CO_2 + 4H_2 \leftrightarrow CH_3COOH + 2H_2O$$

4) Cp. JOU 29

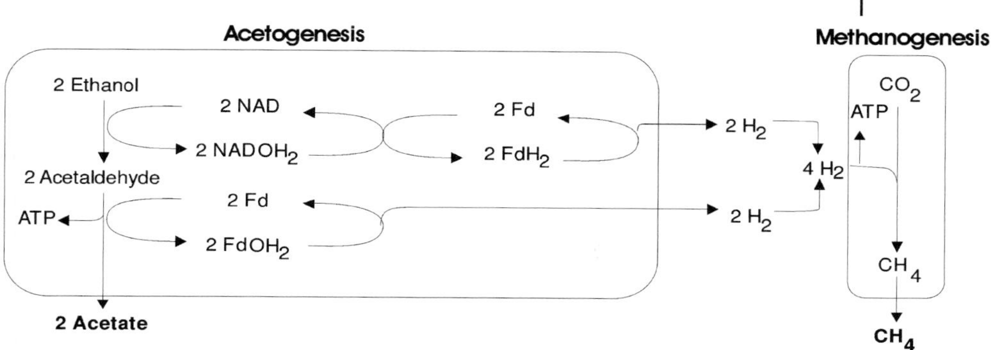

Figure 2.5 "Interspecies hydrogen transfer", as for example in a *Methanobacterium omelanskii* culture.[5)]

Acetogenic bacteria are obligatory H_2 producers. The acetate formation by oxidation of long-chain fatty acids (e.g., propionic or butyric acid) runs on its own and is thus thermodynamically possible only with very low hydrogen partial pressure. Acetogenic bacteria can get the energy necessary for their survival and growth, therefore, only at very low H_2 concentration.

Acetogenic and methane-producing microorganisms must therefore live in symbiosis. Methanogenic organisms can survive only with higher hydrogen partial pressure. They constantly remove the products of metabolism of the acetogenic bacteria from the substrate and so keep the hydrogen partial pressure, p_{H2}, at a low level suitable for the acetogenic bacteria.

When the hydrogen partial pressure is low, H_2, CO_2, and acetate are predominantly formed by the acetogenic bacteria. When the hydrogen partial pressure is higher, predominantly butyric, capronic, propionic, and valeric acids and ethanol are formed. From these products, the methanogenic microorganisms can process only acetate, H_2, and CO_2.

About 30% of the entire CH_4 production in the anaerobic sludge can be attributed to the reduction of CO_2 by H_2, but only 5–6% of the entire methane formation can be attributed to the dissolved hydrogen. This is to be explained by the "interspecies hydrogen transfer" (Figure 2.5), by which the hydrogen moves directly from the acetogenic microorganisms to the methanogenics, without being dissolved in the substrate.

The anaerobic conversion of fatty acids and alcohols goes energetically at the expense of the methanogenics, where these, however, in return, receive the substrates (H_2, CO_2, acetic acid) needed for growth from the acetogenic bacteria.

The acetogenic phase limits the rate of degradation in the final stage. From the quantity and the composition of the biogas, a conclusion can be drawn about the activity of the acetogenic bacteria.

5) Cp. JOU 6

At the same time, organic nitrogen and sulfur compounds can be mineralized to hydrogenic sulfur by producing ammonia.

The reduction of sulfate follows for example the stoichiometric equations below. Sulfate-reducing bacteria such as *Desulfovibrio, Desulfuromonas, Desulfobulbus, Desulfobacter, Desulfococcus, Desulfosarcina, Desulfonema* and *Desulfotomaculum* participate in the process, which uses the energy released by the exergonic reaction.

$$SO_4^{2-} + CH_3COOH \rightarrow HS^- + CO_2 + HCO_3^- + H_2O$$

$$SO_4^{2-} + 2CH_3CHOHCOOH \rightarrow HS^- + 2CH_3COOH + CO_2 + HCO_3^- + H_2O$$

2.1.4
Methanogenic phase

In the fourth stage, the methane formation takes place under strictly anaerobic conditions. This reaction is categorically exergonic. As follows from the description of the methanogenic microorganisms, all methanogenic species do not degrade all substrates. One can divide substrates acceptable for methanogenesis into the following three groups:[6)]

CO_2 type: CO_2, $HCOO^-$, CO
Methyl type: CH_3OH, CH_3NH_3, $(CH_3)_2NH_2^+$, $(CH_3)_3NH^+$, CH_3SH, $(CH_3)_2S$
Acetate type: CH_3COO^-

The reactions[7)] are shown in Table 2.2.

The pathway for the formation of methane from acetate and/or CO_2 in microorganisms is to be seen in Figure 2.6. Long-chain hydrocarbons are involved such as methanofuranes (e.g. R-$C_{24}H_{26}N_4O_8$) and H_4TMP (tetrahydromethanopterin) as

Table 2.2 Methanogenic degradation.

Substrate type	*Chemical reaction*	$\Delta G_f'$ *(kJ mol⁻¹)*	*Methanogic species*
CO_2-Type	$4H_2 + HCO_3^- + H^+ \rightarrow CH_4 + 3H_2O$	−135.4	All species
	$CO_2 + 4\,H_2 \rightarrow CH_4 + 2H_2O$	−131.0	
CO_2-Type	$4HCOO^- + H_2O + H^+ \rightarrow CH_4 + 3HCO_3^-$	−130.4	Many species
Acetate	$CH_3COO^- + H_2O \rightarrow CH_4 + HCO_3$	−30.9	Some species
Methyl type	$4CH_3OH \rightarrow 3CH_4 + HCO_3^- + H^+ + H_2O$	−314.3	One species
Methyl type	$CH_3OH + H_2 \rightarrow CH_4 + H_2O$	−113.0	
e.g. Methyl type: ethanol	$2CH_3CH_2OH + CO_2 \rightarrow CH_4 + 2CH_3COOH$	−116.3	

6) Cp. WEB 74

7) Cp. JOU 5

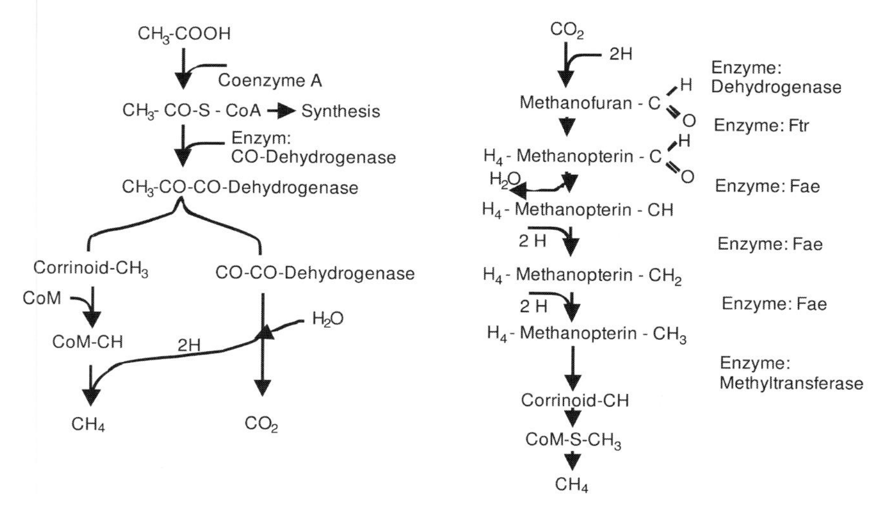

Figure 2.6 Methane formation from acetate[8] (left) and from carbon dioxide (right) (CoA = Coenzyme A, CoM = Coenzyme M).

Co-factors. Corrinoids are molecules which have four reduced pyrrole rings in a large ring and can be represented by the empirical formula $C_{19}H_{22}N_4$.

When the methane formation works, the acetogenic phase also works without problems. When the methane formation is disturbed, overacidification occurs.

Problems can occur when the acetogenic bacteria live in symbiosis instead of with a methanogenic species with other organisms, using H_2. In waste water technology, symbioses can occur with microorganisms which reduce sulfate to hydrogen sulfide. Therefore they need hydrogen and compete with the methanogenics. The methanogenics get less feed and form less methane. Additionally, hydrogen sulfide affects the methanogenics toxically.[9]

All methane-forming reactions have different energy yields.

The oxidation of acetic acid is, in comparison to the reduction of $CO_2 + H_2$, only a little exergonic:

$$CH_3COOH \leftrightarrow CH_4 + CO_2 \text{ at a } \Delta G^0 = -31\,\text{kJ/kmol}$$

$$CO_2 + 4\,\text{BADH}/H^+ \leftrightarrow CH_4 + 2H_2O + 4NAD^+ \text{ at a } \Delta G^0 = -136\,\text{kJ/kmol}$$

Nevertheless, only 27–30% of the methane arises from the reduction, while 70% arises from acetate during methanation. This also is the case in sea sediments.

Acetate-using methanogenics like *Methanosarcina barkeri, Methanobacterium söhngenii*, and *Methanobacterium thermoautotrophicum* grow in acetate very slowly, theoretically with a regeneration time of at least 100 h, whereas CO_2 has turned

8) Cp. BOK 19

9) Cp. BOK 28

Table 2.3 Environmental requirements.

Parameter	*Hydrolysis/acidogenesis*	*Methane formation*
Temperature	25–35 °C	Mesophilic: 32–42 °C Thermophilic: 50–58 °C
pH value	5.2–6.3	6.7–7.5
C:N ratio	10–45	20–30
DM content	<40% DM	<30% DM
Redox potential	+400 to −300 mV	<−250 mV
Required C:N:P:S ratio	500 : 15 : 5 : 3	600 : 15 : 5 : 3
Trace elements	No special requirements	Essential: Ni, Co, Mo, Se

out to be essential for the growth. When a substrate which is rich in energy can be used, as, for example, methanol or methylamine, then the generation time is lower (40 h with *Methanosarcina* on methanol). However, the theoretically given generation times can be substantially longer under real conditions.

2.2 Process parameters

With all biological processes, the constancy of the living conditions is of importance. A temperature change or changes in the substrates or the substrate concentration can lead to shutdown of the gas production. It can last up to three or even more weeks, until the ecological system has adapted to the new conditions and starts biogas production again without any intervention from outside. But in the case of human interference it can take a further three weeks.

The microbial metabolism processes are dependent on many parameters (Table 2.3), so that, for an optimum fermenting process, numerous parameters must be taken into consideration and be controlled. Furthermore, the environmental requirements of the fermentative bacteria, by which the hydrolysis and acidification of the substrates occur, differ from the requirements of the methane-forming microorganisms.

Optimum environmental conditions for all microorganisms involved in the degradation can only be set in a two-stage plant with one stage for hydrolysis/acidification and one stage for acetogenenis/methanation. Provided that the complete degradation process has to take place in the same reaction system (1-stage process), the environmental requirements of the methanogenics must be fulfilled with priority, because these would otherwise have no chance of survival within the mixed culture because of their lower growth rate and higher sensitivity to environmental factors. But the following divergences from this rule are to be taken into consideration.[10]

10) Cp. BOK 36

- With lignocellulose-containing substrates, the hydrolysis limits the process and therefore needs higher priority.
- With protein-containing substates, the pH optima are the same in both stages, so that a single-stage plant is quite sufficient.
- With fats, the hydrolysis proceeds more rapidly with increasing emulsification (bioavailability), so that the acetogenesis is limiting. A thermophilic catabolism of fat is to be preferred.

2.2.1
Parameter: hydrogen partial pressure

A narrow spatial symbiosis is necessary for an undisturbed process between the H_2-producing acetogenic bacteria and the H_2 consuming-methanogenics.

If generally a biological reaction is to take place, the reaction must be exergonic; i.e., the free energy must be negative. The hydrogen concentration should be well balanced: on the one hand methanogenics need enough hydrogen for the methane production. On the other hand the hydrogen partial pressure must be so low that acetogenic bacteria are not surrounded by too much hydrogen and consequently stop the hydrogen production. The maximum acceptable hydrogen partial pressure depends on the species of bacteria and also on the substrates.

For the anaerobic conversion of propionate (salt of the propionic acid) via acetic acid and hydrogen/carbon dioxide to methane, this energetic window is especially small, as Figure 2.7 shows. The degradation of propionic acid can be taken as a measure of the productivity of the plant, because this decomposition is often the limiting factor of an anaerobic fermentation in practice.

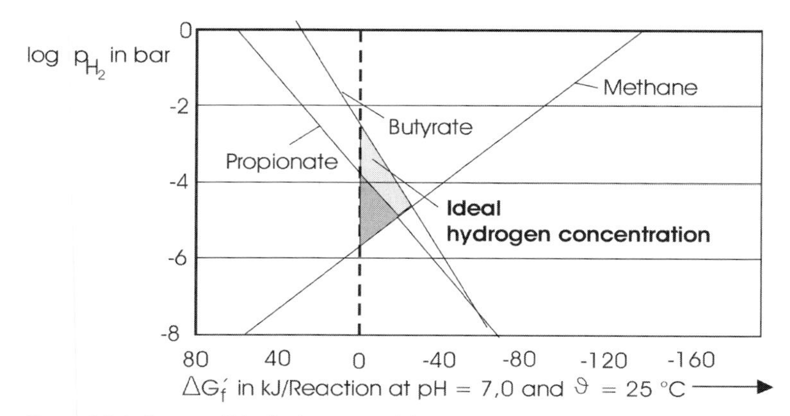

Figure 2.7 Influence of the hydrogen partial pressure p_{H2} on the energy release $\Delta G_f'$ during acetogenesis and methane formation from carbon dioxide and the hydrogen.

Table 2.4 Time of regeneration of different anaerobic microorganism[11] in comparison to aerobic MO

Anaerobic microorganism	*Time of regeneration*
Acidogenic bacteria	
Bacterioids	<24 h
Clostridia	24–36 h
Acetogenic bacteria	80–90 h
Methanogenic bacteria	
Methanosarcina barkeri	5–16 d
Methanococcus	ca. 10 d
Aerobic microorganism	
Escherichia coli	20 min
Active sludge bacteria	2 h
Bacteria living on earth	1–5 h

2.2.2 Parameter: concentration of the microorganisms

Methanogenic microorganisms have a long regeneration time in general (Table 2.4).

To avoid washing out from the reactor, hydraulic residence times must be at least 10–15 days with reactor systems which do not have facilities for retaining and returning biomass.

In comparison with this, the regeneration times of hydrolytic and acid-forming bacteria are significantly shorter, so that with them there is hardly any risk of washout.

The low growth rate of the methanogenics means that for biogas plants a relatively long start-up phase of up to 3 months is required, because the amount of inoculating sludge necessary to start the plant at full capacity immediately is mostly not available and has to be built up in the starting phase.

2.2.3 Parameter: type of substrate

The substrate determines the rate of the anaerobic degradation and must be taken into consideration in the process technology and process operation. If a substrate component of vital importance runs out, the microorganisms stop their metabolism. Therefore, it is often necessary to feed possibly lacking substances (carbohydrates, fat, proteins, mineral substances, and trace elements) as well as the substrate.

11) Cp. BOK 20

Sugar, for example, hydrolyzes and acidifies within very short time. The degradation of cellulose proceeds considerably more slowly, depending on the fraction in the form of lignin. At long residence times of 20 and more days, even medium-heavy and heavy degradable materials hydrolyze and are eventually metabolized to methane.

According to the composition of the substrates, intermediate products of the decomposition can also limit or inhibit the degradation. Thus, for example, the degradation of fats can give rise to fatty acids, which limit the further degradation. With the decomposition of proteins, methane fermentation can be restrained by the formation of ammonia and hydrogen sulfide.

2.2.4 Parameter: specific surface of material

To support a biochemical reaction a material surface[12] as big as possible is necessary. The material surface often varies proportionally to the square of the particle size. In order to increase the material surface, comminution of the biomass is in many cases recommended before fermentation.

Figure 2.8 clearly demonstrates the advantages of comminution in an agitated ball mill (RWKM) for biogas production. The degradation process is accelerated in the first few days as a result of the mechanical treatment and the biogas yield for the whole time of digestion is higher. Already in the first few days the biogas development is more vigorous and the resulting difference of the biogas yield is maintained until the end of the degradation, here after 25 days.

Comminution does not have a big influence on the biogas yield when easily degradable materials (95% or 88% grade of decomposition) are used, which have only a very low content of structural materials (cellulose, lignin, etc.). Such material is easy accessible for the microorganisms (Figure 2.9).

With substrates like hay and foliage, which are rich in structured materials and enable only a grade of decomposition of ca. 50% without comminution, the biogas

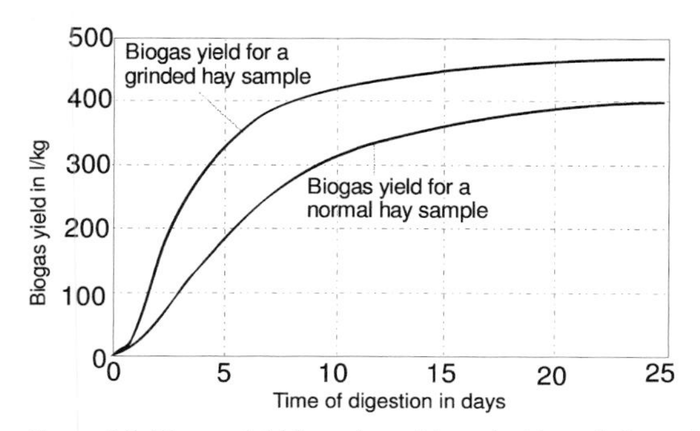

Figure 2.8 Biogas yield from hay with and without being grinded in an agitated ball mill.

12) Cp. JOU 24

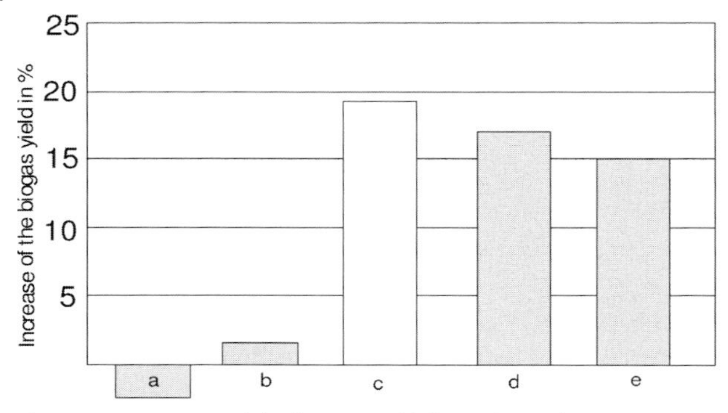

Figure 2.9 Increase of the biogas yield through grinding (a: mix of apples, potatoes and carrots; b: meat; c: sunflower kernels; d: hay; e: leaves).

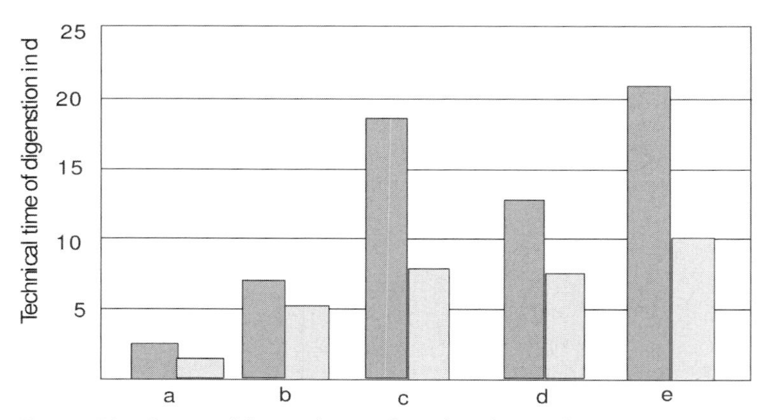

Figure 2.10 Influence of the grinding on the technical time of fermentation for different substrates (a: mix of apples, potatoes, and carrots; b: meat; c: sunflower kernels; d: hay; e: leaves) left columns: without grinding, right columns: with grinding.

yield can be increased by up to 20% depending on the grade of comminution (Figure 2.10). The increase in the particle surface area and the loosening of the fiber structure are the reasons. With sunflower kernels, for example, the destruction of the outer layer of the kernels is decisive.

The technical residence time is the time, at which 80% of the maximum biogas yield is reached. A tendency can be clearly seen, in the case of all substrates, for the technical residence time to be reduced as a result of comminution. The best reduction is achieved with substrates which have the longest technical residence time without comminution. The technical residence time of different substrates can be equalized by comminution.

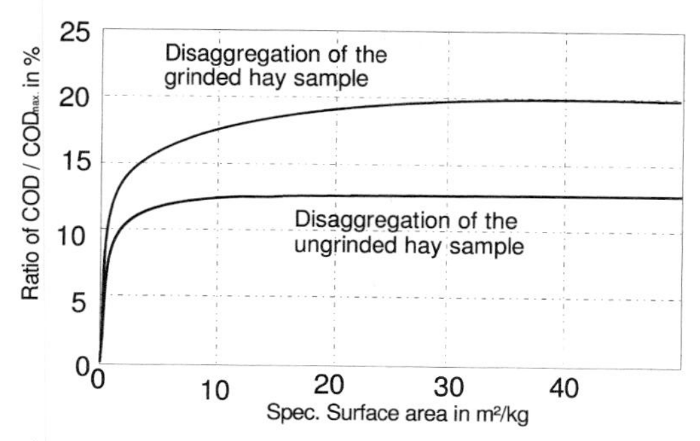

Figure 2.11 Influence of the specific surface area on the release of organic acids within 10 min (given as organic matter measured as COD).

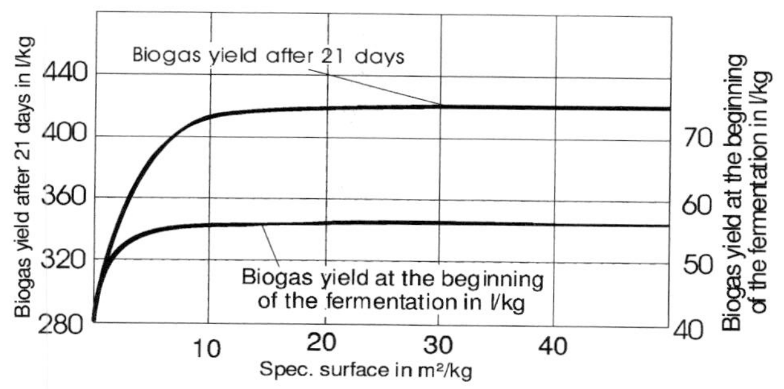

Figure 2.12 Influence of the specific surface on the biogas yield $v_{g,zu}$ after 21 days of fermentation and at the beginning of the fermentation.

The degree of decomposition can be defined as the ratio of the actual chemical oxygen demand (COD) to the maximum achievable COD (COD_{max}), corresponding to the chemical oxygen demand for the entire degradation of all organic components in the sample, as measured by a test in which hay haulms of different length were fermented once comminuted and once not comminuted (Figure 2.11).

It can be seen that, with all hay haulm fractions, there is a relatively high proportion of quickly dissolvable substances. Even with the untreated fraction, more than 10% of the organic material changes into the dissolved form. Figure 2.12 shows the biogas yield at the beginning of the fermentation and after 21 days as a function of the specific surface for hay fractions.

The bigger the specific surface of the biomass, the higher is the biogas yield, but the relationship is not linear. The comminution of fine particles contributes less than the comminution of big particles.

It can be assumed that the increase in the specific surface increases especially the microbiological degradation processes and not the physical solution processes, which would start after a short time.

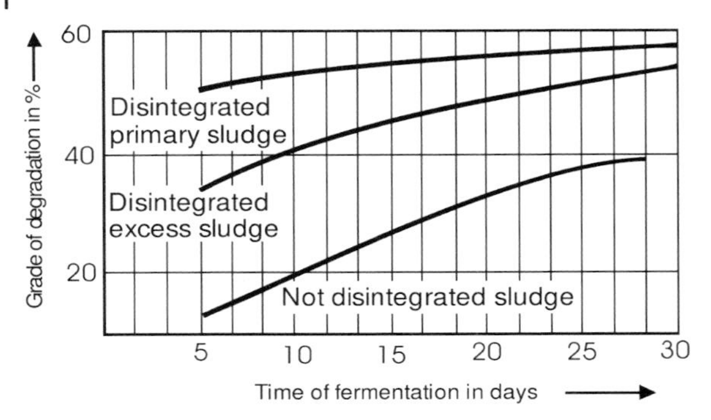

Figure 2.13 Influence of disintegration on the degradability of different sewage sludges.

2.2.5 Parameter: disintegration

The destruction of the cell structure and, with higher energy impact, even of the cell walls, is called "disintegration" or "cell disruption". There are many reasons for adding disintegration devices to a normal biomass fermentation plant, but there are also some reasons against it. Today, th disintegration is mainly recommended for sewage gas production and is applied sporadically.

Reasons for the application of the disintegration are discussed below.

- *Disintegration increases the degree of decomposition and decreases the amount of sewage sludge.*

The disintegration brings, above all, advantages with biomasses that are difficult to destroy (Figure 2.13). For example, excess sludge can be degraded more easily, effectively around 10–30%, with a residence time of 15 days in the bioreactor. This is due to the fact that the time for hydrolysis is shortened as a function of sludge age and the fraction of facultative anaerobic microorganisms. The influence of the disintegration is more important if shorter degradation times must be used.

The dry matter of sewage sludge can be reduced to 50%, because parts of it are diluted. In total, the content of organic dry matter in the sludge is slighty increased.

- *Disintegration increases the biogas yield.*

The biogas yield of sewage sludge can be increased to abt. 350–375 $L\,kg^{-1}$ oDM by disintegration (Figure 2.14).

If degraded sludge is held for 1 h at 70 °C and then further degraded for another 15 days, an increase in the biogas yield of 25% can be achieved. When decomposing at an elevated temperature of 55 °C, after thermal disintegration an increase in the gas yield of up to 50% is even possible.

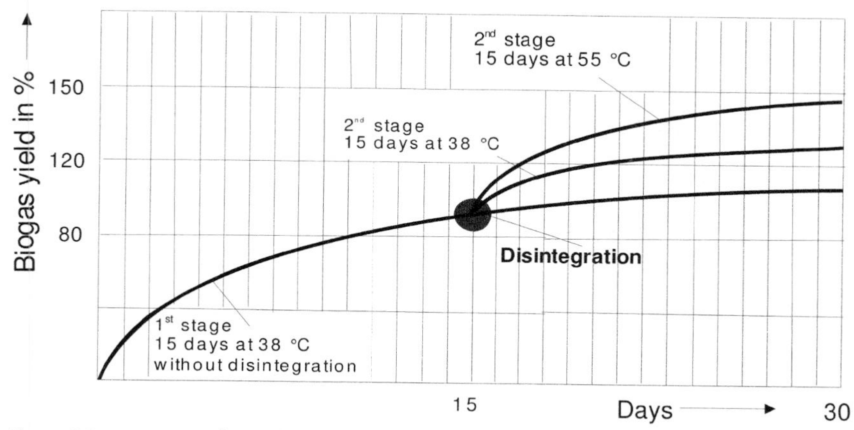

Figure 2.14 Biogas produced from 1 kg organic matter.

Table 2.5 Rate of denitrification of different substrates.

H sources acting as electron donors for denitrification	*Rate of denitrification [Mg NOx-N/h]*
Disintegrated sewage sludge	3.0–4.8
Synthetic waste water according to DEV L24	4.0
Communal waste water	4 0
Acetic acid	3.9

- *The products of disintegration can serve as a hydrogen source or electron donor for the denitrification of waste water.*[13]

With the excess water of disintegrated sewage sludge, the same or even higher velocities of denitrification can be achieved in comparison to using acetate as the hydrogen source (Table 2.5) for waste water treatment.

- *Disintegration lowers the viscosity of the sludge.*

On degradation of the biomass the viscosity decreases of the factor of 30. The heat transfer and the mixing are thereby improved considerably. Less attachment to the heating surfaces is observed.

- *The sedimentation behavior of the sewage sludge is improved.*

The sedimentation behavior of the sludge is improved by up to 70%. Hence, disintegration is to be recommended, especially with bulking sludge, because it sediments with difficulty because of its high content of filamantous microorganisms.

13) Cp. DIS 8

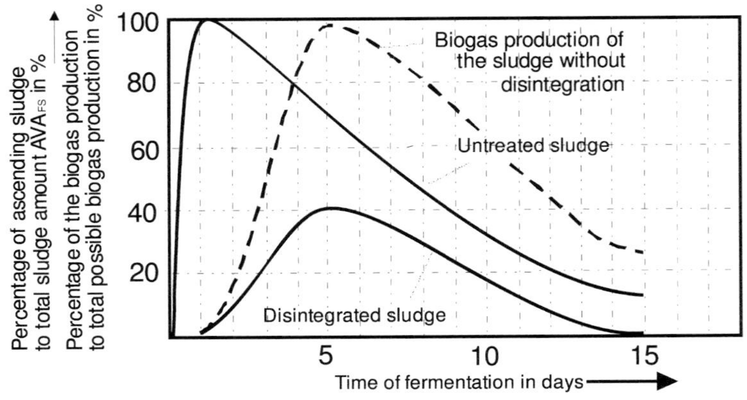

Figure 2.15 Percentage in the bioreactor of ascending sludge AVA_{FS} and biogas production depending on disintegration.

- *The formation of floating sludge can be considerably reduced and even sometimes completely avoided.*

Particularly with the degradation of sewage sludge, it often happens that all the sludge ascends as floating layer to the surface. Then the percentage of the swimming sludge to the total amount of sludge is $AVA_{FS}=100\%$. The ascent of sludge can be prevented by disintegration (Figure 2.15). Then the value rises to only $AVA_{FS}=40\%$ after the first five days after feeding the bioreactor because of strong gas development and then decreases rapidly with decrease of the biogas production.

- *Foaming in the bioreactor (caused by filamentous organisms from the activated sludge tank) can be combated by disintegration.*

If the formation of foam is caused through a high proportion of filamentous microorganisms in the excess sludge, the voluminous flake structure is destroyed by the disintegration, and therefore the possibility that gas bubbles are attached to the flakes is reduced. The amount of floating sludge in the digestion tank can be reduced.

Disadvantages of disintegration are discussed below.

- *Disintegration affects the dehydratability, and the demand on flocculants is thus increased.*

The sludge flakes are chopped up by the disintegration and are more difficult to dehydrate, because, among other things, cell-internal polysaccharides are released which are difficult to degrade because of their complex structure. Therefore, after disintegration and degradation, on average ca. 40% more flocculants are needed for dehydration.

- *Disintegration increases the back load of the water treatment plant with nitrogen, carbon, heavy metals, and phosphates.*

Table 2.6 Influence of the disintegration upon the dehydratability and back load after an anaerobic fermentation.

Sample		*Not disintegrated*	*Mechanically disintegrated*	*Average/average relative change*
Degradability A_s	%	0	43–57	53%
Time of fermentation	d	13–21	13–20	17 d
Degree of degradation	%	28–34	33–49	+19%
Requirement for flocculants (polymers)	g/kg	5.0–6.5	6.7–9.0	+42%
Residual dry matter	g/kg	58–154	61–145	–6%
TKN nitrogen content, after Kjeldahl.	mg L^{-1}	156–680	204–822	+28%
Organic matter after filtration	mg L^{-1}	35–253	46–283	+38%
PO_4-P	mg L^{-1}	24–251	20–208	–9%

As a result of the improvement of the degradation by the disintegration of proteins released from the cells, an increased nitrogen back load of aprox. 30% is to be anticipated with the sludge pumped back to the waste water facilities, because ammonium remains as an end product of the protein degradation in the water (see Table 2.6).

The back load of carbon compounds is raised to more than 40%, and the BOD_5 increases by 30–40%.

Immediately after the disintegration, heavy metals are to be measured in the liquid phase. During the subsequent sludge stabilization the heavy metals are adsorbed again in the matrix of the sludge. The increased degradation of organic matter causes the concentration of the heavy metals in the residue to rise.

The back load on the sewage treatment plant with phosphate is different. As long as enough cations exist for precipitating reactions, there is no increased back load with phosphates in consequence of mechanical disintegration.

- *Disintegration increases the filtration resistance.*

After degradation, the decomposed sludge has to be dehydrated. This can be achieved by belt-type presses or chamber filter presses. Depending on the process of disintegration, the filtration is more or less impaired.

- *Disintegration increases the power consumption of the complete sewage treatment plant.*

In general but not always[14], the power consumption increases with increasing degree of disintegration (Figure 2.16), of course more or less depending on the process of disintegration and particularly on the heat developed in the process. A nearly complete disintegration (>95%) of microorganisms has now been achieved

14) Cp. WEB 1

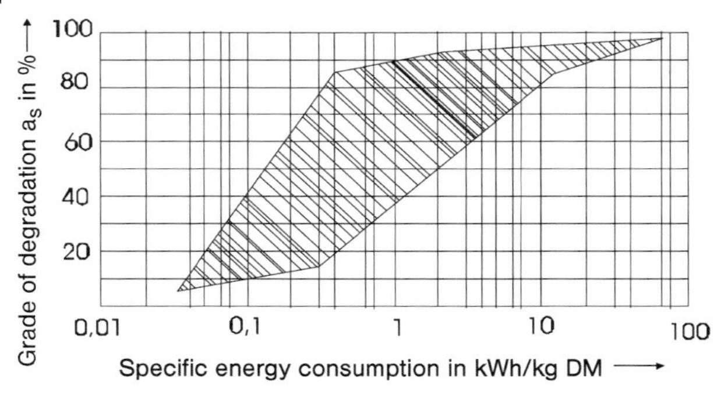

Figure 2.16 Energy consumption during disintegration.[15]

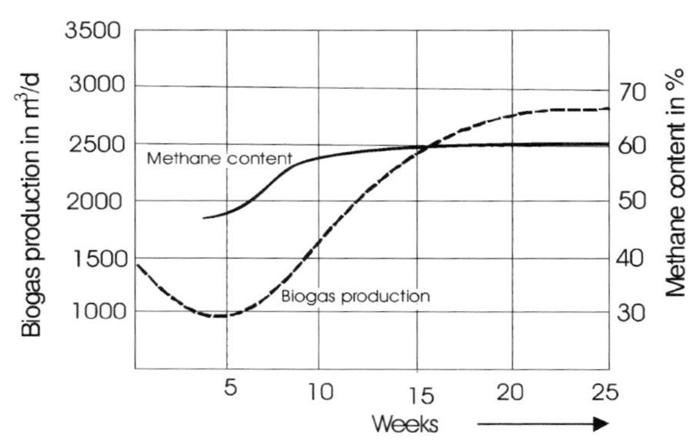

Figure 2.17 Gas production and methane content of the gas during start-up.

by means of agitated ball mills, ultrasonic treatment, and a high-pressure homogenizer.

- *Disintegration leads to considerable erosion or corrosion problems.*

For disintegration, the substrates have to be stressed, either by high temperatures, chemical or enzymatic attacks, high stress factors, or other means. These forces also affect the machines and containers and shorten their life span.

2.2.6
Parameter: cultivation, mixing, and volume load

The start-up phase of an anaerobic plant often lasts 2–4 months (Figure 2.17).

The start-up of the plant usually has a long-term effect on the biocenosis. The start up even can fail completely; i.e., the bioreaction does not work and biogas is not produced. To avoid these problems, the hydrolyzer and the methane reactor

15) Degree of disintegration A_s = decrease in BOD due to breakdown of microorganisms relative to the BOD of untreated sewage sludge

are often inoculated with anaerobic sludge from other fermentation plants. The mixing of the reactor has to be carried out very carefully[16]:

- On the one hand, each individual microorganism must be supplied evenly with nutrients and the metabolism products have to be removed evenly also, which can be achieved by smooth agitation. On the other hand, both movements can be blocked, e.g., by a layer of H_2 around the microorganism, which has to be destroyed by heavy agitation.
- Fresh substrate may have to be mixed with degraded substrate in order to inoculate the fresh substrate with active bacteria.
- The biogas must be removed effectively from the reactor.
- The symbiosis from acetogenic and methanogenic microorganism must not be disturbed.
- The microorganisms are actually stress-sensitive and can be destroyed by too strong agitating.
- Foaming due to a too intensive gassing can be prevented by adequate agitation.
- Temperature gradients in the bioreactor result in lower reaction efficiency.
- Floating and sinking layers have to be destroyed.
- The energy consumption must be minimized
- Solids must be prevented from reaching the discharge, as they must be filtered out afterwards.

Usually, a careful but intensive mixing action is choosen.

The volume load depends particularly on the temperature, the organic dry matter in the substrate, and the residence time (Figure 2.18).[17]

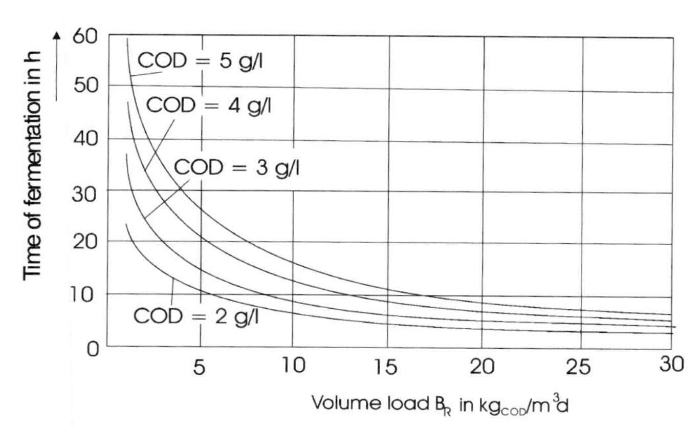

Figure 2.18 Maximum load of the reactor depending on the time of fermentation at a given concentration.

16) Cp. BOK 33

17) Cp. DIS 3

With more than 12% solids in the substrate, gas production is impaired. For economic reasons, the solid content should not exceed 30% DM, because

- a too low water content retards any cell growth
- the material transfer within the substrate becomes a limiting factor
- the biomass cannot be pumped or mixed any longer.

With too low a load, on the other hand, the process works, but it does not work economically, because too much water is passed.

In order to avoid a locally excessive volume load, the bioreactor should be fed frequently, e.g., twice daily or continuously.

For economic reasons, biogas reactors are designed so that 75% of the maximum degradable organic matter is actually decomposed.

2.2.7 Parameter: light

Light is not lethal for methanogenics, but severely inhibits the methanation. The methane formation should therefore take place in absolute darkness.

2.2.8 Parameter: temperature

The temperature shows two optima for acidifying bacteria; a smooth one at abt. 32–42 °C for mesophilic microorganisms and a sharp one at 48–55 °C for thermophilic microorganisms (Figure 2.19).

Most of the methanogenic microorganisms belong to the mesophilics. Only a few are thermophilic. A few others are able to produce methane even at low temperatures (0.6–1.2 °C), e.g., on the surface of permafrost soils. In laboratory tests, methane formation could be proven also with temperatures below freezing, i.e.

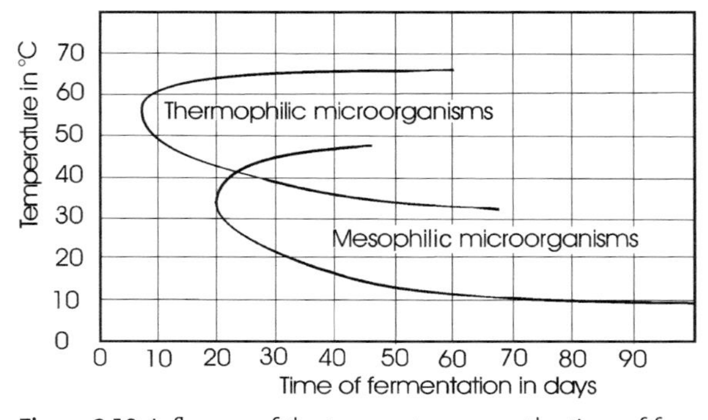

Figure 2.19 Influence of the temperature upon the time of fermentation.

down to −3 °C.[18] In general, the lowest temperature at which microorganisms grow, is −11°C. Below −25 °C, even the enzyme activity succumbs. Methanogenics are sensitive to to rapid changes of temperature. Thermophilic methanogens are more temperature-sensitive than mesophilics. Even small variations in temperature cause a substantial decrease in activity. Therefore, the temperature should be kept exactly within a range of +/−2 °C. Otherwise, gas losses of up to 30% have to be taken in consideration. Particularly critical for mesophilics are temperatures in the range of 40–45 °C, because in that range they lose their activity irreversibly.

Under mesophilic operating conditions, the inhibition of ammonium is reduced because of the lower content of inhibiting free ammonia.

In general, it has to be menshioned that the energy balance is better in the mesophilic range than in the thermophilic range.

The thermophilic mode of operation results in ca. 50% higher rate of degradation, and, particularly with fat-containing materials, a better microbial availability of the substrates and thus a higher biogas yield.

Epidemics and phytopathogenic germs are inactivated by higher process temperatures, so that special hygienic procedures are not necessary when using a temperature >55 °C and a material retention time of >23 h.

Oxygen is less soluble in the thermophilic temperature range, so that the optimal anaerobic operating conditions are reached more quickly.

In many two-stage plants, therefore, different temperatures are applied at the two stages. There are good reasons to drive the methanation thermophilically and the hydrolysis mesophilically. But, depending on the substrate, it can also be favorable to operate the hydrolysis at higher temperatures than the methanation.

2.2.9
Parameter: pH

The equation below gives the relationship between pH and hydrogen ion concentration in mol mL^{-1}:

$$H^{+} = 10^{-pH}$$

Water with a hydrogen ion concentration of $10^{-6}\,mol\,L^{-1}$ or $10^{-4}\,g\,L^{-1}$ has, for instance, a p_H value of 6.

Because of the hydrogen transportation by NAD, different products of fermentation are developed: the H^+ ions isolated from the substrate are carried over to the uncharged NAD. The NAD molecules so charged ($NADH + H^+$) regenerate (oxidize), by forming H_2 molecules:

$$NADH + H^{+} \rightarrow H_2 + NAD^{+} \quad \Delta G^\circ = +18.07\,kJ\,mol^{-1}$$

18) Cp. WEB 37

This reaction occurs independently of the hydrolysis and acidification of hydrocarbon and proteins. Hydrocarbons are easier to acidify, and no pH-buffering ions are released as with the degradation of proteins. Therefore the pH value decreases more easily. With the degradation of carbohydrates, the partial pressure of hydrogen increases more easily, as with other substances. This happens in combination with the formation of reduced acidic intermediate products.

Even when the hydrolysis and the acidification occur in different aparatuses and are separated from the methanation, complete suppression of the methanation is almost impossible.

The pH optimum of the methane-forming microorganism is at pH=6.7–7.5. Therefore, it is important to adjust the pH-value in the second stage higher than that in the first stage of a two-stage biogas plant. Only *Methanosarcina* is able to withstand lower pH values (pH=6.5 and below). With the other bacteria, the metabolism is considerably suppressed at pH<6.7.

If the pH value sinks below pH=6.5, then the production of organic acids leads to a further decrease of the pH value by the hydrolytic bacteria and possibly to cessation of the fermentation. In the reality, the pH-value is held within the neutral range by natural procedures in the fermenter. Two buffering systems ensure this.

A too strong acidification is avoided by the carbon dioxide/hydrogen carbonate/carbonate buffer system. During the fermentation, CO_2 is continuously evolved and escapes into air. With falling pH value, more CO_2 is dissolved in the substrate as uncharged molecules. With rising pH value, the dissolved CO_2 forms carbonic acid, which ionizes. Thus, hydrogen ions are liberated.

$$CO_2 \leftrightarrow H_2CO_3 \leftrightarrow H^+ + HCO_3^- \leftrightarrow 2H^+ + 2CO_3^{2-}$$

At pH=4 all CO_2 is as free molecules; at pH=13 all CO_2 is dissolved in the form of carbonate in the substrate. The center around which the pH value swings with this system is at p_H=6.5. At a concentration of 2.5–5 g L^{-1}, hydrogen carbonate gives particularly strong buffering.

A too weak acidification is avoided by the ammonia-ammonium buffer system (Figure 2.20). With falling pH value, ammonium ions are formed with release of hydroxyl ions. With rising pH value, more free ammonia molecules are formed.

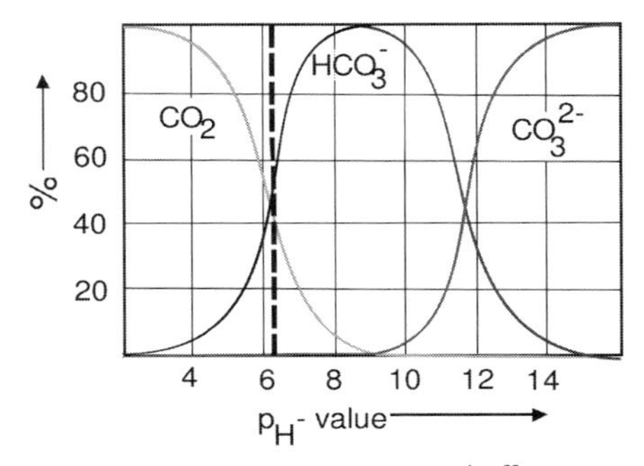

Figure 2.20 Ammonia-ammonium buffer system.

$$NH_3 + H_2O \leftrightarrow NH_4^+ + OH^-$$

$$NH_3 + H^+ \leftrightarrow NH_4^+$$

The center, around which the pH value swings with this system, is at pH = 10.

Both buffering systems can be overloaded by a feed of particularly rapidly acidifying waste water or organic material, by toxic substances, by a decrease in temperature, or by a too high volume load in the bioreactor; e.g., by feeding waste water out of a starch processing plant, which incurs the possibility of acetic acid toxification. Consequences are:[19)]

- increase in the amount of uncharged fatty acid molecules – this leads sometimes to increase in the hydrogen content in the substrate and CH_4 production, sometimes to the detriment of the methanation
- inhibition of the methanation by increase in the proportion of unhydrolyzed inhibitors, e.g., sulfide
- rise of the pH value due to degradation from sulfate to H_2S
- inhibition of reactions by rise in the proportion of free ammonia.

A drop in the pH-value and a rise of the CO_2 in the biogas is an indication of a disturbance of the fermentation process. A first sign of the acidification is the rise of the propionic acid concentration. Measures for the prevention of excessive acidification are:

- Stoppage of the substrate supply, so that the methanogenic bacteria are able to degrade the acid
- Reduction of the organic space load (increase of the residence time)
- Increase of the buffering potential of the substrate by addition of selected co-substrates, in particular if the buffering potential of the substrate is small. It must be taken into consideration that the buffering potential changes because of the removal of the CO_2
- Continuous removal of the acids
- Addition of neutralizing substances: milk of lime (CaO, $Ca(OH)_2$), sodium carbonate (Na_2CO_3), caustic soda solution (NaOH)
- Addition of diluting water
- Emptying and restarting the fermenter.

As a result of the feed of special caustic solutions for adjusting the pH-value or of the addition of cleaning and disinfecting agents, values of pH > 10 can arise in the reactor, which will lead to an irreversible loss of the activity of the bacteria.

19) Cp. BOK 35

Table 2.7 C/N ratio of organic wastes.[20]

Waste	*DM content*	*Organic substances % of DM*	*C/N ratio*
Straw	ca. 70	90	90
Waste from sawmills	20–80	95	511
Paper	85–95	75	173
Waste from housholds	40–60	40	18
Sewage sludge	0.5–5	60	6

Cleaning and disinfecting agents should therefore be tested for their inhibiting potential before their first application in the plant.

2.2.10
Parameter: redox potential

In the bioreactor, low redox potentials are necessary; e.g. monocultures of methanogenics need between −300 and −330 mV as an optimum. In addition, the redox potential can also rise to 0 mVs in the fermenter. In order to keep a low redox potential, few oxidizing agents should be supplied, e.g., no oxygen, sulfates, nitrates, or nitrites.

2.2.11
Parameter: nutrients (C/N/P-ratio)

The C/N-ratio of the substrate should be in the range of 16 : 1–25 : 1 (Table 2.7). But this is only an indication, because nitrogen can also be bound in lignin structures.

The need for nutrients is very low due to the fact that with the anaerobic process not much biomass is developed, so that for methane formation even a nutrient ratio C : N : P : S of 500–1000 : 15–20 : 5 : 3 and/or an organic matter ratio of COD : N : P : S = 800 : 5 : 1 : 0.5 is sufficient.

Substrates with a too low C/N ratio lead to increased ammonia production and inhibition of methane production. A too high C/N ratio means lack of nitrogen, from which negative consequences for protein formation and thus the energy and structural material metabolism of the microorganisms result. A balanced composition is absolutely necessary; e.g. the mixture of rice straw and latrine waste as usual in China of the co-fermentation of elephant dung with human waste as done in Nepal.

2.2.12
Parameter: trace elements

For survival, microorganisms need certain minimum concentrations of trace elements Fe, Co, Ni, Se, W, and Mg (see Section 3.2.2.15).

20) Cp. JOU 4

2.2.13
Parameter: precipitants (calcium carbonate, MAP, apatite)

At abt. 100 mg L^{-1} Ca^{2+}, calcium carbonate ($CaCO_3$) begins to precipitate from aqueous suspensions.[21,22)]

At concentrations up to 150 mg L^{-1} Ca^{2+}, the formation of readily-sedimenting sludge flakes (pellets) is promoted.

When the calcium carbonate concentration exceeds 500 mg L^{-1} Ca^{2+}, on the one hand (positively) the formation of biofilms and biomass growth is supported, but on the other hand (negatively), some types of bioreactors can be overgrown too fast and blocked by lime deposits.

With calcium concentrations of over 1000 mg L^{-1}, as they occur, e.g., in the German sugar industry, separation of the lime is recommended, e.g., in a hydrocyclone or a sedimentation apparatus or with the aid of carriers, which can than be cleaned from the lime residues and fed back into the reactor.

In the paper and milk industries, various phosphate compounds precipitate, e.g., apatite, because of the higher Ca concentrations (400–1000 mg L^{-1} Ca^{2+}) and the high concentrations of phosphoric compounds (40–100 mg L^{-1} P). In order to avoid calcium carbonate linings, permanent magnets are used.

Magnesium ammonium phosphate (MAP), struvite ($MgNH_4PO_4 \cdot 6H_2O$) precipitates if the three components in combination are present together, e.g. with waste water from the wheat and potato industry. MAP precipitates mainly by turbulent flow as in pumps or by exceeding a certain pH value, depending on the waste water, frequently starting at pH = 7 and temperatures >30 °C.

2.2.14
Parameter: biogas removal

The removal of the produced gases[23)] (methane, carbon dioxide, hydrogen sulfide) from the substrate has a considerable influence on the biological reactions. The attainable microorganism concentration can be increased by a factor of 12 in the case of extremely thermophilic methanogenics when produced gases are removed to a sufficient extent.

In particular, the degassing behavior of the substrate plays a tremendous role whenever the produced gases inhibit the metabolism.

The material transfer from the liquid phase into the gaseous phase can be effected by technical means. By feeding back the produced biogas into the reaction system, the $k_L.a$[24)] value, characteristic of the material transfer between liquid and gaseous phase, can be increased by 50% and the inhibition of the microorganisms by H_2S can be reduced.

21) Cp. BOK 73
22) Cp. BOK 74
23) Cp. WEB 71
24) Material transfer number $k_L \cdot a$ = product of material transfer coefficient k_L and surface a

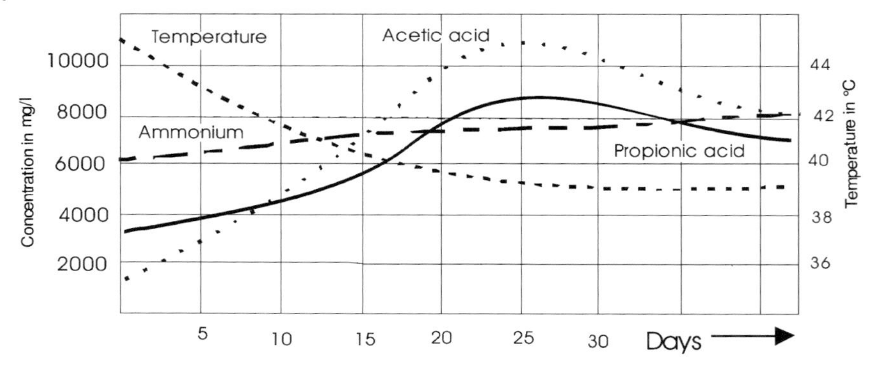

Figure 2.21 Progression of an inhibition.

2.2.15
Parameter: inhibitors

When planning and operating a biogas plant, it has to be borne in mind that some compounds which are formed, even to a limited extent, as products of the metabolism of the anaerobic disgradation, inhibit the biocenosis and can even be toxic at higher concentrations (Figure 2.21).

The inhibition depends on the concentration of the inhibitors, the composition of the substrate, and the adaptation of the bacteria to the inhibitor. Anaerobic bacteria need a low concentration of the inhibitors as trace elements. They degrade a high percentage of the inhibitor.

About inhibitors and their effect on the anaerobic process, many and sometimes very contradictory comments have been published. This is because

- microorganisms adapt to their environment. How quickly the adaptation takes place depends on the composition of the biocenosis.
- inhibitors have different effects, depending upon whether they flow intermittently or continuously. In the case of a single brief addition, irreversible damage develops only if the exposure time is too long and the concentrations are too high.
- the inhibitors are not able to penetrate very fast into the biofilm if fermenters with retained biomass are used in preference to conventional fermenters with lower biomass concentration or immobilized biomass.
- the effect of different inhibitors is affected by other components; thus, for example, the restraining effect of heavy metals is affected by any anions present at the same time, since metals can be precipitated, e.g., by hydrogen sulfide, or can be bound in complexes. On the other hand, a too high sulfide ion concentration can even have a toxic effect – depending upon temperature and pH value. The

reduction of a pollutant by addition of a material working against it can therefore be problematical because of different interactions.

Usually, however, anaerobic processes seem to be relatively insensitive to inhibitors and mostly adaptable, even to concentrations that are toxic under other circumstances.

2.2.15.1 **Oxygen**

The majority of the acidifying bacteria are facultatively anaerobic, so that the exclusion of oxygen is not absolutely necessary for acidification. But methanogenics are obligatorily anaerobic – the inhibition begins at $0.1\,mg\,L^{-1}$ O_2. They take nitrate, sulfate, or carbonate as the oxygen replacement, i.e. as the hydrogen acceptor.

Since under operational conditions the methanogenics always grow in the presence of facultatively anaerobic acidifying bacteria, which consume available oxygen immediately, anaerobic conditions can be maintained in closed reactors. The small amount of air, which is injected frequently for the biotechnological desulfurization of the biogas, usually has no inhibiting effect on the methane formation.

2.2.15.2 **Sulfur compounds**

Wastes and industrial waste water can contain a high concentration of sulfur compounds, e.g., waste water from the production of yeast, viscose rayon, cartons, citric acid, or fiber board. Sulfur compounds can be present

- as sulfate (in industrial waste water in high concentrations)
- as sulfide
- as hydrogen sulfide in the gas
- as undissociated hydrogen sulfide in the liquid, and (toxically)
- in dissociated form HS^-, S^-.

Sulfate can be problematical, because H_2S develops from it in a stage before methane formation. H_2S can be inhibiting to the process (see below).

$$SO_4^{2-} + 4H_2 \rightarrow H_2S + 2H_2O + 2OH^-$$

$$SO_4^{2-} + CH_3COOH \rightarrow H_2S + 2HCO_3^-$$

Sulfate can also inhibite the methane formation, because sulfate-degrading microorganisms are dominant as they need less energy and/or do not need a symbiosis partner: for sulfate degradation, a free energy of $-154\,kJ\,mol^{-1}$ and/or $-43\,kJ\,mol^{-1}$ is needed, while for the methanation of carbon dioxide and hydrogen or acetic acid only $-135.4\,kJ\,mol^{-1}$ and/or $-30.9\,kJ\,mol^{-1}$ is needed.

Dissolved sulfide in the waste water up to $50\,mg\,L^{-1}$ does not lead to problems for methane formation if the pH is less than 6.8 in the bioreactor. Microorganisms can adapt themselves to sulfide and can survive at up to $600\,mg\,L^{-1}$ Na_2S and $1000\,mg\,L^{-1}$ H_2S.

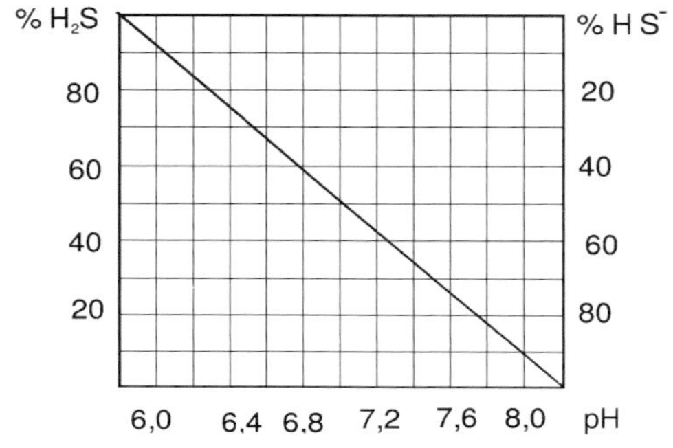

Figure 2.22 Dissociation of hydrogen sulfide.

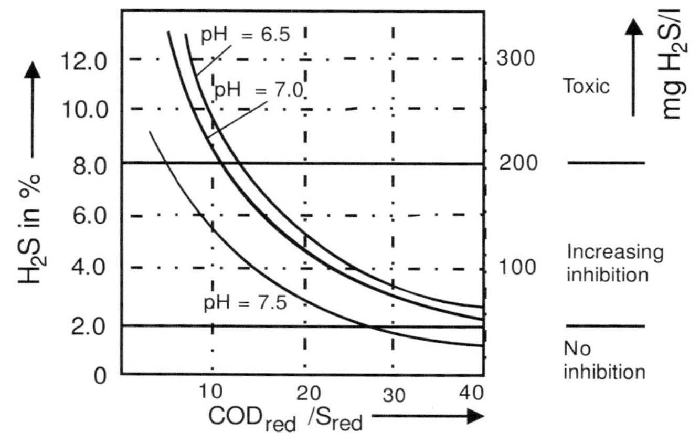

Figure 2.23 Inhibition through hydrogen sulfide at 30% CO_2, 38 °C, 10 g/L, organic matter as COD_{red}.

H_2S develops as by-product with nearly all substrates. It escapes with the biogas and is dissolved in undissociated and dissociated form in the substrate as a weak acid, whereby hydrogen sulfide ions and sulfide ions develop.

$$H_2S \leftrightarrow HS^- + H^+ \leftrightarrow S^{2-} + 2H^+$$

The chemical equilibrium between the undissociated and dissociated form depends on the pH value (Figure 2.22).

With decreasing pH value, the proportion of dissolved undissociated hydrogen sulfide rises. In dissolved form it works directly as cellular poison even at a concentration of ca. $50\,mg\,L^{-1}$ (Figure 2.23). Above that, H_2S can cause process inhibition indirectly by the precipitation of essential trace elements as insoluble sulfides.

With organic matter as COD > 100, basically no problems are to be expected through H_2S. With organic matter as COD < 15, methane production is possible only in special cases.[25)]

25) Cp. BOK 59

However, the temperature is also important: with rising temperature the toxicity of hydrogen sulfide increases.

In some cases, hydrogen sulfide in the substrate is favorable, particularly if the substrate contains heavy metal ions at a level at which they must be precipitated.

Continuous monitoring of the H_2S content of the biogas is recommended. An unexpected rise should be dealt with by

- increasing the pH value by addition of caustic soda
- admitting iron salt, which works as an H_2S scavenger
- lowering the volume load.

When high sulfur content in the waste water is to be expected, a two-stage process should be prefered, so that the removal of the sulfur compounds can be achieved in the first stage.

2.2.15.3 **Organic acids (fatty acids and amino acids)**

Normally, organic acids are present in the substrate. These are decomposed during methanation. They exist partly in the undissociated and partly in the dissociated form (Figure 2.24).

Especially undissociated acids have an inhibiting effect, because they penetrate as lipophilics into cells, where they denaturate the cell proteins.

If too much organic acid is fed to the bioreactor within a short time and/or their degradation is hindered, a too strong acidification takes place. Therefore the load of the methane reactor may be increased only very slowly, fermenting easily acidifiable substrates or ensilaged energy plants in particular.

The inhibiting effect is additionally intensified by the drop in the pH-value (Figure 2.25). At pH<7, the inhibiting threshold is up to $1000\,mg\,L^{-1}$ acetic acid. For *iso*-butyric acid or *iso*-valeric acid, the inhibiting threshold can be as low as a concentration of $50\,mg\,L^{-1}$ undissociated fatty acid without adaptation. Propionic acid is strongly inhibiting even at a concentration of $5\,mg\,L^{-1}$ (corresponding to ca. $700\,mg\,L^{-1}$ undissociated acid at pH=7).

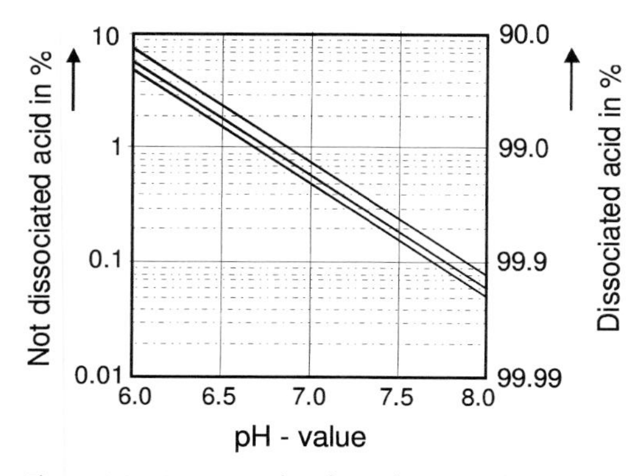

Figure 2.24 Dissociated and not dissociated acid.

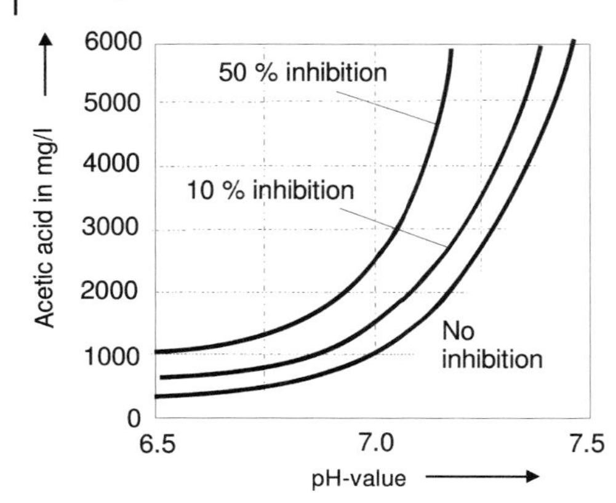

Figure 2.25 Inhibition through acetic acid.

Table 2.8 Limits for sulfur compounds and fatty acids.

Substance	*Minimum amount required as trace element [mg L^{-1}]*	*Concentration at which inhibition starts [mg L^{-1}]*	*Toxicicity [mg L^{-1}] for adopted MO*
Sulfur compounds	Organic matter as COD : S ratio at 800 : 0.5	H_2S: 50	1000
		S: 100	
		Na_2S: 150	600
Iso-Butyric acid	n.a.	50	n.a.
Long-chain fatty acids	n.a.	1.2 mM C12 and C18	n.a.
Petrochemical products	n.a.	0.1 mM hydrocarbonates, aromatic halogenic products	n.a.

Aromatic amino acids probably lead to inhibition, e.g., with waste water from potato processing (Table 2.8).

2.2.15.4 Nitrate (NO_3^-)

Nitrate is denitrified in the first stage of decomposition, in any case before the methanation.

Inhibition of the methanation is only possible with substrates with high nitrate content NO_3-N > 50 mg Mg^{-1}. This is the case if the denitrification does not work properly. Therefore, before nitrate-rich sludges can be fermented, the oxygen bound in the nitrate/nitrite must be split off, e.g., in a denitrification stage. On denitrification in the anaerobic reactor,

- the redox-potential becomes more negative, in favoring the anaerobic process.
- the gas quality decreases because of the higher nitrogen content of the biogas.

- carbon is consumed and is no longer available for methanation, which is noticeable in the output of the plant.

2.2.15.5 Ammonium (NH_4^+) and ammonia (NH_3)

Ammonia and ammonium result from the anaerobic biological degradation of nitrogen compounds. Ammonia forms ammonium ions in the substrate, the extent of this depending on the pH value. Ammonia has an inhibiting effect, and with larger concentrations can even be toxic, while ammonium is innocuous. It can only have an inhibiting effect at concentrations NH_4^+-N > 1500 mg Mg^{-1}, where its inhibiting effect is predominantly due to the species whose concentration depends on the pH value. Ammonium leads to potassium loss of methanogenic microorganisms and can show reciprocal effects with Ca^{2+} or Na^+.

The inhibition by ammonium increases with rising pH value (Figure 2.26) at constant concentration: e.g., the ratio ammonium to ammonia is 99 : 1 at pH = 7, and 70 : 30 at pH = 9. Often, even small pH value variations are sufficient to affect the inhibiting effect.

The equilibrium between ammonium and the inhibiting ammonia (Table 2.9) is temperature-dependent: with rising temperature the equilibrium between NH_4 and NH_3 is shifted in favor of NH_3, so that the inhibition increases with rising temperature.

In communal plants, waste water from the anaerobic plant is recycled to the denitrification tank of the sewage treatment plant. It has to be borne in mind that

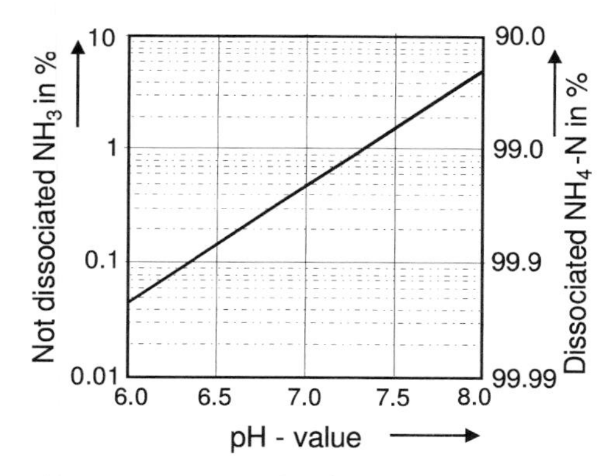

Figure 2.26 Dissociated and not dissociated ammonium.

Table 2.9 Limit values of ammonium and ammonia.

Substance	*Regulation/interaction with*	*Concentration at which inhibition starts [mg/L]*	*Toxicicity [mg/L] for adopted MO*
Ammonium	Ca^{2+}, Na^+	1 500–10 000	30 000
Ammonia	–	80	150

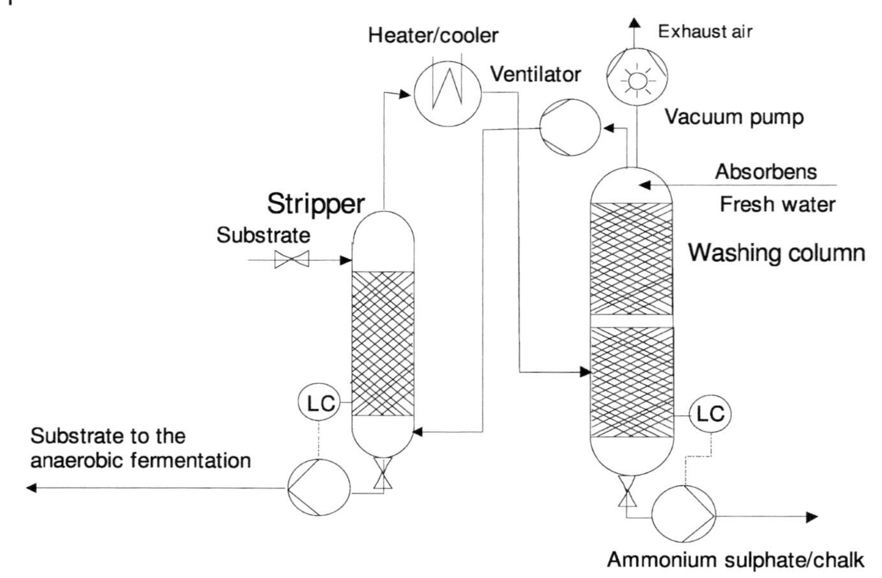

Figure 2.27 ANAStrip®-process.

the ratio of organic matter (as COD) to nitrogen content should be at 5–6 : 1 and has often to be adjusted by adding raw waste water to the waste water from the anaerobic plant.

Ammonia can be removed from the substrate by the ANAStrip-process, system GNS[26] (see Figure 2.27).

The plant is operated with a circulated gas. The circulated gas ascends in the substrate in the stripper at 80 °C (max.) and low vacuum and leaves the stripper ammonia-enriched. The circulated gas then flows through a scrubber in which ammonia reacts with an absorbent (REA gypsum) in aqueous solution with formation of ammonium sulfate and lime fertilizer.

$$CaSO_4 + 2NH_3 + CO_2 + H_2O \rightarrow CaCO_3 + (NH_4)_2SO_4$$

The ammonium sulfate salt is very water soluble and can be taken off as 40% fertilizer concentrate solution from the scrubber. The ammonia-free circulated gas from the scrubber is injected again into the stripper by a blower. Within 2 h, the circulated gas takes off at least 70% of the bound nitrogen of the substrate. The special advantages of the procedure are:

- no consumption of caustic soda solution or other alkalis to extract ammonia from the liquid manure
- no neutralization of the liquid manure necessary after stripping

26) Cp. WEB 39

- no increase of the salt concentration in the liquid manure
- low operating temperature (80 °C max.) when stripping, as the waste heat supply can be used
- no need for stripping steam, since the procedure works under low vacuum
- minimum water distillation when stripping, because the process works without steam injection. The fertilizer from the plant does not become dehydrated.
- direct formation of ammonium sulfate from the vaporized ammonia without formation of concentrated ammonium solution as intermediate
- no odors from the concentrated ammonia solution possible
- no expensive safety devices necessary for the storage and dosage of ammonia solution, caustic soda solution, and sulfuric acid
- no environmental pollution or danger to personnel from aggressive chemicals

2.2.15.6 Heavy metals

Heavy metals, which act as trace elements at low concentrations stimulating the activity of the bacteria, can have toxic effects at higher concentrations (Table 2.10). In particular lead, cadmium, copper, zinc, nickel, and chromium can lead to disturbances in biogas plants, but so also can sodium, potassium, calcium, and magnesium.

The content of lead, cadmium, chromium, copper, nickel, mercury, and zinc in the biomass rises during fermentation, but only slightly. The high zinc contents of the pig liquid manure originate from the piglet fodder, which often contains a zinc additive as an antibiotic.

Heavy metals can be bound or precipitated by 1–2 $mg\,L^{-1}$ sulfides. Copper, cadmium, lead, and zinc can be eliminated by polyphosphates (complexing agents). They are then no longer biologically available.

2.2.15.7 Tannins[27]

Condensed tannins – secondary plant compounds which are to be found in many legumes – can specifically and dose-dependently inhibit methane formation. Data on the inhibiting threshold are not available.

2.2.15.8 Other inhibiting thresholds

Disinfectants (hospitals, industry), herbicides, and insecticides (agriculture, market gardens, households), surfactants (households), and antibiotics can often flow with the substrate into the methanation and can there cause nonspecific inhibition (Table 2.11).

Chlorinated hydrocarbons such as chloroform at levels above 40 $mg\,L^{-1}$ and/or chlorinated fluorocarbons also have toxic effects. The swelling concentration,

27) Cp. WEB 44

Table 2.10 Inhibiting or toxic concentrations of different metals in solution in the reactor.[28),29)] The inhibition depends on whether the metals are there as ions or as carbonates.

Substances	***Minimum amount required as trace element [mg/L]***	***Parameters affected***	***Concentration at which inhibition starts [mg/L]***		***Toxicity [mg/L] for adopted MO***
			Free ions	***As carbonate***	
Cr	0.005–50	–	28–300	530	500
Fe	1–10	–	n.a	1750	n.a
Ni	0.005–0.5	–	10–300		30–1000
Cu	Essentially with acetogenic MO	–	5–300	170	170–300
Zn	Essentially with acetogenic MO	–	3–400	160	250–600
Cd	n.a	–	70–600	180	20–600
Pb	0.02–200	–	8–340	n.a.	340
Na	n.a.	pH – Wert	5000–30 000	n.a.	60 000
K	n.a.	Osmosis of the methane formers	2500–5 000	n.a.	n.a
Ca	n.a.	Long-chain fatty acids	2500–7 000	n.a.	n.a.
Mg	Essentially with acetogenic MO	Fatty acids	1000–2 400	n.a.	n.a.
Co	0.06	–	n.a.	n.a.	n.a.
Mo	0.05	–	n.a.	n.a.	n.a.
Se	0.008	–	n.a.	n.a.	n.a.
Mn	0.005–50	–	1500	n.a.	n.a.
HCN	0.0	–	5–30	n.a.	n.a.
C_6H_6O	Inhibiting until the microrganisms are adapted. Then it is completely degraded.				

Table 2.11 Inhibiting or toxic concentrations of different inhibitors.

Substance	***Concentration at which inhibition starts [mg/L]***	***Toxicity [mg/L] for adapted MO***
Chloroform	40	n.a.
Chlorofluorohydrocarbons etc.	50	n.a.
Formaldehyde	100	1200
Ethene and terpenes	1–50	n.a.
Disinfectants and antibiotics	1–100 (not necessarily)	n.a.

28) Cp. BOK 12

29) Cp. JOU 25

starting from which an inhibition begins, depends frequently on the time of adaptation during which the bacteria are exposed to the inhibitor. Usually the inhibition decreases with increasing time of adaptation. Therefore, sudden changes of concentration of inhibitor have to be avoided.

2.2.16
Parameter: degree of decomposition

Complete degradation of the organic dry matter is theorically only possible if no lignin from wood-containing biomasses is present. The normal degree of decomposition varies within the range 27–76% and is usually abt. 43.5%.

2.2.17
Parameter: foaming

Growing filamentuous microorganisms produce foam.[30] They arrive in the digestion tower with the secondary sludge. There they become poorly degraded and hinder the gas discharge, so that the anaerobic sludge attains a foamy consistency and can foam over in the digestion towers. The foam-forming microorganisms develop in the denitrification zone and in the aeration basin of the water treatment plant on oxygen impact during agitation, via the recirculated sludge and over the inlet to the water treatment plant, since low winter temperatures strongly reduce the oxygen consumption, so that the quantity of dissolved oxygen increases. The filamentuous microorganisms need only a small oxygen concentration.

Foam can be combated through

- mechanical foam destroyers and cutting up of the foam blisters
- addition of precipitants (e.g., two-material dosage of polymers and iron salts)
- frequent removal of the floating sludge
- decrease of the oxygen entry into the denitrification zone in the water treatment plant; e.g., by a scum on the basins, in which the denitrification runs off.
- by an air-tight housing over the denitrification zone in the water treatment plant, so that no contact with the ambient air occurs.

2.2.18
Parameter: scum

The foaming and the scum[31] formation are interdependent.

The flotation of the activated sludge, which leads to the formation of a scum, is effected by the occurence of filamentous bacteria in large quantities. A low

30) Cp. WEB 47

31) Cp. WEB 34

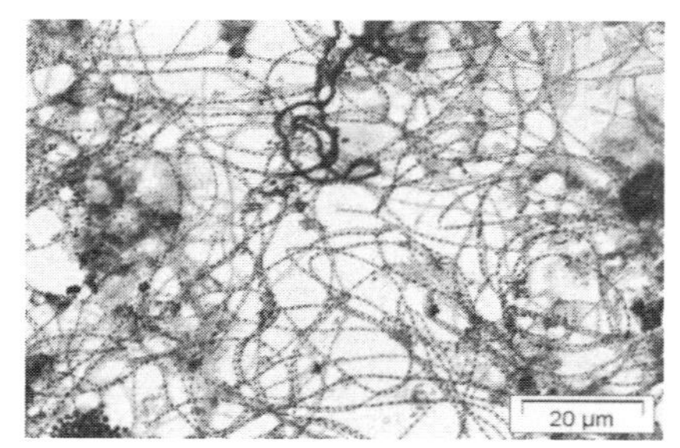

Figure 2.28 Scum (left), agitated by a RECK-Blizzard (right).

substrate concentration in nutrient-limited plants supports filamentous bacteria more than flake-forming bacteria.

Nowadays, approximately 30 different filamentous microorganisms are known, which can cause both scum and bulking sludge (Figure 2.28). These microorganisms can be roughly divided into three groups: sulfur bacteria, Gram-negative bacteria from high-load plants, and Gram-positive bacteria from low-load plants. Most frequently, the bacterium *Microthrix parvicella* is responsible for the filament formation in the activated sludge of communal plants.

Microthrix parvicella, an unusual member of the actinomycete phylum, is an aerobic organism with a high affinity for oxygen. The bacterium prefers low temperatures with a pH optimum around pH=8. *Microthrix parvicella* can be suppressed by long-chain fatty acids, reduced nitrogen and sulfur bonds, and a low sludge load with relatively high sludge age. *Microthrix parvicella* can also be inactivated by aerobic, anaerobic, and anoxidic contact zones (selectors), but cannot be banished completely from the system.

3
Bacteria participating in the process of degradation

The species of microorganisms vary depending upon the materials which are to be degraded; e.g., *Clostridium* degrades butyricum, *Cl. Pasteurianum* and *Citrobacter freundii* degrade particularly hexadecimal chlorine cyclohexane, *Micrococcus, Aerobacter, Alcaligenes, Flavobakterium*, and *Pseudomonas* decompose the alkylsulfonate of detergents.

Alcohols, fatty acids, and aromatic bonds can be degraded by microorganisms with anaerobic respiration. They use, among other nutrients, nitrate (*Paracoccus denitrificans, Pseudomonas stutzerii*), sulfur (*Desulfuromonas acetoxidans, Pyrodictium occultum*), sulfate (*Desulfovibrio desulfuricans, Desulfonema limicola*) carbonate (*Acetobacterium woodi, Clostridium aceticum, Methanobacterium thermoautotrophicum*), fumarate (*Escherichia coli, Wolinella succinogenes*) or Fe(III) (*Alteromonas putrefaciens*) as electron acceptors, so they are called nitrate reducer, sulfate reducer etc. accordingly.[32),33)] However other microorganisms also compete around nitrate as electron acceptors, so that nitrate is rapidly reduced to ammonium and the nitrate reducer plays a subordinate role in fermentation processes. But the sulfate reducers are deeply involved in the degradation of low-oxygen compounds such as lactates and ethanol. The microorganisms form a culture from obligatory or facultative anaerobic bacteria at a concentration of 10^8–10^9 fermenting bacteria per mL. In the first and second phase of degradation, at least 128 orders are involved of 58 species and 18 genera. *Clostridium, Ruminococcus, Eubacterium* and *Bacteroide* are the species mainly occurring.[34)] In the third and fourth phase of degradation, methane bacteria are in the majority – so far, 81 species from 23 genera, 10 families, and 4 orders have been specified. Also, there are many microorganisms which belong to the ecological system of a bioreactor and which are indirectly involved in the degradation. *Staphylococcus*, which can cause health dangers for personnel, should be mentioned.

At all four phases of degradation, the species *Acetobacter* and *Eubakterium* are involved in nearly equal amounts (Table 3.1). Some kinds are homoacetogenic,

32) Cp. WEB 108
33) Cp. WEB 109
34) Cp. DIS 4

Biogas from Waste and Renewable Resources. An Introduction.
Dieter Deublein and Angelika Steinhauser

ISBN: 978-3-527-31841-4

Table 3.1 Bacteria participating in the fermentation process in all four phases.

Taxonomy	*Species*	*Description*	*Metabolism*
Genus: *Acetobakterium*	*A. woodii* *A. paludosum*	The genus *Acetobacter* contains homacetogenic rod-shaped bacteria.	They reduce autotrophic polymer compounds, oligomers and monomers or CO_2, with hydrogen as the electron source. They serve as hydrogen-fed partners and make possible the decomposition of fatty acids and aromatic compounds.
Genus: *Eubacterium*	*E. rectale* *E. siraeum* *E. plautii* *E. cylindroides* *E. brachy* *E. desmolans* (produces desmolase, which can degrade the steroids), *E. callandrei* *E. limosum*	The genus *Eubacterium* consists of obligate anaerobic Gram-positive bacteria, which do not generate endospores.	Most of the saccharolytic *Eubakteria* produce butyrate as the metabolism product. Many species are able to decompose complex substrates via special pathways. Some species grow autotrophically and are therefore able to fulfill special challenges in anaerobic decomposition, e.g., as hydrogen-consuming partners in the acetogenesis of fatty acids. *E. limosum* grows autotrophically and metabolizes methoxyl groups of substituted aromatics.

i.e., they convert CO_2 and/or monomers of carbonhydrates via the acetyl CoA pathway. Since CO_2 is constantly delivered by the metabolic processes and therefore CO_2 is available in the reactor in unlimited quantity, some of carbohydrates are degraded by homoacetogenesis. There are some signs, that homoacetogenic bacteria are involved in other metabolic processes also, e.g., in the splitting of the aromatic ring system of substituted aromatics and/or in the dissolution of structures similar to lignin.

Other *Eubakterium* species appear in bioreactors but apparently do not participate in the decomposition.

3.1
Hydrolyzing genera[35)]

Microorganisms of many different genera are responsible for hydrolysis (Table 3.2).

3.2
Acidogenic genera

Nearly all acidogenic microorganisms also participate in hydrolysis.

The genera *Clostridium, Paenibacillus* and *Ruminococcus* appear in all phases of the fermentation process but are dominant in the acidogenic phase.

The cluster *Cytophaga-Flavobacterium-Bacteroides* is the second largest group of microorganisms during the first two phases of decomposition, but represents in the methanogenic phase less than 5% of all microorganisms. This is an indication that species of these groups are mainly responsible for the degradation of the monomeric compounds.

Genus: Clostridium (Figure 3.1)
The genus *Clostridium* features a large phenotypical and genotypical diversity (Table 3.3). It is therefore difficult to describe special functions of clostridia.

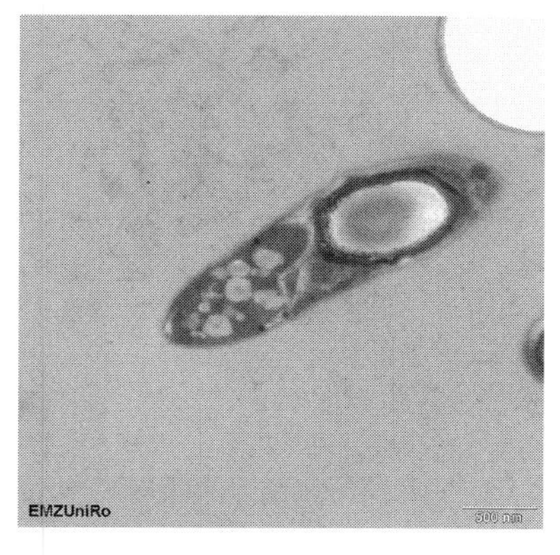

Figure 3.1 *Clostridium.*

35) Cp. BOK 28

Table 3.2 Bacteria participating in the fermentation process in the hydrolysis phase.

Taxonomy	*Species*	*Description*	*Metabolism*
Genus: *Bacteroides*	*B. uniformis* *B. acidifaciens* *B. vulgatus* *B. splanchnicus* *B. ruminicola*	The genus *Bacteroides* consists of immobile, Gram-negative rods.	They take as substrate carbohydrates, peptones, and metabolic products of other microorganisms like sugar, amino acids, and organic acids. The metabolic products of the *Bacteroides* are succinate, acetate, formate, lactate, and propionate. Butyrate is mostly not a main product of the fermentation of carbohydrates and occurs normally with isobutyrate and isovalerate.
Genus: *Lactobacillus*	*L. pentosus* *L. plantarum* *L. agilis* *L. aviarius* *L. lindneri*	The genus *Lactobacillus* consists of Gram-positive, catalase-negative rods, which do not generate endospores. They are normally immobile.	They ferment glucose to lactate and other organic acids either homofermentatively or heterofermentatively.
			Lactobacilli are known for their need of additional nutrients like vitamins, amino acids, purines, and pyrimidines.
Genus: *Propioni-bacterium*	*P. microaerophilum* *P. granulosum* *P. lymphophilum* *P. acnes* *P. avidum*	They are immobile Gram-positive rods, which do not form spores.	Propionibacteria are catalase-positive.
	P. propionicus *P. combesii* *P. thoenii* *P. freudenreichii* *P. cyclohexanicum*	Propionibacterium	They are chemoorganotrophic and produce much propionate and acetate during fermentation of carbohydrates.

Table 3.2 *Continued*

Taxonomy	*Species*	*Description*	*Metabolism*
			Byproducts of the fermentation are isovalerate, formate, succinate, lactate, and CO_2.
Genus: *Sphingomonas*	*S. aromaticivorans* *S. subterranea* *S. stygia*	They occur in deep sediments.	Sphingomonas are able to degrade aerobically a wide spectrum of substituted aromatics.
		Sphingomonas on Xanthos	They can utilize anaerobically the methoxyl groups of trimethoxybenzoate without splitting the aromatic ring.
Genus: *Sporobacterium*	*Sp. olearium*		*Sporobacterium* is able to degrade stoichiometrically trimethoxybenzoat to acetate and butyrate by splitting the aromatic ring.
Genus: *Megasphaera*	*M. elsdenii*	These occur in the rumen.	The *Megasphaera* use the acrylate pathway.
Genus: *Bifidobacterium*			*Bifidobacteria* ferment glucose to lactate and acetate. The decomposition of hexoses occurs via a special pathway.

Genus: Ruminococcus

Species of the genus *Ruminococcus* (Table 3.4) are anaerobic chemoorganotrophic cocci, which ferment carbohydrates to acetate, formate, succinate, lactate, ethanol, H_2, and CO_2.

Genus: Paenibacillus

At least one species of the genus *Paenibacillus* produces lactate, formate, acetate, and propionate from polymeric compounds of molasses. It is assumed that *Paenibacillus* belongs to the primary decomposers in the biomass fermentation.

Some species of the genus *Paenibacillus* are able to respirate nitrate, a by-process in methanation.

Table 3.3 Species of *Clostridium*.

Microorganism	*Substrate*
C. celerecrescens	Cellulose
C. aerotolerans	Xylan, glucose is decomposed to succinate
C. butyricum	Polymers like starch and pectin, and also metabolic products of other organisms, e.g., lactate
C. tyrobutyricum	Monosaccharides, but not disaccharides or polymers
C. propionicum	Different amino acids, lactate, and acrylate. The substrates are fermented to fatty acids such as propionate, butyrate, *iso*butyrate, *iso*valerate, and acetate on the acrylate pathway.
C. clostridiiformis	Carbohydrates to acetate, lactate, and formate
C. methylpentosum	Grows only on pentoses
C. viride	Different fatty acids (valerate, crotonate)
C. spiroforme	Carbohydrates to acetate
C. piliforme	Causes the illness Tyzzers[36]
C. propionicum	Lactate via the acryloyl pathway

Table 3.4 Species of Ruminococcus.

Microorganism	*Substrate*
R. hydrogenotrophicus	*H_2/CO_2; degrades short-chain fatty acids and aromatics in symbiosis with obligate syntrophic organism*
R. gnavus	*Sugar to ethanol; starch*
R. bromii	*Sugar and starch to ethanol*
R. flavefaciens	*Cellulose*
R. callidus	*Different sugars to succinate*
R. albus	*Cellulose and different sugars to ethanol and formate*

3.3
Acetogenic genera

Acetogenic genera (Table 3.5) can only survive in symbiosis with hydrogen-consuming genera.

All acetogenic microrganisms have a long period of regeneration up to 84 h. An enrichment of butyric acid by a butyric acid-degrading acetogenic bacterium in symbiosis with *Methanobacterium hungatii* takes 120 h at 35 °C.

36) Infection of the liver

Table 3.5 Species of acetogenic bacteria.

Taxonomy	*Species*	*Metabolism*
Genus: *Desulfovibrio*	*D. desulfuricans* *D. termitidis.*	*Desulfovibrio* oxidizes organic acids and alcohols to acetate and transfers the released electrons to sulfate. This pathway offers a higher energy yield than fermentation. Through this, the high number of sulfate-reducing bacteria in the reactor can be explained, although this genus shows only a small diversity. According to experience, the number of sulfate-reducing bacteria decreases at the end of fermentation.
Genus: *Aminobacterium*	*A. colombiens*	*Aminobacteria* ferment aminoacids and produce acetate.
Genus: *Acidaminococcus*		Species of the genus *Acidaminococcus* ferment aminoacids, *trans*-aconitate and citrate to acetate, CO_2 and H_2.

Sulfate-reducing acetogenic bacteria are able to degrade lactate and ethanol, but not fatty acids and aromatic compounds.

3.4 Methanogenics

The last phase of the anaerobic decomposition is dominated by a special group of microorganisms, the methanogenic Archaea.[37] These are characterized through the co-factor F_{420}, which acts in the presence of hydrogenase as a carrier for H_2, appears only in methanogenics, and can be detected by its autofluorescence in an optical microscope.

Active methanogenics appear in the second phase of fermentation, the acidogenic phase, but the number of methanogenic Archaeae obviously increases in the methanogenic phase.

Methanobacterium, Methanospirillum hungatii, and *Methanosarcina* are the main species. When the acetate concentration is low, the filamentous *Methanosaeta* is able to stand up to the coccus *Methanosarcina*, because of its considerably lower substrate saturation efficiency and its higher substrate affinity. Especially in flow-through systems, *Methanosaeta* has advantages over other methanogenics because of its ability to grow on hydrophobic surfaces and hence to withstand washing effects.

37) Cp. BOK 26

The domain Archaea[38] is polymorphic, so that it can taxonometrically be distinguished from the other domains only by the sequence signature of the 16S rRNA.[39]

The cell walls of all Archaea contain phytanyl ether lipids but no muramic acid. Gram-positive species have cell walls of pseudomurein with incorporated C_{20} – isopranyl diether and C_{40} – isopranyl tetraether, methanochondroitin, or heteropolysaccharides. Gram-negative organisms have only (glyco)-protein cell walls.[40]

The domain Archaea is divided into four phyla: Crenarchaeota, Euryarchaeota, Korarchaeota, and Nanoarcheota. The methanogenic Archaea belong to the phylum Euryarchaeota (Figure 3.2).

The Euryarchaeota are a physiologically deverse phylum with many extremophilic organisms. Extremely halophilic, methanogenic, and hyperthermophilic genera and species belong to it. But also some species of the hyperthermophilic Archaea are methane forming. The acidophilic group called Thermoplasma, which does not have cell walls, should be mentioned.

How the phylum Euryarcheota is best subdivided down to the species is still matter of research. Often, the taxonomy corresponds to Figure 3.30.

Methanogenic Archea exist in all possible shapes: rods, short lancet-shaped rods, cocci, irregular plates, or spirilli (Figure 3.3).

Methanogenic Euryarchaeota are strictly anaerobic and live off H_2/CO_2, H_2/CO, formate, acetate, methanol, methanol/H_2, methylamines, or dimethyl sulfides. Some species use primary and secondary alcohols as electron donors. Nickel is obligate for euryarcheota. They produce methane from methyl-CoM catalyzed by the enzyme methyl-CoM reductase in presence of the cofactors F_{420}, F_{430}, methanopterin, and CoM.

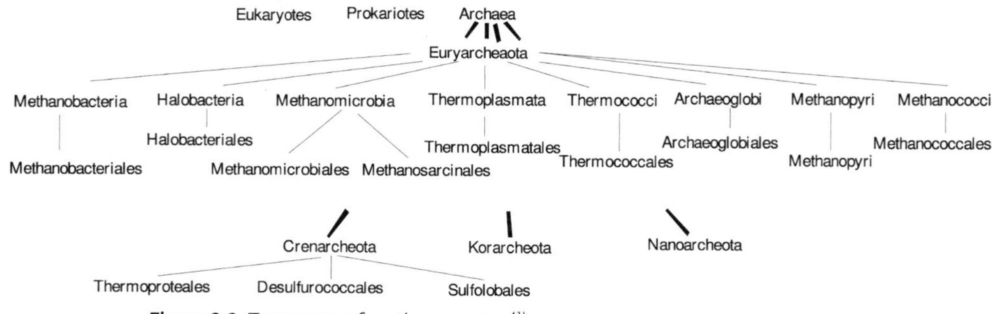

Figure 3.2 Taxonomy of methanogenics.[41]

38) Cp. BOK 26

39) 16S rRNA belongs to the main parts of the procaryotic ribosome. It consists of 1540 base pairs and plays a significant role during the initiation of the translation. By formation of base pairs the 3′-end of the 16S rRNA binds to the mRNA. In this way, the starting codon of the mRNA is transferred to the correct position in the ribosome

40) Glycoproteins are macromolecules, which consist of a protein to which carbohydrates are covalently bound. The fraction of carbohydrates in such proteins can be up to 85%

41) Cp. BOK 29

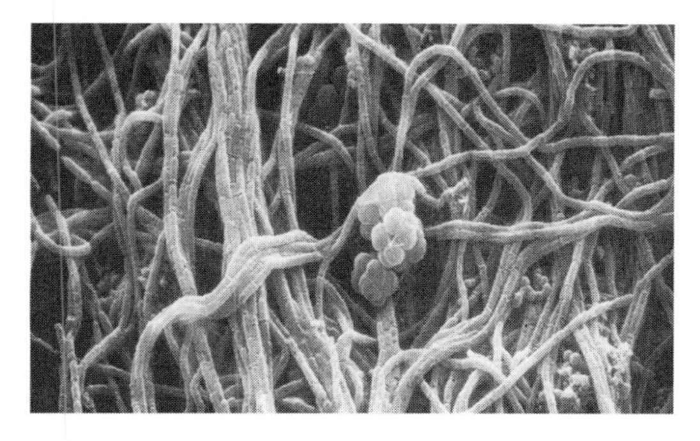

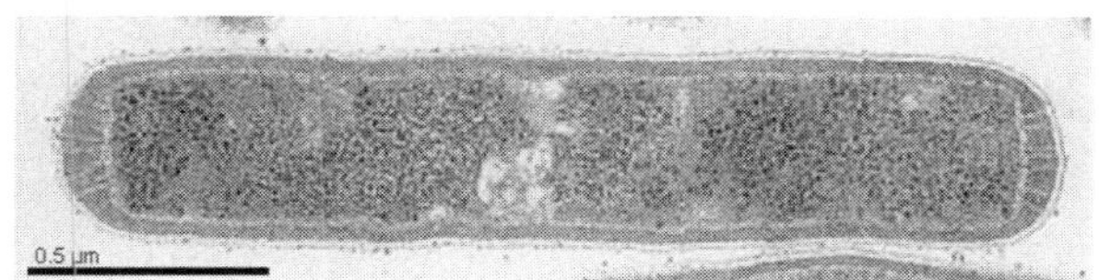

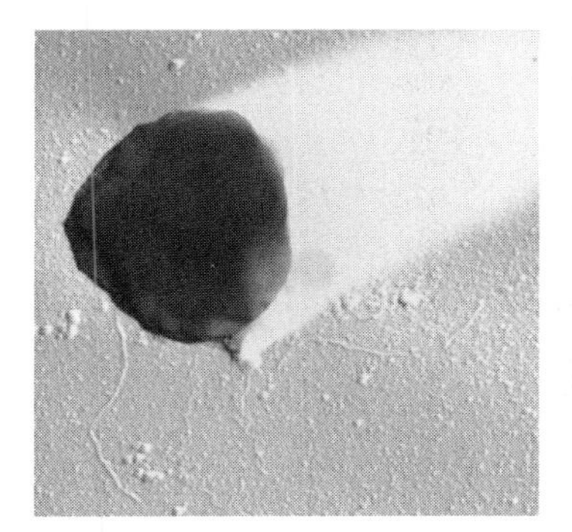

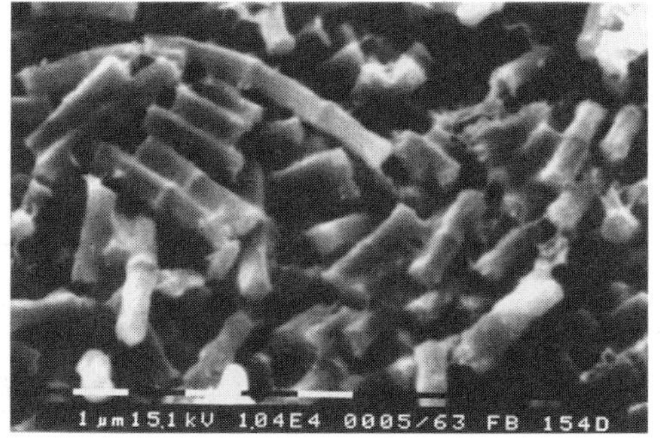

Figure 3.3 Methanogenics: Methanosaeta (filament) and Methanosarcina (cocci) (top left), Methanothermus (top right), Methanotorris igneus (bottom left), Methanosaeta on a solid-state material after 154 days. The length of the white bar equals 1 µm (bottom right).

The orders, families, genera and species listed in Table 3.6 are able to produce biogas. Many of them do not occur in biogas plants of today and/or in the rumen of cattles, but could perhaps be used in special applications to produce biogas.

3.5 Methanotropic species

Methanotropic species (methane-consuming microorganisms) (Figure 3.4) are everywhere. They are undesired in biogas plants, but very importend for our climate. Most of them are aerobics. They use oxygen to split methane and to get their energy. The metabolic products are water and carbon dioxide.

Table 3.6 Methanogenic species.

Taxonomy	*Species*	*Description*	*Metabolism*
Order: Methanobacteriales		Methanobacteriales are Gram-positive organisms.	The cell division occurs by binary fission.
Order: Methanobacteriales	*M. Defluvii*	Methanobacteriaceae are rod-shaped and filament-forming microorganisms.	Their growth optimum lies between 37 °C and 70 °C.
Family: Methanobacteriaceae	*M. Oryzae*	Their cell walls do not contain an S layer.	They take their energy from H_2 or formate. Six species grow on formate, three species on 2-propanol/CO_2
	M. Thermoflexum	The DNA base composition is 29–62 mol% G + C (guanine + cytosine).	
Order: Methanobacteriales	*Mbr. Arboriphilus*	Methanobrevibacter are short, lancet-shaped rods, which prefer to live in pairs or in chains. Some species have one flagellum.	Cofactors are obligate for their growth.
	Mbr. Ruminantium (dominant type in stomachs of cattle)		
Family: Methanobacteriaceae	*Mbr. Smithii (sewage sludge and mammalian colons)*	Two species contain threonine or ornithine within their peptides; one species has exclusively GalN instead of GlcN in the glycan strand of the pseudomurein.	They are able to use formate. They grow mesophilically.
Genus: Methanobrevibacter	*Mbr. Curvatus*	DNA base composition is 27–32 mol% G + C.	
	Mbr. Cuticularis		
	Mbr. Filiformis		
	Mbr. Oralis		
Order: Methanobacteriales	*Mb. Alcaliphilum*	Methanobacterium are straight, long, sometimes irregular rods. Some species live in filaments. They do not have a locomotor system.	They grow in mesophilic to thermophilic conditions. Some species prefer an alkaline environment.
	Mb. Bryantii		
Family: Methanobacteriaceae	*Mb. Espanolae*		
	Mb. Formicicum		

Genus: Methanobacterium	*Mb. Ivanovii* *Mb. Palustre* *Mb. Thermoaggregans* *Mb. Uliginosum* *Mb. Subterraneum* *Mb. Thermoautotrophicum*	DNA base composition is 33–61 mol% G+C.	
Order: Methanobacteriales Family: Methanobacteriaceae Genus: Methanosphaera	*Msp. Cuniculi* *Msp. Stadtmanae*	Methanosphaera are coccoid cells, living singly or in clusters. They live in the intestinal tract of human beings and rabbits. The Pseudomurein in their cell walls contain Ser. The DNA base composition is 23–26 mol% G+C.	They produce methane from methanol and H_2 and not from H_2, CO_2, or formiate, because they lack a CO_2 reductase and also a methyltransferase complex (N_5-methyl-tetrahydromethanopterin: coenzyme M). CO_2 and acetate serve as carbon sources.
Order: Methanobacteriales Family: Methanobacteriaceae Genus: Methanothermobacter	*Met. thermoautotrophicus* *Met. Wolfeii* *Met. Marburgensis*	Very similar to the genus *Methanobacterium*	
Order: Methanobacteriales Family: Methanothermaceae		To be found in volcanic springs	
Order: Methanobacteriales	*Mt. Fervidus* *Mt. Sociabilis*	Methanothermus are long rods with a double-layered cell wall composed of pseudomurein and a glycoprotein layer. They have a bipolar polytrichous flagellation.	They grow optimally above 80 °C. Maximal growth temperature is 97 °C.
Family: Methanothermacea Genus: Methanothermus		Their DNA base composition is 33–34 mol% G+C. Methanothermus	They live solely from H_2 and CO_2.

Table 3.6 *Continued*

Taxonomy	*Species*	*Description*	*Metabolism*
Order: Methanococcales			
Order: Methanococcales Family: Methanococcaceae		Methanococcaceae are regular to irregular formed cocci, have cell walls of protein and move through a polar flagellum.	Growth is stimulated by selenium. All species take H_2 and formate as electron donors and are prototropic[42] except methanocaldococcus and methanoignis igneus.
Order: Methanococcales Family: Methanococcaceae Genus: Methanococcus	*Mc. Deltae* *Mc. Thermolithothrophicus* *Mc. Vannielii** *Mc. Voltae* *Mc. Aeolicus*	Methanococcus are Gram-negative cocci. Their surface layer is composed of non-glycosylated protein subunits – only one species has C_{20}– isopranylglycerolether in its cell walls. Their DNA base composition is 30–41 mol% G+C. *Methanococcus janaschii*[43]	They grow mesophilically to extremely thermophilically. The energy sources are H_2/CO_2 and formiate.
Order: Methanococcales Family: Methanococcaceae Genus: Methanothermococcus	*M. thermolithotrophicus*		
Order: Methanococcales Family: Methanocaldococcaceae	*Methanocaldococcus jannaschii (Ost-Pazifik-Rücken)* *M. Fervens* *M. Vulcanius* *Methanoignis igneus*	Methanocaldococcaceae are similar to Methanococcaceae. The DNA base composition is 31 to –33 mol% G+C.	The genus *Methanocaldococcus jannaschii* grows faster than all methanogens with a regeneration period of 30 min.

42) Proton-containing solvents similar to water are prototropic

43) Cp. WEB 74

Order: Methanomicrobiales			They grow only below 60 °C
Order: Methanomicrobiales Family: Methanomicrobiaceae		Methanomicrobiaceae are Gram-negative, irregular cocci, rods and spirilli. Cell walls are of proteins and the lipids contain C_{20} and C_{40} isopranyl glycerol ether. DNA base composition is 39–50 mol% G + C.	Nearly all species take H_2 and formate as substrate, some secondary alcohol.
Order: Methanomicrobiales Family: Methanomicrobiaceae Genus: Methanomicrobium	*Mm. Mobile*	Methanomicrobium is short and rod-shaped, with a monotrichous flagellation. Their DNA base composition is 49 mol% G + C.	Acetate is required as a carbon source. H_2 and formate serve as energy source. In the rumen a special growth factor could be found.
Order: Methanomicrobiales Family: Methanomicrobiaceae Genus: Methanolacinia	*Ml. Paynteri**	Methanolacinia are short, irregular rods or coccoid to lobe-shaped cells without flagellation. Their cell wall is of polyamines and lipid patterns. DNA base composition is 45 mol% G + C.	They produce methane from H_2/CO_2, 2-propanol/CO_2, 2-butanol/CO_2 and cyclopentanol/CO_2. Acetate is required. Formate cannot be used.
Order: Methanomicrobiales Family: Methanomicrobiaceae Genus: Methanogenium	*Mg. Cariaci* *Mg. liminatans* *Mg. Organophilum* *Mg. Tationis* *Mg. Frittonii* *Mg. Frigidum (optimum temperature 15 °C)*	Methanogenium are highly irregular cocci with polytrichous or monotrichous flagellation. Their cell wand is out of proteins. Their DNA base composition is 47–52 mol% G + C.	They live from H_2/CO_2, formate, and sometimes alcohols. Most strains require acetate as carbon source. Two species use secondary alcohols. They tolerate high salt concentrations and grow best at around 1 M Na^+. They need growth factors.
Order: Methanomicrobiales Family: Methanomicrobiaceae Genus: Methanoculleus	*Mcl. Bourgense** *Mcl. Marisnigri* *Mcl. Olentangyi* *Mcl. Thermophilicus*	Methanoculleus are irregular formed coccoids and Gram-negative. The DNA base composition is 49–62 mol% G + C.	Most of the species need H_2/CO_2, formate and some secondary alcohols as substrates. Their best growth rate is at an Na^+ concentration of 0.1–0.4 M.

Table 3.6 *Continued*

Taxonomy	***Species***	***Description***	***Metabolism***
Order: Methanomicrobiales Family: Methanomicrobiaceae Genus: Methanoplanus	*Mp. Endosymbiosus, Mp. Limicola, Mp. Petrolearius*	Methanoplanus is Gram-negative, plate-shaped with sharp edges and with polar tuft of flagella. The cell wall consists of a glycoprotein S layer. One species lives together with marine ciliates and is responsible for the methane production in sea sediments. DNA base composition is 38–48 mol% G + C.	Acetate is absolutely necessary as a carbon source. H_2 and formate serve as electron donors.
Order: Methanomicrobiales Family: Methanomicrobiaceae Genus: Methanofollis	*M. Tationis* *M. Liminatans*	Their DNA base composition is 54–60 mol% G + C.	Methanofollis needs formate as substrate.
Order: Methanomicrobiales Family: Methanomicrobiaceae Genus: Methanocalculus	*M. halotolerans (Offshore Ölquelle)*		Methanocalculus grows in 5% NaCl and tolerates up to 12% NaCl – a maximum for methanogenic MO.
Order: Methanomicrobiales Family: Methanocorpusculaceae			

Order: Methanomicrobiales Family: Methanocorpusculaceae Genus: Methanocorpusculum	*Mcp. Aggregans*, *Mcp. Bavaricum*, *Mcp. Labreanum*, *Mcp. Parvum**, *Mcp. Sinense*	Methanocorpusculum are Gram-negative small cocci (<1 µm) with polar flagella. The DNA base composition is 48–52 mol% G + C.	They take H_2/CO_2, formate, and some species, also 2-propanol/CO_2. Acetate, yeast extract, and tungstate are utilized.
Order: Methanomicrobiales Family: Methanospirillaceae			
Order: Methanomicrobiales Family: Methanospirillaceae Genus: Methanospirillum	*Msp. Hungatei*	Methanospirillum create rods forming long filaments in which the single cells with polar flagellation are separated by spacers. Several cells are surrounded by an SDS-resistant protein sheath. The cell wall contains 70% amino acids, 11% lipids and 6.6% carbohydrates. C_{20} isopranyl diether and C_{40} isopranyl tetraether are present in the cell wall. DNA base composition is 45–50 mol% G + C.	All species take H_2/CO_2 and formate, some species 2-propanol and 2-butanol as hydrogen donors for the methanogenesis of CO_2.
Order: Methanosarcinales			
Order: Methanosarcinales Family: Methanosarcinaceae		Methanosarcinaceae are Gram-positive or Gram-negative cocci occurring singly or in large clusters.	Methanol and methylamines serve as carbon and energy sources. Some species also use H_2/CO_2, or acetate. Two genera containing rods use only acetate as substrate.
Order: Methanosarcinales Family: Methanosarcinaceae	*Ms. Acetivorans* *Ms. Barkeri** *Ms. Frisia* *Ms. Mazei*	Most species are Gram-positive cocci and contain methanochondroitin cell walls. Cytochromes are present. Their cell walls consist of *N*-acetyl-D-galactosamin and D-glucuronic- or D-galacturonic acid in a molecular ration of 2 : 1, as well as some D-glucose and traces of D-mannose. Some species contain only C_{20}–isopranyl glycerol ether.	The demand on nutrients varies widely. Methanosarcina can grow on acetate, methanol, methylamines, and most species, also on H_2/CO_2. Formate is not used.

Table 3.6 *Continued*

Taxonomy	*Species*	*Description*	*Metabolism*
Genus: Methanosarcina	*Ms. Methanica* *Ms. Thermophila* *Ms. Vacuolata* *Ms. Siciliae*	Their DNA base composition is 40–51 mol% G + C. *Methanosarcina Barkeri*[44]	They dominate many anaerobic ecosystems.
Order: Methanosarcinales Family: Methanosarcinaceae Genus: Methanolobus	*Mlb. Sicilae* *Mlb. Tindarius* *Mlb. Vulcani* *Mb. Oregonensis*	Methanolobus have a monotrichous flagellation or no flagella. Cytochromes are present. Their DNA base composition is 39–46 mol% G + C.	Methanolobus are obligate methylotrophs. They can only use methanol and methylamines as energy and carbon sources. They are mainly mesophilic and prefer salt concentrations of 0.5–1.5 M.
Order: Methanosarcinales Family: Methanosarcinaceae Genus: Methanococcoides	*Mcc. Methylutens* *Mcc. Burtonii (from Antarctica)*	Methanococcoides are very similar to methanolobus. Their cytoplasmic membrane contains only C_{20}-isopranyl diether. DNA base composition is 40–42 mol% G + C.	Preferred salt (NaCl) concentration is 0.2–0.6 M. Magnesium (50 mM) is essential.
Order: Methanosarcinales Family: Methanosarcinaceae Genus: Methanohalophilus	*Mh. Halophilus* *Mh. Mahii* *Mh. Oregonense* *Mh. Euhalobius*	Methanohalophilus are irregularly formed cocci. The DNA base composition is 38–49 mol% G + C.	They are moderately halophilic – the optimum salt concentration is 0.6–2.5 M NaCI. The cell wall is degradable with 0.05% sodium dodecyl sulfate. They take methylamine or methanol as substrate.
Order: Methanosarcinales Family: Methanosarcinaceae Genus: Methanosalsus	*Ms. Zhilinae*	Their DNA base composition is 38 mol% G + C.	Methanosalsus are able to catabolyze dimethyl sulfide.

44) Cp. WEB 74

Order: Methanosarcinales Family: Methanosarcinaceae Genus: Methanohalobium	*Mhm. evestigatus**	Methanohalobium are irregular flat cells. The cell walls are not degraded in sodium dodecyl sulfate (0.05%).	They are extremely halophilic: the optimum salinity is 2.5–4.3 M NaCI (25%) at a temperature of 50 °C. Only methylamine, not even methanol, serves as substrate.
Order: Methanosarcinales Family: Methanosarcinaceae Genus: Halomethanococcus	*Hmc. Doii**	Halomethanococcus are differently formed cocci. The DNA base composition is 43 mol% G + C.	They prefer temperatures of 5–45 °C with an optimum temperature of 35 °C. Their minimal salt content is 1.8 M NaCl, the optimum saltinity is 3.0 M NaCl. They take methylamine and methanol as substrates. Acetate and liquid of the rumen are essential for their growth.
Order: Methanosarcinales Family: Methanosaetaceae			
Order: Methanosarcinales Family: Methanosaetaceae Genus: Methanosaeta	*Mst. Concilii (statt früher Methanotrix soehngenii), Mst. Thermophila (statt früher Mst. Thermoacetophila)*	Methanosaeta on a solid-state material after 154 days.[45] Methanosaeta are Gram-negative cells of about 0.8–3.3 µm size with a typical bamboo-like rod shape and flat ends. Conglomerations mostly of many hundreds of single cells are surrounded by a layer of mucilage consisting of fibrous glycol proteins.	Methanosaeta takes only acetate as energy source, which is degraded to methane and CO_2. Methanosaeta has a high affinity and a very low threshold concentration to usual substrates. Therefore the microorganism can be found especially at places with low acetate concentrations.

45) Cp. DIS 6

Table 3.6 *Continued*

Taxonomy	***Species***	***Description***	***Metabolism***
		The DNA base composition is 50–61 mol% G + C.	In the anaerobic stages of the municipal sewage plants with retention times of more than 15 days at 35 °C, Methanosaeta is dominant. The genus has a low growth rate and a doubling time of 4–7 days at 37 °C.
Order: Methanopyrales			
Order: Methanopyrales	*Methanopyrus Kandleri*	Methanopyraceae are Gram-positive with a cell wall of pseudomurein containing ornithine and lysine, but not *N*-acetylglucosamine. The main lipid consists of phytanyl diether.	Methanopyrales grow at 110 °C and a salinity of 4%.
Family: Methanopyraceae		The DNA base composition is 60 mol% G + C.	In presence of sulfur they form H_2S.

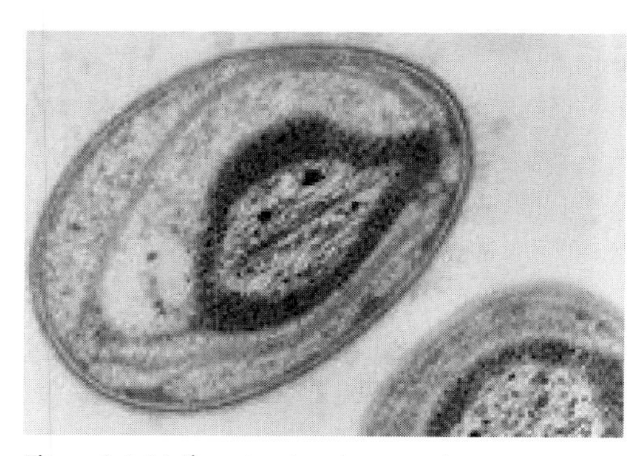

Figure 3.4 Methanotropic microorganism.

Aerobic methanotropics degrade about 17% of all methane in the atmosphere. Besides these, another group of methanotropics exists which is able to consume methane without needing oxygen. They are mostly to be found in sea sediments.

Methanotropic microorganisms build up their lipids based on methane.

Part IV
Laws and guidelines concerning biogas plants

In Europe, particularly in Germany,[1] many state laws, regulations, norms, and guidelines of branch institutions were issued in order to maintain a safe and smooth operation of biogas plants. These cover their installation, operation, supply, and waste management. Many offices are involved in an administrative decision for the construction of a plant, e.g., the planning department and building control office, the public order office, the water regulatory authority, the natural conservation authority, the authority for nutrition, the authority of agriculture, the office for veterinary matters, food control, the office for technology and plant safety, the office for noise control-traffic-energy-climatic protection, and a lot more besides.

In the following, some of these regulations, focusing mainly on Germany, are described in more detail, as similar regulations relating to the operation of biogas plants are likely to be issued in many other countries.

1) Cp. WEB 19

Biogas from Waste and Renewable Resources. An Introduction.
Dieter Deublein and Angelika Steinhauser

ISBN: 978-3-527-31841-4

1
Guidelines and regulations

The regulations and laws are aimed at the elimination or minimization[2),3)] of all safety hazards affecting both people and the environment, especially those due to biogas plants:

- Danger to life and health caused by suffocation and toxification in tanks, silos, and pits filled with gases like H_2S, CH_4, and CO_2. Some of these gases are heavier than air and tend to remain in the vessels.
  The following gases are e.g. dangerous even in low concentration:
  Hydrogen sulfide (H_2S)[4)] = 10 ppm
  Carbon monoxide (CO) = 50 ppm
  Chlorine (Cl_2) = 0.5 ppm
  Carbon dioxide (CO_2) = 5000 ppm
  Danger of being poisoned by co-fermentation products
- Emission of pollutants into air, groundwater and surface water, especially when disposing residues
- Danger of explosions caused by flammable gas/air mixtures. All flammable gases, vapors, and dust are able to form explosive mixtures.
- Severe falls for workers
- Injuries caused by machinery and plant parts, e.g., rotating machinery, electromagnetic forces, electrostatic charging, or burning by hot surfaces
- Weather, e.g., flooding or lightning
- Damages by fire
- Freezing of gas or substrate pipes, e.g., by condensation of water in the pipes leading to blocking and finally to bursting of the pipe
- Corrosion caused by aggressive substances like ammonium nitrate, H_2S, and many others

2) Cp. LAW 16
3) Cp. WEB 17
4) TLV-value is the maximum concentration of toxic gases at the workplace = PEL.

Biogas from Waste and Renewable Resources. An Introduction.
Dieter Deublein and Angelika Steinhauser

ISBN: 978-3-527-31841-4

- Plugging of tubes by solid components
- Noise, especially in the areas of the CHP.

Some regulations concerning the installation of biogas plants are listed in Table 1.1.

The principle behind all regulations concerning health and safety at work is:

Constructive/technical measures better than organizational better than personal measures

1.1
Construction of plants

1.1.1
Corresponding regulations

According to constructive laws, biogas plants are considered as commercial plants, which are located in industrial zones. The location of a biogas plant outside an industrial area is legal if it does not conflict with public interest. This is not the case, when

it serves an agricultural enterprise whose ground is mostly used for other purposes.

it is installed away from settlements. This is desired because of its demand on agricultural products, its requirement to remove the residue.

it provides for the energetical use of biomass (plants and animal products).

it cannot cause environmental damage and is not affected by natural climate influences.

it serves the public supply of electricity, gas, heat and water.

it does not interfere with land use planning, the landscape plan, or any other plan, especially for the utilization of water and the avoidance of emissions.

no uneconomic costs are incurred for the infrastructure (roads and so on) for the supply and the waste removal of the plant or for security and health procurement.

it does not interfere with existing areas whose soil, monuments, and relaxation value are environmentally protected.

it does not conflict with existing measures to improve the agricultural structure or the water situation.

Table 1.1 Guidelines from the EU (RL), laws (G), regulations (VO), and technical requirements (TA).

Statutory instrument	***Subject matter***
Regulations concerning the biogas plant	
German law of Construction and Building	Approval, authorization to present building documents, design drawings, and clearance
German law of emissions and dedicated regulations	Emissions (air pollution, noise, vibrations)
German law of environmental sustainablity	
German law of soil protection	
German law of nature conservation	
German law of air pollution	Limit values for exhaust air
German law of noise protection	Limit values for noise
German law of industrial plants handling material hazardous to waters	
Regulations concerning the biomass and residual	
German law of biomass	Governmental support when certain substrates are fermented
German law of waste-recycling	Treatment of wastes
German law of cadaver removal	Treatment of cadavers
German law of epizootic	Hygiene
German law of fertilizers	Residues permitted to be used as fertilizer
German law of fertilizing	Basics for good fertilizing practice
German law of waste registration	Bio wastes as fertilizers
German law of bio waste	
German law of determination and supervision of hazardous wastes	
German law of on the removal of animal excrement	
European guideline on removal of nitrate (91/676/P.E.)	Protection of waters from contamination with nitrate from agricultural sources
German law of water supply	Treatment of sewage sludge
German law of waste water	Treatment of waste water
General Administrative Regulation on the Classification in Hazard Classes of Substances Hazardous to Waters	
German law of sewage sludge	Sewage sludge as fertilizer
German law of mineral oil tax	Addition of diesel oil to biogas
Regulations concerning the use of biogas	
German law of Electric power to be fed into the network	Promotion of the use of renewable resources
European Guideline on natural gas (2003/55/EG)	Equalization of natural gas and biogas

Table 1.1 *Continued*

Statutory instrument	***Subject matter***
Regulations of the DVGW[5]	Features of biogas to be fed into the natural gas network
Regulations concerning work	
German law of work protection	Health and safety at work
German law of workplaces	Workplaces
German law of safety at workplaces	Safety and health with tools and in industrial plants
Safety regulations of the employer's liability insurance: association of gasworks and waterworks	

In general, all parts of the plant have to be inspected and approved by the authorities. This includes installations like bioreactors, gasholders, combined heat and power stations (CHP), ignition oil tanks, stores and tanks for liquid manure.

In biogas plants, the formation of explosive gas mixtures can occur. Therefore, a system for plant security must exist relating to installation and operation of electrical devices in areas with danger of explosions. These areas are

- Enclosed spaces which serve as gasholders or where gases are produced, or other enclosed spaces which are connected, e.g., by pipes, to these spaces
- Channels and storage tanks
- Spaces inside machinery, tanks, and pipes, which contain biogas
- Spaces around machinery where gas is discharged, e.g., overpressure security valves at the bioreactor or the gasholder.

Buildings which are used for gas consumption (heaters or engines) and are not connected to bioreactors or gasholders are excluded from the above-mentioned prescriptions. In these buildings, the installation of gas pipes and electrical wires should be easy to keep under surveillance. Ideally, the gas pipes and the electrical wires should be installed on different walls.

1.1.2 Checklist of regulations concerning the plant

The following checklist refers to regulations concerning the design, organization, and control of the plant.

5) Deutsche Vereinigung des Gas- und Wasserfachs e.V.

Regulations relating to the design

Are all parts of the plant protected against lightning according recommendations (DIN V VDE 0185)? Is the lightning protection in accord with the special demands of explosive areas? Is the lightning protection regularly checked and are these checks documented?

Is it ensured that unfermented substances do not come in contact with fermented substances? Arriving vehicles must be cleaned and disinfected at the plant entrance, departing vehicles after leaving the plant area.[6)]

Is the plant (i.e.all parts of it) situated at least 300 m away from the closest planned or existing settlement and are sufficient measures taken to protect the settlements from smell?

Are all recommendations[7),8),9),10),11),12),13),14),15)] from local authorities for the liquid manure storage tanks and bioreactors fulfilled?

Is ensured that the vibrations of the machinery do not exceed the allowed limits (DIN 4150)? Were measures taken to diminish the vibrations?

Regulations relating to organization

Are unauthorizied persons prevented from entering, e.g., by closed access doors, appropriate design features, and a fence at least 1.50 m high around flammable material.

Regulations relating to control

Are all technical installations regularly checked by authorized experts? Special attention should be given to air supply, chimneys, automatic fire extinguishers fed from the main water supply, compressors, fire warning systems and alarms, and power supplies for safety devices.

1.2 Utilized biomass

The German law of bio waste regulates the circulation of bio wastes in agriculture, in forestry, and in horticulturally used ground (Table 1.2).

Bio waste recycling includes all plants which ferment biological waste, even if the fermented substances are only partly of biological origin. The residue is called bio waste and is the subject of the bio waste regulations. Bio waste is, for example, the content of the grease removal tank in catering establishments, mash from

6) Cp. WEB 86
7) Cp. LAW 18
8) Cp. LAW 19
9) Cp. JOU 33
10) Cp. JOU 34
11) Cp. JOU 35
12) Cp. LAW 20
13) Cp. LAW 21
14) Cp. LAW 22
15) Cp. LAW 23

Table 1.2 Restrictions when bio wastes are to be fermented in biogas plants.

Different wastes according to Bioabf VO	*Usable types of waste*	*Comments*
Animal excrement Liquid manure Muck (incl. spoiled straw) Waste water Bark and cork wastes Bark and waste wood Wastes from forestry	Poultry excrement Pig and cattle liquid manure Muck Old straw Bark (wood processing) Bark Wood, wood resudues	These wastes are subject to the regulations only when they are traded. Infectious muck (LAGA code 137 05) is generally not allowed to be utilized. Separately collected bark except bark of trees and bushes, grass cuttings from edges of roads or from industrial sites, and vegetable components of flotsam is excluded from the obligations of examinations and special treatments.
Biologically degradable wastes	Garden/park wastes, wood residues Vegetable components of flotsam	Bark of trees and bushes of edges of road may be supplied to a utilization only if it is stated that the heavy metal contents specified in the regulation are not exceeded.
Wastes	Mushroom substrate residues	Inactivation of the cultures by steaming prescribed.
For consumption or processing unsuitable materials Wastes from animal tissue Sludge from waste water treatment Biologically degradable kitchen and cafeteria wastes Vegetable oils and fats	Fat wastes (meat and fish processing) Bristle and horn wastes Contents of grease removal tanks and flotate (meat and fish processing)	Utilization allowed only in so far as it is not against the law of cadaver removal or epizootic. Fat wastes may be used only in biogas plants. Residuals may be used, also as a component of a mixture, as fertilizer for grassland only after pasteurization (70 °C; >1 h).
Wastes	Sludge from the gel production Feathers Stomach and intestine contents	Residues, also as a component of a mixture, may be used as fertilizer on grassland.
Sludges from washing, cleaning, peeling, centrifugation, and separation processes	Other suspended wastes from food production Starch sludge	Sludges may be used only if they are not mixed with waste water or sludges from other origins.

Table 1.2 *Continued*

Different wastes according to Bioabf VO	*Usable types of waste*	*Comments*
For consumption or processing unsuitable materials	Overlaid food and luxury articles	
	Wastes from production of tobacco, coffee, tea, cocoa, canned goods	
	Oil seed residuals	
Wastes	Sludge from seed processing	
	Bleaching earth (deoiled)	
	Seasoning residues molasses residues	
	Residues from starch production from potato, corn or rice	
	Whey	
	Wastes from beer, hop, and malt preparation	
	Mud, sludge, mash from wine preparation	
	Yeast and yeast-similar residues	
	Moorland mud and medical earth	
Wastes from alcohol distillation	Fruit and potato mash	
	Mash from distillery	
Sludges from waste water treatment	(food and luxury article production)	
Wastes from vegetable tissue	Husks, dust from cereals	
	Wastes from fodder	
Wastes from laundry, cleaning and mechanical cutting up of raw materials	Used filter and absorption masses (diatomite), activated carbon	Dry diatomite may not be distributed on land. It has to be incorporated in the soil immediately after distribution.
Other solid wastes	Mash of medical plants mycelia	Mycelium from pharmaceutical production may be further used only after special approval.
	Mushroom substrate residues	
	Protein wastes	
Solid wastes from filtration and sieving	Materials from skimming, cutting, and raking	Only cut grass may be used in biogas plants.
Paper and pasteboard	Recovered paper	Waste paper may only be added to bio wastes or compost in small quantities (ca. 10%). Addition of art paper and wallpaper is not allowed.

Table 1.2 *Continued*

Different wastes according to Bioabf VO	***Usable types of waste***	***Comments***
Wastes from housing estates	Bio wastes from households and small businesses	
Market wastes	Lime, bentonite, clay	Only separately collected biologically degradable fraction is allowed to be utilized.
	Powdered mineral, stone abrasion dust, sand	
	Biologically degradable products from renewable resources	Degradability must be demonstrated on the basis of a technical standard
Unidentifiable calcium carbonate sludge	Sludge from carbonization (sugar beet processing)	Residues may be added to bio wastes, which are used as fertilizer for grassland.
Sludges from decarbonization	Sludge from water purification	

distilleries, waste from kitchens, waste from landscape conservation, bio waste from garbage, and cut grass from the roadsides.

For all bio wastes, the origin, treatment, and disposal of the residues must be documented.

The residues of the fermentation will be reused as fertilizer according to the German law of waste recycling. The German law of bio waste prescribes the hygienization of bio waste before reuse. According to the regulation, the operator of the plant must verify the effect of his hygienization process and compliance with the necessary process temperatures as well as the hygienic harmlessness.

In the fermentation process, the heavy metal concentration increases because of the volume reduction of the substrate. According to the law of bio waste, residues from bio waste fermentation may only be distributed in agriculture at certain rates, when the heavy metal content of the soil is found to be low.

Max. 30 Mg DM/ha in 3 years if the heavy metal concentration is low
Max. 20 Mg DM/ha in 3 years if the heavy metal concentration is high

The heavy metal content in soil must not exceed the values given in Table 1.3.

The plant operator must analyze the residues from fermentation every three months according to the documentation required. Nevertheless, the heavy metal concentration, pH value, salt concentration, and content of dry matter and organic dry matter must always be determined when 2000 Mg biomass is supplied to the biogas plant.

Biogas plants processing animal byproducts have to be admitted in general according to the EU guideline 1774/2000. The exact documentation of the product

Table 1.3 Maximum permitted heavy metal amount to be distributed in Mg DM/ha within 3 years according to Germans law.[16)]

	20 Mg DM/ha	*30 Mg DM/ha*
Lead	150	100
Cadmium	1.5	1
Chromium	100	70
Copper	100	70
Nickel	50	35
Mercury	1	0.7
Zinc	400	300

flow rates as well as the measures for cleaning and disinfection is obligatory. The EU guideline regulates exposure to special substrates, dividing all by-products into three categories.

Category 1

Bodies of diseased animals, bodies of pets, and specific wastes from kitchen and the food industry.

These products are not allowed to be processed in biogas plants.

Category 2

Bodies of the livestock husbandry, unusable slaughterhouse disposals, liquid manure, gastrointestinal contents, milk.

These products are allowed to be processed with specific permission and after a certain preparation process (pressure sterilization).

Category 3

Slaughterhouse disposals of animals, wastes from kitchen and food industry.

These products are allowed to be processed in biogas plants after pasteurization (hygienization).

In agricultural biogas plants, wastes are not normally processed with co-ferments produced for that purpose and with animal by-products.

1.3 Biomass to be used preferentially

A high price for the electricity produced in a biogas plant is guaranteed by the German EEG (Renewable Energy Law) only if very specific, approved biomass is used for fermentation and specific technical regulations are met.

16) Cp. DIS 3

The German law of bio waste biomass[17] specifies the biomasses and the technical processes supported. Approved biomass consists of pure herbaceous products. Not approved are fossil fuels, mixed wastes, mud, sewage sludge, and port sludge.

1.4 Distribution of the residues

Residues from biogas plants are subject to the German law of fertilizers, including when domestic waste water is fermented or when it is mixed with agricultural substrates.

The German law of fertilizers regulates single parameters such as temporary permissions to deploy fertilizer, the determination of fertilizing needs, the upper limits for the deployed amount of fertilizer, techniques to deploy, and a lot more. These regulations are called the "principles of the good and professional execution of fertilizing". In Germany, no liquid manure or liquid secondary raw material fertilizer are allowed to be deployed from November 15th to January 15th. This is because the soil is frozen and the fermentation product cannot penetrate the soil.

The deployment of residues from biogas plants is regulated also. Not more than 210 kg/ha.a total nitrogen from industrial fertilizer are allowed on pasture land and not more than 170 kg/ha.a on ground used agriculturally. For phosphates, the limits are set to 120 kg/ha.a on pasture as well as on ground used agriculturally, for calcium the limit is restricted to 360 kg/ha.a. Farms with more than 10 ha of ground used agriculturally are obliged to prepare a written fertilizer balance to keep track of their usage.

The German law of fertilizers prescribes that all raw materials incorporated in fertilizers must be of plant cultivational, productional, or technological (improvement of application properties) use. Additionally, all raw materials to produce an organic fertilizer have to be listed in the attachment of the law and may not interfere with the mixing prohibitions (e.g. bio waste to sewage sludge).

As soon as domestic waste water is co-fermented in agricultural biogas plants, the residues of the fermentation become subject to the German law of sewage sludge. This law also applies to sewage sludges from small water treatment plants ($<8\ m^3 d^{-1}$).

With the employment of sewage sludge from waste water treatment plants, both soil analyses and sewage sludge analyses are to be performed by the operator. The complexity of the prescribed analyses depends on the size of the plant. The precise expenditure on this is to be agreed with the responsible authorities.

17) Cp. WEB 9

1.5
Feeding biogas to the gas network

In the EU guideline for the equalization of biogas and natural gas, it is stipulated that biogas must be allowed to be fed to the gas network without restriction,[18] provided that the addition is possible without technical problems and impairment of security (Figure 1.1). The relevant agreement[19] between the German natural gas suppliers concerning the biogas supply to the natural gas network has become of marginal importance, because nowadays the gas suppliers have the function just to transport gases.[20]

The biogas producer himself has to look for a customer who would like to purchase his biogas. The preconditions for feeding biogas into the natural gas network are in Germany considered to be fulfilled when they correspond to the technical regulations of the authorities.[21] Therefore, the biogas must meet the specifications such that at the point of discharge to the customers no special treatment of the gas is required and the gas is sufficiently pressurized.

1.6
Risk of explosion

To avoid the risk of explosion, the guidelines for explosion prevention, the regulations concerning electrical systems in spaces where there is a high risk of

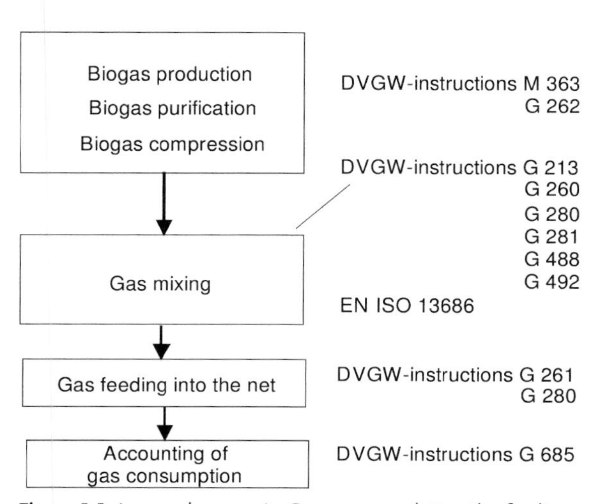

Figure 1.1 Law and norms in Germany regulating the feeding of biogas into natural gas networks.[22],[23],[24]

18) Cp. JOU 11
19) Cp. LAW 14
20) Cp. BOK 63
21) Cp. LAW 7
22) Cp. JOU 11
23) Cp. LAW 4
24) Cp. LAW 5

explosion, the technical regulations concerning flammable liquids and the advices contained in these regulations, and the regulations[25] of the Employer's Liability Insurance Association have to be followed.

The prescribed tests before start-up and the subsequent routine tests are derived from the ElexVO (regulation on electrical systems in spaces where there is a high risk of explosion).

Since explosive gas/air mixtures are to be expected in the proximity of gasholders and bioreactors, so-called ex-zones have to be declared.

1.6.1 Explosion-endangered areas – ex-zones

The explosion limits for ambient temperature can be seen in the triangular diagram (Figure 1.2). They are very dependent on temperature.

Spaces with risk of explosion are graded in zones according to the probability of the occurrence of a dangerous explosive atmosphere.[26] If a dangerous explosive atmosphere can occur in a space, the entire space is to be regarded as highly explosive (Figure 1.3).

Zone 0 covers spaces with a constant, long-term, or frequent (most of the time) dangerous explosive atmosphere which consists of a mixture of air and gases, vapors, or mists.

In biogas plants, the gasholder, the air intake of the combustion engine, the combustion chamber of the gas flare, and under special operating conditions the bioreactor itself belong to zone 0. A special operating condition is given when air enters the interior of the bioreactor. Under normal operation conditions, a small positive pressure prevents the penetration of air into the bioreactor.

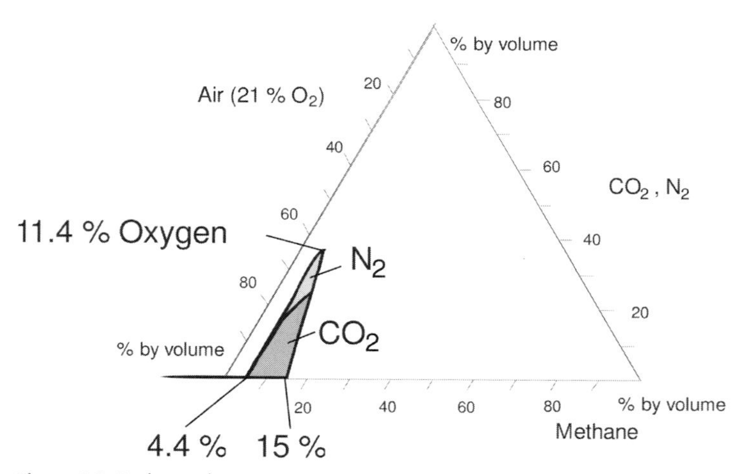

Figure 1.2 Explosion limit at 25 °C.

25) Cp. LAW 24

26) Cp. LAW 16

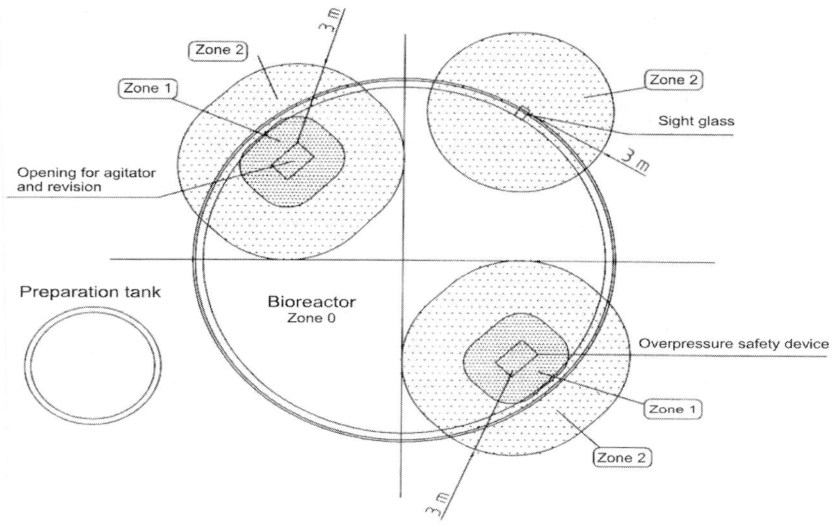

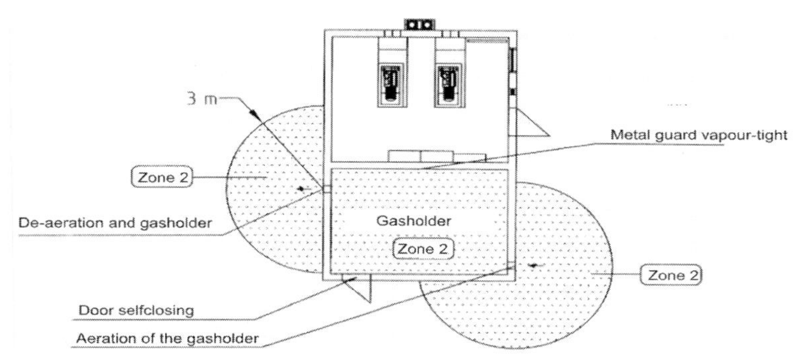

Figure 1.3 Explosive areas in biogas plants.

In the air intake of the combustion engine or in the combustion chamber of the gas flare there is constantly an explosive mixture. The engine and the gas flare must be separated from the other gas system by a flame trap as safety device.

Zone 1 covers spaces with the occasional occurrence of an explosive atmosphere which consists of a mixture of air and gases, vapors, or mists.

In conditions of good ventilation, zone 1 is to be assumed in a region within 1 m from components of the plant, items of equipment, connections, sight glasses, grommets, and service openings at the gasholder and at the bioreactor, but only if leakage of biogas is technically possible. Likewise the space around the mouth of exhaust pipes, positive pressure safety devices, and gas flares is considered to be zone 1. In closed spaces, the endangered space is extended to a periphery of 4.5 m.

Enclosed spaces or pits through which anaerobic sludge flows belong to zone 1.

Zone 2 covers spaces where the occurrence of dangerous atmospheres from gas mixtures cannot be assumed, but if it does occur it is seldom and only for a short time. Zone 2 is in the region 1–3 m from components of the plant technically classified as leak-tight, items of equipment, connections, grommets, service openings, and bursting disks. Open pits (e.g., pits for pumps for anaerobic sludge) or basins, enclosed spaces wherein gas pipes are installed and which do not have ventilation, are zone 2.

The radius of 1–3 m is given for good ventilation. Closed spaces are considered to belong to zone 2 in their entirety. More details about the explosive zones in biogas plants are stipulated[27] by the Employer's Liability Insurance Association.

1.6.2 Checklist of measures for explosion protection

The following checklists are to be considered during the planning stage and, at the latest, before start-up of the plant.

Measures related to the design

Are all parts of the biogas plant readily accessible, sufficiently stable, above ground so that no heavy noxious gas can settle, and well ventilated? (In case of doubt, forced ventilation is particularly necessary before entering the space.)

Were sealed electrical components and electrical appliances installed, which are certified for the respective explosion protection zone and characterized accordingly? Electrical systems and operating equipment in explosion-endangered spaces must in Europe correspond to EN 60079-0, EN 60079-14, and EN 60079-17 and all non-electrical devices to EN 13463?

Were all process control equipment classified according to the European standard of EN 61508/61511?

If, exceptionally, electrical devices which do not correspond to zone 0 (e.g., engines of submerged agitators) have to be installed inside the bioreactor, then explosion prevention has to be assured, e.g., by redundantly implemented float switches which are certified for zone 0 and which are installed above the electrical equipment so that they switch off all loads before the equipment is emerged. Or is safety guaranteed by a monitoring of the minimum positive pressure in the gas space of the bioreactor (to prevent the penetration of air)?

Are all electrically conductive components connected to each other and provided with protective grounding (also known as earthing)?

27) Example: a plastic bag in a wooden housing is assigned to explosion proof zone 2; i.e. in a space with a radius of 3 m around it no CHP may be installed; but if the plastic bag is located in a building made of concrete or steel only the bag itself belongs to zone 2 and the space within a radius of 3 m around the doors and other openings. The CHP could be installed just outside this space.

Are the gravel pots designed appropriate to explosion prevention for a test positive pressure of 1 bar? Does the half cross section of the gravel pot correspond to at least three times the cross section of the associated gas pipe? Is the partition plate between the entrance and the withdrawal side welded gas-tight? Is a seal cap installed made of material stable toward methane gas? Is there a nozzle to connect rinse waters as well as differential pressure measuring devices?

Have the gas filters a pressure loss of 3 mbar max. at nominal flow rate with a new filter body and 6 mbar max. under operation? Are filter bodies installed only of ceramic material or equivalent? Is replacement of the filter cartridges easily possible? Is there a nozzle to connect a differential pressure measuring device?

Are foam traps provided with a water pipe for cleaning purposes? It would be best if the foaming is already dealt with in the gas zone of the bioreactor by a suitable spraying device.

Are condensate traps for draining of gas pipes and gasholders equipped with two three-way armatures, which are locked mechanically, so that gas cannot escape? The manipulation of the armatures can be done either manually or automatically (e.g., with pneumatic drive of the armatures controlled by a pneumatic level control).

Are the gas suction pipes as well as the upstream gas collectors under negative pressure, so that with a leakage in the pipe or in the pipe connection air is drawn into the pipe? If this is the case, are all the prescribed steps taken to prevent explosion in the gas transportation system?

Are all ventilators perfectly sealed against the maximum gas pressure?

Are all covers installed in such a way that no sparking is possible when opening or closing?

Is the pipeline just before a compressor furnished with a pressure-monitoring device which switches off the compressor automatically when the negative pressure drops below the limit, or is it furnished with a negative pressure protection device?

Are explosion-endangered spaces, in particular in direct proximity to the filling mechanism, marked with plates indicating the danger by black writing on a yellow background?

Does the gas alarm system really sound an alarm in case of emergency? Does it switch on the forced ventilation automatically?

Does the substrate supply discharge into the bioreactor at least 1 m below the lowest liquid level?

Is it certain that the following are avoided: transportation vehicles in the plant, screw conveyors with screws that scrape the walls, agitators that scrape the walls of the bioreactor, welding and electrical installation work in the plant during operation?

Are gas-containing pipes and other parts of the biogas plant coming into contact with gas protected against chemical influences and from mechanical damage (e.g., protection against impacts where vehicles move) as well as against overheating?

Is it guaranteed that in a bioreactor with gas hood no negative pressure develops under any circumstances? When discharging residue, is fresh or excess sludge fed into the bioreactor at the same rate? Is a gas pressure measuring device installed on the gas hood of the reactor? Is the water filling of all pressure safety devices daily controlled and is the correct water level maintained? Is the water-filled pressure safety valve installed at a non-freezing point?

Are all closed tanks, in which fermentation can occur, provided with safety devices which prevent positive or negative pressure? Does the sealing liquid in the device stay at low or high pressure and does it flow back to its original position when normal pressure is achieved again? Is the liquid non-freezing? Did the manufacturer certify in writing that the safety device is suitable for the special purpose?

Is no shut-off possibility/armature in the pipe to the positive or negative pressure safety device?

Is it certain that a separate negative pressure indicator in the gas system or an equivalent measurement before responding to the negative pressure safety device will ensure that the gas utilization is switched off?

Does the liquid storage tank withstand five times the maximum pressure which can occur in the bioreactor, especially when it is used as pressure protection and additionally as a condensate trap?

Are measures taken to prevent foaming?

Do the positive and negative pressure protection devices discharge at least 3 m above the ground and/or 1 m above the roof or the edge of the tank or at least 5 m away from buildings and roads? Was it taken in consideration that the space of 1 m around the discharge belongs to zone 1? Was the suitability of the safety device proven by a comprehensible computation, a description of function, and a validation of the construction unit?

Is it possible, that the biogas rate exceeds the flow rate of $20\,m^3\,h^{-1}$ when biogas has to be blown off due to failures of the gas consumption equipment? Is it certain in this case that the flow rate can be decreased to that value within 48 h, e.g., by a flare or a second gas consumption equipment or throttling of the substrate feed? Do the exhaust gases from the flare discharge at least 3 m above ground and at least 5 m away from buildings and roads?

Can the gas production of the plant be reduced by suitable measures when disturbances at the gas consumption equipment are observed, so that the amount of blown-off gas can be kept at a minimum? Suitable measures for the reduction of gas production are, e.g., stoppage of the substrate supply or discontinuation of heating the bioreactor.

Is it certain that all gasholders contain the prescribed minimum gas content, in order to avoid negative pressure in the gasholder? Is the gasholder gas-tight, pressure resistant and media-, UV-, temperature- and weather-proof? The following requirements have to be met by the materials of gasholders, especially when they are constructed of plastics.

- Ultimate tensile strength: 500 N/5 cm (possibly 3000 N/5 cm)
- Tensile strength: at least 250 N/5 cm

- Gas permeability related to methane: <1000 cm^3/m^2.d.bar
- Temperature resistant from −30 °C to + 50 °C (occasionally[28] 70 °C)
- Bleeder resistor less than 3×10^9 Ohm.

Are gasholders and related equipment, e.g., cushion gasholders and bioreactor hoods (which are of elastic material), protected from mechanical damage. This requirement is fulfilled, e.g., with a fence around cushion gasholders. When the fence is closer than 850 mm to the gasholder, it must have a narrow mesh size. The fence must be at least 1.50 m high.

Are all gasholders equipped with positive and negative pressure protection devices?

Are enclosed spaces in which gasholders and/or gas consumption equipment are set up ventilated diagonally (transverse ventilation)? Is it guaranteed that the ventilation equipment will extract the exhaust air below the ceiling? In this case it is immaterial whether the supply air opening is near the floor or the ceiling. Is the supply air opening arranged near the floor and the discharge opening in the opposite wall below the ceiling when natural ventilation is applied? Do the supply air and discharge openings have the minimum cross sections given in Table 1.4?

Where the free minimum cross section of each air opening A calculated according to the equation:

$$A = 10P + 175$$

A = free cross section in cm^2
P = maximum electrical power of the generator in kW.

Is the free cross-sectional area at least 400 cm^2 for each opening? Do all gas discharge pipes lead into the open air?

Are condensate traps and discharge valves installed at the lowest point of the gas pipe? Do the condensate drain pipes flow as closed systems? Automatic condensate traps with discharges into enclosed spaces is not allowed.

Table 1.4 Minimum cross-sectional area for ventilation openings.

Gasholder volume [m^3]	*Cross-sectional area according to German regulations [cm^2]*	*Cross-sectional area according to Austrian regulations [cm^2]*
Up to 100	700	1000
Up to 200	1000	1500
Over 200	2000	2000

28) According to Austrian reulations

Are the condensates disposed off according to the local waste water regulations?

Are two stop valves which close automatically with stoppage of the engine fitted in the gas pipe before each power generator? Is the tightness of the space between the two values regularly checked? Is the pressure in the space between the two values steadily monitored, as the feeding pipe to the engine is constantly under a pressure of >5 mbar even when the engine is out of operation?

Are the two emergency quick-acting shut-down valves, which are located in the safety pipe between bioreactor and gas consumption equipment, controlled, so that gas cannot flow before the engine is running and so that the gas flow is interrupted when, in the gas consumption equipment

- the nominal rotation speed exceeds
- the gas pressure drops to a minimum
- the gas pressure exceeds the maximum gas pressure
- the temperature in the cooling water circle exceeds a given value
- the emergency switch is actuated
- the control energy fails
- the gas-warning and fire alarm systems are actuated
- the temperature monitoring of the enclosed space is actuated
- the ventilation system fails?

Were all technical regulations considered in the areas where the gas consumption equipment is installed?

Is the CHP accessible from three sides?

Are floor drains furnished with oil traps and/or is a catch pan which can take the entire engine oil quantity installed below each generator?

Is the CHP at all times disconnectible by an illuminated switch within and outside of the room where it is installed? Are the switches clearly visible and durably marked "emergency circuit breaker CHP"?

Can the gas feed pipe to the CHP be closed from outside the enclosed space containing it? Are the open and closed positions of the armature indicated?

If engines are used in which the air/gas mixture is compressed by a turbo charger, were the following precautions taken for the prevention of explosive mixtures in the case of accidents?

Air monitoring of the installation area with design-certified equipment and automatic disconnection of the turbo charger and electrical system

or

air monitoring of the installation area with design-certified equipment and automatic disconnection of the turbo charger and simultaneous switch-on of a forced ventilation plant, which is so designed that no flammable mixtures can occur or

forced ventilation of the installation area to give air exchange at a rate exceeding a specified minimum, which guarantees a sufficient dilution of gas escaping at the maximum possible rate. The necessary minimum air exchange amounts to $35\,m^3h^{-1}$ air per 1 kW installed electrical power. The maximum gas concentration is then abt. 1.5% by volume and corresponds thereby to about 25% of the lower gas explosion limit (the limits for biogas are 6–12% by volume.

Does the ventilator work when the shut-down valves are open, and is its function monitored by a flow rate registration?

Are enclosed spaces aerated, into which gas can penetrate and which have to be regularly accessible for maintaining the operation of the plant. Does the aeration prevent dangerous gas mixtures? Is it possible to leave the enclosed spaces without entering the CHP area?

Is explosion-protection provided in the case that the enclosed spaces are not aerated adequately? Are the TLV values for air pollution always below the limiting values?

Are all walls, columns, and ceilings above and below the installation area fire-resistant and made from fire-proof materials if the gas consumption equipment is installed in an inhabited building? Are also the linings and insulations of walls, ceilings, and columns of non-flammable materials?

Do ventilation ducts and other pipes through walls and ceilings not transmit fire and are precautions taken in case of such transmissions, e.g., certified cable bulkhead, fire protection flaps?

Are gaps in the break-throughs filled with non-combustible dimensionally stable materials?

Is the methane content of the exhaust gas from the engines reduced by technical measures?

Do all exhaust gases from the engine released into the open air flow over exhaust stacks/chimneys in accordance with the local regulations?[29] Is the prescribed minimum height of 10 m complied with?

Are the chimneys and ventilating outlets without roofing? As a protection against rain, deflectors can be installed.

Measures relating to the organization

Is it certain that the operation and the maintenance of the biogas plant is done only by reliable and qualified persons?

Did the operators receive work contracts which prohibit smoking on the plant area? At prominent places are there signs reading “Fires, naked flames, and smoking are forbidden!”?

29) Cp. LAW 8

Is it certain that the gas pipe is closed before the bioreactor cover is opened?

Is it certain that the bioreactor is gas-free before entering it?

Are the operators obliged to stay out of the fire-break around the gasholder unless it is essential to be in it?

Were the operators instructed not to use machinery or carry out work which can endanger the gasholder, e.g., welding and cutting? If such work cannot be avoided, other measures have to be taken, e.g., fire prevention, fire protection, fire fighting.

Is a journal obligatory in which all daily measurements, controls, and maintenance work as well as disturbances are documented?

Are the engines maintained according to the timetable given by the manufacturer and if necessary maintained and/or functionally checked by a specialized company?

Is pressurized water and not compressed air used for rinsing sludge pipes?

Is it certain that an operational manual is available before any work is done? Did a specialized craftsman start up the plant? Was the operational manual diligently read by the operators? Is the manual at hand when working inside the plant?

Is it stipulated in the manual that safety devices have to be checked at least once a week and after any disturbance?

Are operating instructions readily available, easy to see, and easy to read for the operators during their work?

Measures relating to the inspections

Is the explosion prevention document[30] conscientiously filled out in order to avoid dangers of explosion due to gases in agricultural biogas plants?

Are measuring instruments installed at suitable places for safety reasons, e.g.,

- Gas measuring instruments (based on the principle of heat coloring or heat conductivity) to evaluate dangers of ignition
- Measuring instruments indicating the composition of the gas or the concentration of components in the gas. The explosiveness of sewage gas can be determined, e.g., with a carbon dioxide test tube (when no carbon dioxide is measured, no methane can be present)?

Is there a plan indicating the explosion protection zones?

Are safety devices and armatures that are installed in the gas-containing areas, the central emergency stop system and automatic emergency devices, e.g., emergency ventilation, and electrical locking mechanisms of the submerged agitators with floating switches examined annually? Regulations specify that emissions must be measured (3 individual measurements) and evaluated with respect to limit values; other parameters have to be determined, e.g. electrical power (kW_{el}), air number (lambda) of the respective engine, sulfur content of the biogas, oxygen content of the engine exhaust gas.

30) Cp. LAW 25

Are all parts of the biogas plant containing a gas flow regularly checked and submitted to a documented pressure test at least every 3 years?

Are emission measurements performed every 3 years by a neutral organization and are the results documented?

1.7 Risk of fire

1.7.1 Fire protection sectors

In order to reduce the fire risk, the plant is to be divided into fire protection sectors, e.g., the bioreactor and gasholder, the gas consumption equipment, and the gas compressor. Certain distances must be maintained between the fire protection sectors. Depending upon how much space is avoilable in between, the material of the external walls of buildings containing equipment or of protecting walls has to be chosen.

Protection distances around above-ground fixed gasholders (Table 1.5)

Fixed gasholders include plastic bags in fixed containers or in enclosed spaces, e.g., in former silage tanks.

The protection distance can be smaller if endangered spaces are covered with earth or a suitable metal guard or fire protection insulation is installed.

Protection distances around underground and earth-covered gasholders (Table 1.6)

For such gasholders the following protection distances are to be provided around armatures and openings.

Table 1.5 Metal guards.

Gas volume per tank [m³]	*Up to 300*	*300 to 1500*	*1500 to 5000*	*over 5000*	*Material of the walls*
Fire break in m	6	10	15	20	Other materials, class B
Fire break in m	3	6	10	15	Noncombustible,
Fire break in m	3	3	6	10	class A, fire-retardant (F 30[31]), vapor tight

31) For the fire resistance classes F 30 and F90 according to DIN 4102 the resistance lasts a minimum of 30 min and 90 min respectively

Table 1.6 Fire breaks (underground gasholders etc.).

Gas volume per tank [m^3]	*Up to 300*	*300 to 1500*	*1500 to 5000*	*over 5000*
Fire break in m	3	6	10	15

Table 1.7 Fire breaks (balloon gasholders etc.).

Gas volume per tank [m^3]	*Up to 300*	*300 to 1500*	*1500 to 5000*	*over 5000*
Fire break in m	4.5	10	15	20

As far as earthcovered containers are possibly impassable, they must be marked and provided with gates.

Protection distances around cushions and balloon gasholders as well as around foil hoods for gas holding above liquid manure storage tanks or bioreactors (Table 1.7)

1.7.2
Checklist for fire protection measures

The following checklist for fire protection measures may be used.

Measures relating to the design

Are the prescribed protection distances realized? Are the fire protection zones free of buildings? Are metal guards correctly dimensioned?

Are all openings in fire walls provided with fire-resistant and self-closing covers? Are the doors in the fire walls at least fire-retardant and self-closing and/or lockable if they do not lead into the open air?

Is it certain that no gases can settle at any place in the factory?

Are gas pipes in all areas insulated to give protection against continuous fire and provided with fire protection flaps?

Are gasholders from flammable materials in fire-protected areas shielded against radiation. The shield should be made of non-flammable materials.

Do escape doors open to the outside?

Are certified flame traps installed as safety devices in all pipes to and from the gasholder close to the consumer according to the prescriptions of the manufacturer? The flame traps must be easily cleaned and correspond to the standards. The often used gravel pots must be design-examined.

Are there adequate and well-marked routes for fire brigade vehicles? Are the roads strongly constructed such that they can be safely used by fire brigade vehicles up to 14 Mg total weight? Are the routes always free?

Are enough fire extinguishers on plant site? At least 12 portable units of suitable extinguishing agent should be availabe per plant or per fire protection sector.

Is a minimum of one portable fire extinguisher available at the gas consumption equipment building, easily seen, easy to reach in case of fire, and always working? Are the extinguishing foams stable in contact with alcohol?

Are there hydrants available which are capable of delivering 1600 L min^{-1} water for a period of 2 hours?

Are all areas clearly marked showing their use?

Measures relating to organization

Are smoking, naked flames, and storing of flammable materials forbidden in the entire area of the plant?

Are fire-protection posts set up and suitable fire extinguishers made available when work involving a risk of fire is carried out such as welding, cutting, abrasive cutting, soldering, etc., and use of a naked flame?

Is storage of flammable materials, flammable liquids, and gases limited to small amounts inside a building? Is no more than 200 kg of engine oil, waste oil, and other flammable materials stored inside the CHP plant area.

Are all hazardous areas and safety areas marked, e.g., entrances to gasholders?

Was a responsible person designated for all fire protection measures?

Are fire protection exercises regularly carried out?

Is the local fire brigade informed about the entire plant in detail and is a fire brigade plan available in accordance with local regulations, e.g., DIN 14096?

1.8 Harmful exhaust gases

From biogas plants, climate-relevant gases and gases harmful to humans can escape, such as ammonia, methane, nitrous oxide, and others. The amount of leakage depends on the applied technology and the substrates.

1.8.1 Prescriptions and guidelines

The German law of Immissions protects humans, the environment, and property against immissions by prohibiting emissions (Table 1.8) from industrial facilities. According to this law, biogas plants need special permission[32] if

- more than 10 Mg d^{-1} wastes are co-fermented, if these are wastes which have not to be specially monitored and which are listed in the attachment to the law.
- more than 1 Mg d^{-1} wastes are co-fermented, if these are wastes for which special monitoring is prescribed.
- the volume of stored wastes exceeds 2500 m^3.

32) Cp. BOK 17

Table 1.8 Emissions from liquid manure (without measures to minimize emissions such as scum).

Emissions	*Possible sources of emissions*	*Compared to unfermented liquid manure [%]*	*Remarks*
Germs	Preparation tank, storage tank		When the plant is well operated, the number of germs is decreased (e.g., by hygienization, intermediate pH value decrease, ammonia content). A germ increase is possible if co-ferments with inappropriate pretreatment are used.
Odors	Storage, feeder, preparation tank, storage tank, distribution	(−31)	Odor emissions tend to be decreased by the degradation of smell-causing substances in the liquid manure. They can be increased, however, by the use of co-ferments, improper equipment technology, and impurities.
Ammonia	Preparation tank, storage tank, distribution	+21 to +64	Ammonia emissions increase due to an increase of the pH value and the temperature in the storage tank corresponding to a change of the nitrogen compounds in the ammonium nitrate
Methane	Storage tank	−67 to +87	
Nitrous oxide (N_2O)	Preparation tank, storage tank, distribution	−36 to −72	

- engines of more than 350 kW of combustion heat performance (FWL) of the CHP are installed or the total combustion heat performance is higher than 1 MW. The obligation to carry out a detailed examination of environmental effects exists starting from a FWL of 50 MW.[33] Small plants may possibly have to pass a simplified process for approval.
- gasholders with more than 3 Mg gas content are to be set up.

33) Combustion heat performance (FWL) is the maximum fuel rate necessary for the maximum load of the plant (lower calorific value multiplied by the fuel rate per hour)

- the biogas plant is to be installed close to an animal husbandry plant for which permission is obligatory.

An environmental compatibility examination may be necessary. Limiting values for emissions are given in local regulations.

If, in the plant, more than 10 000 kg biogas, equal to 8333 m^3 of biogas (60% CH_4), is stored, other regulations might be relevant.

In plants with a throughput of more than 10 Mg d^{-1} of wastes, emission-reducing measures, e.g., a computation of the smell propagation, must be taken or a minimum distance to settlements must be maintained. This must be

- 300 m with encapsulated biogas plants (shelter, fermentation, rotting),
- 500 m with open biogas plants.

1.8.1.1 Germs

The microorganisms taking part in the fermentation process are mainly to be assigned to the group 1 risks and to a small extent to the group 2[34] risks. Therefore no special protection devices are necessary. But substrates and also residues often contain organisms which do not participate in the process, like viruses and parasites, which can have severe effects on the health of humans as well as on the environment (Table 1.9). Therefore certain measures for protection are necessary.

For all who work in biogas plants, the danger of an increased ingestion of endotoxins exists. Endotoxins are metabolic products of microorganisms. They are taken up from humans via skin cracks. At low concentrations (<200 EU[35] m^{-2}), endotoxins can induce fever, and at high concentrations (>200 EU m^{-2}) a stimulation of the mucous membrane; respiratory diseases up to chronic inflammations of the respiratory system can be evoked. Increased endotoxin concentrations (>50 EU m^{-2}) could be found with increased aerosol formation, e.g., at the exhaust of a capsulated surface blower of an activated sludge tank, when cleaning a chamber filter press, when cleaning pump pits, etc.

People living in the neighborhood are not endangered in general, since the germ concentration in the air is low and independent on the weather conditions.

1.8.1.2 Emissions of smells

Smell-intensive materials in biogas plants include particularly ammonia, organic acids, phenol, and hydrogen sulfide (H_2S). Smells are often causes of annoyance and/or complaints obout biogas plants. One differentiates between

perceptible smells	reasonable in general
perceptible and strongly perceptible smells	reasonable up to 8% of the yearly hours
strongly perceptible smells	reasonable up to 3% of the yearly hours

34) Cp. LAW 26

35) EU = Endotoxin-units

Table 1.9 Infectious agents for humans (mainly present in sewage sludge (1)[36] or in liquid manure and organic wastes (2)) (R = Class of risk according to the European guideline 2000/54/EG).

Infectious agent	*R*	*Sources*	*Uptake and illnesses provoked*	*1*	*2*
Bacteria					
Leptospira sp.	2	Waste water (urine of rats), livestock husbandry, slaughterhouse	Uptake via the softened and injured skin. Leptospirose often not discovered, rare but severe illness (fever, ague, meningitis)	X	X
Escherichia coli	2	Stomach-intestine bacterium, which is excreted.	Oral uptake A possible reason for stomach-intestinal illnesses, diarrhea	X	X
Salmonella sp., Shigella sp., S. typhii	3	Releaser of salmonellae, water, soil, plants	Smear infection, oral uptake Low risk of infection; only when epidemic, higher risk: diarrhea with vomiting, fever, heart desease	X	X
Yersinia enterocolitica	2	In stables, especially in piggeries	Oral uptake after contact with meat of pigs or waste water Fever, diarrhea	X	X
Klebsiella pneumoniae	2	In the stomach-intestine tract, soil, water, cereals	Oral uptake Only with low immunity are urinary passage and airway problems caused.	X	X
Pseudomonas aeruginosa	2	Widely spread on soil and in water	Contact Infection of the skin, urinary passage, and meningitis	X	
Staphylococcus sp.	2	Ubiquituous, not specifically in biogas plants	Injuries and oral uptake Acute inflammation, sanies, food poisoning	X	X
Streptococcus sp.	2	Livestock (pigs, horses)	Via injuries by direct contact Inflammation of the endocardium, meningitis, arthritis	X	X
Trichinella spiralis, Trichinella pseudospiralis	2	Domestic and wild animals: e.g., pigs, fox, marten, bear, wolf, rats)	Oral uptake, smear infection Fever above 39 °C, face edema		X
Clostridium tetani, C. botulinum	2	Ubiquituous, not specifically in biogas plants	Injuries of the skin Tetanus	X	X
Bacillus anthracis	3	Cattle, sheep, goats, horses, buffaloes, camels, reindeer, mink, very seldom pigs and carnivores	Percutaneous through injuries, aerogenic and oral uptake Danger especially when working with cadavers and fats Anthrax		X

36) Cp. JOU 9

Table 1.9 *Continued*

Infectious agent	*R*	*Sources*	*Uptake and illnesses provoked*	*1*	*2*
Brucella sp.	3	Cows, sheep, goats, pigs, dogs	Injuries of the skin and mucosa through dust expulsion from infected animals into the air. Bacteria-containing milk and milk products Brucellosis, Malta fever	X	X
Mycobacterium sp., Mycobacterium tuberculosis	3	Soil, water	Aerogenic, smear infection (sputum, milk, urine, and excrement of infected animals, oral uptake, injuries.	X	X
		In the lungs of humans	Abscess of the skin, tuberculosis		
Enterobacter	2	Ubiquituous	Infection of the urinary passage and airways, meningitis	X	X
Erysipelothrix rhusiopathiae	2	Pigs, poultry, crustaceans, and fishes	Via injuries to the skin in contact with infectious material. redness (red murrain)		X
Fungi					
Candida sp.,	1–2	Material from rakes	They are able to cause infections and allergies	X	
Aspergillus sp. (fumigatus)	2		Depending on the species, different ways of infection possible Risk of infection very low		
Parasites					
Round worm	2	Because of their weight, protozoans and worm eggs sink to the bottom of the bioreactor and are enriched in the residue, where they can survive a long time.	Mainly oral uptake		X
Gastrointenstinal worm			Are able to provoke stomach-intestinal illnesses		X
Hookworm			Low risk of infection		X
Distome					X
Liver fluke					X
Lung worm					X
Stomach worm					
Tapeworm				X	X
Nematode				X	X
Viruses					
Polio Virus	2	Excreted	Oral uptake Diarrhea, polio, meningitis Extremely low risk of infection	X	

Table 1.9 *Continued*

Infectious agent	*R*	*Sources*	*Uptake and illnesses provoked*	*1*	*2*
Hep.-A Virus Hep.-B Virus	2 3	High concentration in the waste water in autumn due to travellers coming home	Oral uptake Icterus Certain risk of infection	X	
Enteroviruses – Coxsackie-A Virus – Coxsackie-B Virus – Echo-Virus	2	Proliferating in the intestinal mucosa	Transfer from human to human via droplets or dirt Fever, flu, Meningitis, encephalitis, hepatitis, pneumonia, myo- and pericarditis	X	X
HIV Virus	3	Virus is not stable in waste water	Transfer through blood or other body liquids AIDS Very low risk of infection	X	
Rota Virus	2	In excrement	Oral uptake	X	X
Noro Virus	2		Diarrhea with vomiting	X	
Picorna Virus	2	In saliva, nasal discharge, sperms, and milk of infected animals	Transfer through contact with animals, persons, vehicles via the air Virus of aftosa The virus is inactivated by temperatures above 56 °C or acids		X
Swine fever		In meat of pigs, even in cured and frosted meat.	Sputum, secretion of the eyes, breath Fever, cramps, palsy, bleeding, cardio-vascular desease		X
Paramyxovirus	2	Poultry	Avian plague (Newcastle Illness), seldom transferred via air to owner and laboratory personal Trachoma, sometimes flu		X
Protozoans		Waste water, sewage sludge	Oral uptake Epidemic (malaria, sleeping sickness, amebiasis)	X	X
Entamoeba histolytica	2		diarrhea	X	
Giardia lamblia	2		diarrhea, fever		
Toxoplasma gondii					

Table 1.10 Toxicity of hydrogen sulfide.[37]

Concentration (in air)	*Effect*
0.03–0.15 ppm[38]	Threshold of detectability (odor of bad egg)
15–75 ppm	Irritation of the eyes and the airways, nausea, vomiting, headache, unconsciousness
150–300 ppm (0.015–0.03%)	Palsy of the olfactory nerve
>375 ppm (0.038%)	Death through poisoning (after some hours)
>750 ppm (0.075%)	Unconsciousness and death by apnea within 30–60 min
Above 1000 ppm (0.1%)	Sudden death by apnea within a few minutes

The odor emissions depend particularly on the composition of the materials used (liquid manure, co-ferments). The composition of the liquid manure varies depending upon animal species, stable technology, feeding, and water requirement for the cleaning of the stable (dilution).

Hydrogen sulfide Hydrogen sulfide, a mostly inevitable component of biogas, not only smells, but can have a toxic effect in higher concentrations (Table 1.10).

Ammonia Ammonia (NH_3) has a strong smell, induces eutrophy, and is indirectly relevant to the climatic situation, since ammonia is partly converted to nitrous oxide (N_2O) in the soil.

The degradation of organic substances results in nitrogen in the form of ammonium in the substrate. Since during the degradation the pH value increases about 1 unit due to the decomposition of acids and the temperature in the residue storage tank is high due to the high processing temperature in the bioreactor, a lot of volatile ammonia is liberated.

Nitrous oxide (laughing gas) Nitrous oxide (N_2O) smells slightly sweet. It can cause cramp-like laughter, hallucinations, and intoxication. Nitrous oxide is climatically relevant (CO_2 equivalent = 310).

Nitrous oxide can develop in higher concentrations, when nitrite and nitrate are decomposed under anaerobic conditions together with easily degradable organic compounds (denitrification) or when ammonium is oxidized under aerobic conditions.

1.8.2 Checklist for immission prevention measures

Harmful gases and smells have to be prevented in all parts of the plant.[39]

37) Cp. BOK 61

38) ppm = parts per million (1 ppm = 0.0001%)

39) Cp. JOU 9

Measures relating to design

Are the roadways sufficiently finished?

Are they cleaned frequently and sufficiently, especially in the delivery and the storage areas? Are closed containers used for transportation and storage?

Are there deflectors for gusts of wind and guards, e.g., compact plantings?

Is the biogas plant sufficiently ventilated (Table 1.11)?[40]

Are there places for washing boots and protective clothing, several facilities for washing hands, mechanisms for drying of wetted work, and protective clothing?

Do pits and pits underground have forced ventilation or a guard for warnings of dangerous gases?

Are there exhaust-gas decontamination plants when the throughput of the plant exceeds $30\,Mg\,d^{-1}$ wastes, so that emissions of smell-intensive materials in the exhaust gas are limited to $500\,GE\,m^{-3}$? Are the local prescriptions considered?

Is it certain that no leakage can occur from the delivery of substrates from the bioreactors, storage tanks, gasholders, pipework, or vehicles which deliver the residue?

Are measures taken for reduction of emissions arising from the use of dusty materials, e.g., husks, bonemeal, dry chicken excrement, milk powder, or whey powder?

Are there measures for odor emission reduction when feeding particularly smell-intensive materials like wet chicken excrement, wastes from the bio waste collection, kitchen and cafeteria wastes, and putrescent vegetable wastes?

Are stinking materials such as chicken excrement stored in a closed building and not in the open air? Are rats and vermin consequently kept away?

Are the storage tanks and the preparation tank provided with gas-tight covers?

Are the receiving conveyors and the lorry sluices in a closed building? Can the air inside the closed building be sufficiently aspirated off, especially when unloading refuse collectors? Is the aspirated air decontaminated, e.g., in a biofilter or exhaust gas scrubber?

Is the composition of the substrate in the substrate storage tank adjusted so that only a little methane is formed? The composition depends on the animal husbandry, fodder composition, and techniques in the stable. The liquid manure of

Table 1.11 Air renewal.

Plant sector	***Air renewals per hour***
Truck sluice	3–5 times
Storage	3–5 times
Pretreatment	2–5 times
Engine hall	1–1.5 times
Rotting hall	1 time

40) Cp. BOK 68

ruminant animals, for example, is very well inoculated with methane-forming bacteria. In contrast, pig liquid manure must develop a biocenosis for methane-forming microorganisms. The fermentation is promoted or retarded depending on whether the plant is operated with or without pre-fermentation and with complete or partial emptying of the storage tank.

Is unfermented or only partly degraded raw material prevented from reaching the residue storage tank, even during disturbances like biological change of the plant?

Is storage temperature kept low?

Is the substrate supply spatially closed? In the case of open feeding, considerable odors are released because of increased aerosol formation.

Are substrate fats from the grease removal tanks and flotate fats sanitized as a precaution?

Is all dust-containing air from the substrate preparation plant (comminution, sieving, changing storage place) collected and passed to a decontamination plant, so that the mass concentration in the exhaust air does not exceed $10\,mg\,m^{-3}$?

Is the pH value controlled? The pH value can vary considerably due to the composition of the substrates and thus influence the living conditions of the microorganisms.

Are suitable measures taken to prevent the escape of gases when the positive-pressure safety device opens?

Is the gas regularly analyzed for its H_2S content?

Are the emissions of the biogas combustion engines below the limits listed in Table 1.12?

Are there samplers in the exhaust pipe of the CHP which are large enough, easily accessible, and so constructed that an emission measurement representative

Table 1.12 Limits for the emissions of combustion engine related to dry exhaust gases, (273.15 K, 101.3 kPa and 5% oxygen-content).[41)]

Capacity	*<1 MW*		*>1 MW*	
	Gas motor	*Ignition oil Diesel engine**	*Gas motor*	*Ignition oil Diesel engine**
Dust	–	50 mg/m³	–	20 mg/m³
Nitrogen oxides like NO_2:	0.5 g/m³	1.5 g/m³	0.50 g/m³	1.0 g/m³
Carbon monoxides	1.0 g/m³	2.0 g/m³	0.65 g/m³	2.0 g/m³
sulfur oxides like SO_2:			0.31 g/m³	0.31 g/m³
Formaldehyde			60 mg/m³	60 mg/m³

41) For ignition oil Diesel engines the percentage of ignition oil has to be kept as low as possible (guidance figure: 10%), and the consumption of ignition oil has to be continuously documented.

Table 1.13 Limits for substances in exhaust air according local regulations (30. BImSchG).

Harmful substance	*Long term value*	*Short term value*
Aerosol	0.15 $mg\,m^{-3}$ exhaust	0.3 $mg\,m^{-3}$ exhaust
Carbon monoxide	10 $mg\,m^{-3}$ exhaust	30 $mg\,m^{-3}$ exhaust
Sulfur dioxide	0.14 $mg\,m^{-3}$ exhaust	0.4 $mg\,m^{-3}$ exhaust
Nitrogen dioxide	0.08 $mg\,m^{-3}$ exhaust	0.2 $mg\,m^{-3}$ exhaust

Table 1.14 Limits for substances in exhaust air according local regulations (30. BImSchG).

Average value	*Over one day*	*Over half an hour*	*Over one month*	*Of a sample*	*Over the time of sampling*
Dust	10 $mg\,m^{-3}$	30 $mg\,m^{-3}$			
Organic matter (TOC)	20 $mg\,m^{-3}$	40 $mg\,m^{-3}$	55 $g\,Mg^{-1}$		
Nitrous oxide			100 $g\,Mg^{-1}$		
Odors				500 $GE\,m^{-3}$	
Dioxins/furans					0.1 $ng\,m^{-3}$

Table 1.15 Limits for ammonia and organic matter.

Parameter	*Discharge rate [kg/h]*	*Average value over one day [mg/Nm³]*
Ammonia	≤0.15	≤30
Organic matter as C_{org}	≤0.50	≤50

of the emissions from the plant is obtained? Does the location of the sampler comply with the local regulations?

Is it possible to remove all emissions to the extent that the analysis figures fall below the limit values given in Table 1.13?

Were all pollutants removed in the decontamination plants in such a way that the limiting values[42] (see Table 1.14) are adhered to?

Is ammonia removed in the decontamination plants such that the values go below the values in Table 1.15?

Is the percentage of the still degradable material in the residue for disposal low, and is the germ concentration thereby lowered in the exhaust air because of the hygienization effect, provided there is normal operation?

Is the duration of storage of the residue as short as possible? Otherwise difficultly degradable fiber-rich substrate components degrade in the storage tank, whereby emissions are released.

42) Cp. WEB 41

Is the residue tank emptied without creating and/or releasing aerosols?
Is the residue transported in closed containers?
Is the fermented liquid manure distributed for agriculture according to the rules of good technical fertilization practice? Dragging-shoe and/or dragging-hose injection procedures are to be preferred to procedures with deflector plates from the point of view of air pollution.

Measures relating to the organization

Are all necessary personal hygiene measures given in local regulations considered, mainly from the Employer's Liability Insurance Association, e.g., TRBA 500 in Germany.
Were the operators instructed, on the basis of written instructions, to wear protective clothing such as work clothes for skin protection, gloves to protect against mechanical injury and against chemicals and microorganisms, safety shoes, eye protection against splashing, and respiratory protection against airborne germs?
Are the operators inoculated, in accordance with the recommendations of the Employer's Liability Insurance Association?
Is cleaning work carried out from safe locations and/or from the downwind side?
Are weather conditions considered when work is planned?
Are dirty implements and vehicle cabs properly cleaned?
Are rats and other vermin effectively conrolled and eliminated?

Measures relating to control

Are biogas plants, in which dangerous co-ferments are sometimes fermented, frequently checked to determine whether the regulation is met, that germ reduction should be kept around 4 powers of ten?
Are gas detectors with test tubes installed for the evaluation of health dangers? (In these detectors, gas is pumped through a test medium.)
Are there written procedures for pest control, cleaning, hygiene control, and sampling at the necessary time intervals, and is the necessary equipment provided?

1.9 Noise protection

Noise is defined to be disturbing sound. Any location where sound does not cause disturbance, even it is very loud, does not incur restrictions.
The area in a biogas plant where noise is most intense is near the gas engine. Near CHP plants, the limiting value 80dB (A) for workplaces is far exceeded. The noise radiates through the exhaust pipe and the ventilation openings of the plant area.

Transportation of the co-ferments, pumps, compressors, and emergency cooling systems also causes noise.

1.9.1 Regulations and guidelines

The German law of immissions protects humans from both emissions and noise. Noise must be avoided in principle, with special attention to the following:

- Sounds from the biogas plant, including sounds of traffic, have to be below the limiting values of Table 1.16. The sound has to be measured at the nearest place of the imission. The limiting values depend on the time of day and the usual noise in the area.
- On-off short noises may not exceed the indicated values during day time by more than 30 dB (A) and at night by more than 20 dB (A). In specified areas the noise has to be kept 6 dB (A) below the indicated value.
- The operation of engines, machines, and plant must correspond to the state of the art of noise protection.
- Impact sound-radiating plants must be decoupled from airborne sound-radiating buildings and components.
- In exhaust gas pipes and/or in openings for ventilating enclosed spaces, sound absorbers have to be installed.
- Doors, gates, and windows of the generator house must be closed when the engine is under load.
- The space close to the generator house must be noise-protected by sound-damping measures according to local regulations.
- The vehicular traffic as well as the operation of wheeled loaders or other transportation machines must be limited to the period of 06.00–22.00 hours, excluding vehicular traffic in the course of the harvesting raw materials and distributing residues.

The relevant recommendations of local organizations can be taken as guideline.

Table 1.16 Limits for noise according to local regulations (in Germany: TA-Lärm).

	Time	*Housing area*	*Housing and commercial area*	*Commercial area*	*Industrial area*
Day	06:00–22:00	50 dB(A)	60 dB(A)	65 dB(A)	70 dB(A)
Night	22:00–06:00	35 dB(A)	45 dB(A)	50 dB(A)	70 dB(A)

Figure 1.4 Sound-absorbing wall.

1.9.2
Checklist for noise protection measures

The following questions facilitate the evaluation of noise protection measures:

Do all workers and operators wear ear protection correctly? Is the area of the CHP marked by an appropriate sign?

Are all pipes and piping break-throughs insulated against noise?

Does the plant stand well away from settlements?

Are the plant areas of the CHP and the CHP itself sufficiently sound-insulated, e.g., by insulation of the area with an additional brick lining with transverse clay bricks (Figure 1.4), which break the acoustic waves due to their large surface, so that a sound level of less than 55dB (A) is to be measured outside the wall?

1.10
Prevention of injuries

For protection from injuries there is an abundance of regulations and recommendations, which cannot all be specified.

Some risks of injuries are avoidable by use of the following check list of preventive measures.

Measures relating to design

Are the machine safety regulations (MSV) applied? If so, the CE sign has to be marked in Europe on all machinery together with a declaration of conformity.

Are precautions taken to prevent falls during operational, maintenance, and control work?

Are all components sufficiently illuminated?

Measures relating to organization

Are there operating instructions for all machines and have all operators been adequately instructed?

Are instructions given concerning operation, monitoring, behavior in special situations such as damage and disturbances, required tests, responsibilities, and authorization for access to the electrical installations.

Measures relating to control

Have all electrical devices been validated by an expert?

1.11 Protection from water

A big danger to the environment arises when water, e.g., from pressing plants or contaminated precipitation water, penetrate into the soil or, even worse, reach the groundwater. Main causes can be

- insufficiently tightened ground within the plant site
- cracks in tanks and/or in the crankshaft housing
- corrosion of pipework.

1.11.1 Regulations and guidelines

In biogas plants, materials capable of polluting water are treated. In nearly all countries it is prohibited by law to contaminate water. Best available measures must be taken in biogas plants to protect water from contamination and even more from unfavorable changes of water at any place.[43)] Recommendations are given by the authorities.[44)]

The distribution of residue in agriculture must be done according to prescribed methods.

1.11.2 Checklist for water protection measures

Choice of location

Is the biogas plant planned outside an area liable to flooding? A biogas plant may not be built within an area which was flooded within the last 30 years. If it is constructed in an area which was flooded within the last 100 years, it must be guaranteed that the plant cannot be damaged by the water pressure or by flotsam and that the safety devices operate even at the highest water level.

Is the distance to any surface water at least 20 m?

43) Cp. BOK 27

44) Cp. LAW 11

Is the distance from existing house wells which provide private potable water at least 50 m?

Are all requirements fulfilled for occasionally flooded areas, areas liable to flooding, and diked areas?

Are the requirements fulfilled for all parts of the plant which reach into the groundwater layer when the groundwater level is high?

Measures relating to the design

Are the plants robust and able to withstand the loads which can be expected? The watertightness of the plants must be reliable and easily monitored. Leakage or overflowing of the substrate and its penetration into groundwater, surface waters, and drains must be reliably prevented.

Are joints and connections of prefabricated parts durably sealed? The joints must have an adequate certificate.

Are the parts of the plant containing highly flammable, flammable, and potentially water-polluting materials provided with high-quality sealing systems, e.g., at least double mechanical seals with sealing liquid for pumps, comb-shaped seals for flange connections, and dry state-of-the-art couplings for hose connections?

Are the tanks made from certified reinforced concrete with high water penetration resistance and high resistance against strong chemical attacks? Is cracking of the concrete unlikely?

Are the inner surfaces of the walls of concrete blocks and a 0.5 m broad strip at the bottom protected by a suitable, permanently elastic and crack-bridging coating or lining? Bioreactors and tanks may only be made from concrete blocks or concrete laid in situ, if installed above ground. The suitability of this coating or lining is to be proven.

Are all tanks and reactors which are made from corrodable material, e.g., unalloyed steel, corrosion protected by coating or lining at the zone of the liquid level of the manure and above in the gas dome and around the inlet pipes?

Are tank bottoms made from reinforced concrete without joints?

Is protection from impact installed where necessary at a sufficient distance from the tanks and from above-ground pipework?

Are all tanks inserted so that they cannot lift?

Are underground tanks double-walled or installed in a trough with an alarm device for leaks? Single-walled tanks should not be installed, even with leak detection equipment. If this cannot be avoided, a plastic leakage recognition mat at least 1 mm thick must be placed below the tank, so that even small leakages can be recognized.

Is there a catchment area next to above-ground bioreactors for leaking substrate? The catchment area has to be waterproof. For the calculation of the volume of such an area, the free volumes of other tanks should be included, e.g., the residue tank. The precautions in case of cracks must be documented.

Is there a trough below the biogas utilization equipment that is adequate to the volume of the liquid, which could run onto the floor in the worst case?

Is there a leakage catching device and a drainage layer below the bioreactor and relevant tanks? Is this layer above groundwater level? Are the pits for catching leakages protected against rain?

Are slidegate valves installed that are easily accessible in a water-tight pit? Are all return pipes according to the recommendations of local authorities (DIN 11832 for agricultural technology, armatures for liquid manure, slidegate valves for static pressures up to 1 bar)? Are pumps installed easily accessible? Are the preparation tanks, the pump sump of the pumping station, and open or covered channels manufactured in a water-tight condition? If their volumes are more than 50 m^3 the same requirements have to be fulfilled as those placed on tanks.

Are the co-ferments protected against rainfall? Are they stored on water-tight ground? Are the liquids from the ground sealings correctly disposed of, e.g., into the bioreactor?

Does the sealing layer for the sealing of the underground consist of a welded plastic, e.g., polyethylene sealing foil with a thickness of at least 0.8 mm or of a mineral seal?

Are all pipes above ground, if possible? Single-walled pipes should be avoided. In water protection zones such pipes are not allowed. Here they have to be double-walled or must be fitted inside a water-tight protective pipe with visible discharge to the control pit.

Are all pipe grommets or connections to the tanks durable, tight-fitting, firm, and flexible, even the heating pipes?

Are all pipes of corrosion-resistant material? Is the nominal pressure of all pipes higher than the maximum pump pressure? Are all pipes without seams or welds? Was the soil well consolidated before installing pipework? Are all pipe connections to the tanks below the liquid level in the tank? Are the pipes to the tank closable by two slidegate valves? Is one of the two valves an emergency shut-down a slide-gate valve? Are underground pipes easily controllable? Are above ground pipes adequately fixed by mounting plates?

Measures relating to control

Was the tightness of bioreactors, tanks, and pipework checked before-start up of the plant? Were water level examinations and pressure tests applied? Were all tests conducted by qualified persons?

Are these tests to be repeated regularly? Are the test procedures and inspection reports checked by qualified persons before and after the tests?

Are the foundations regularly checked?

Were the recommended inspections performed before start-up?

Was self-monitoring regulated, e.g., standard operation procedures formulated, annual reviews (inspection, operation control) organized of all accessible components of the plant, daily control established of the regulated level of the bioreactor, the documentation of self-monitoring in a journal stipulated, the immediate informing of the authorities provided for in case of possible leakages?

Has a design qualification, operational qualification, and performance qualification taken place before start-up?

2
Building a biogas plant

The procedure for the planning and construction of a biogas plant is illustrated in Figure 2.1.

For the planning and constructional phase of a biogas plant, work has to be done by external companies. Contracts in conformity with local laws and regulations have to be arranged.[45]

2.1
Feasibility study

The decision to build a biogas plant is made based on a feasibility study.[46] The market for products, the supply of raw materials, the infrastructure, and the recruitment of qualified operators are taken into consideration. The final result is a cost-benefit calculation based on assumed values.

2.2
Preliminary planning

In the preliminary planning stage, the preconditions for the construction of a biogas plant are investigated. Engineering consultants are often contracted for this purpose. Current planning aids for engineering are often available from associations and authorities.

Planning includes, among other things

- confirmation of all parameters of the feasibility study, e.g., by fermentation trials and/or simulation calculations
- selection of the plant location
- assembling documents on such topics as expert opinion about building ground, official layout plans, development plans, permission and contracts for the supply of electrical power

45) Cp. DIS 2

46) Cp. WEB 110

Biogas from Waste and Renewable Resources. An Introduction.
Dieter Deublein and Angelika Steinhauser

ISBN: 978-3-527-31841-4

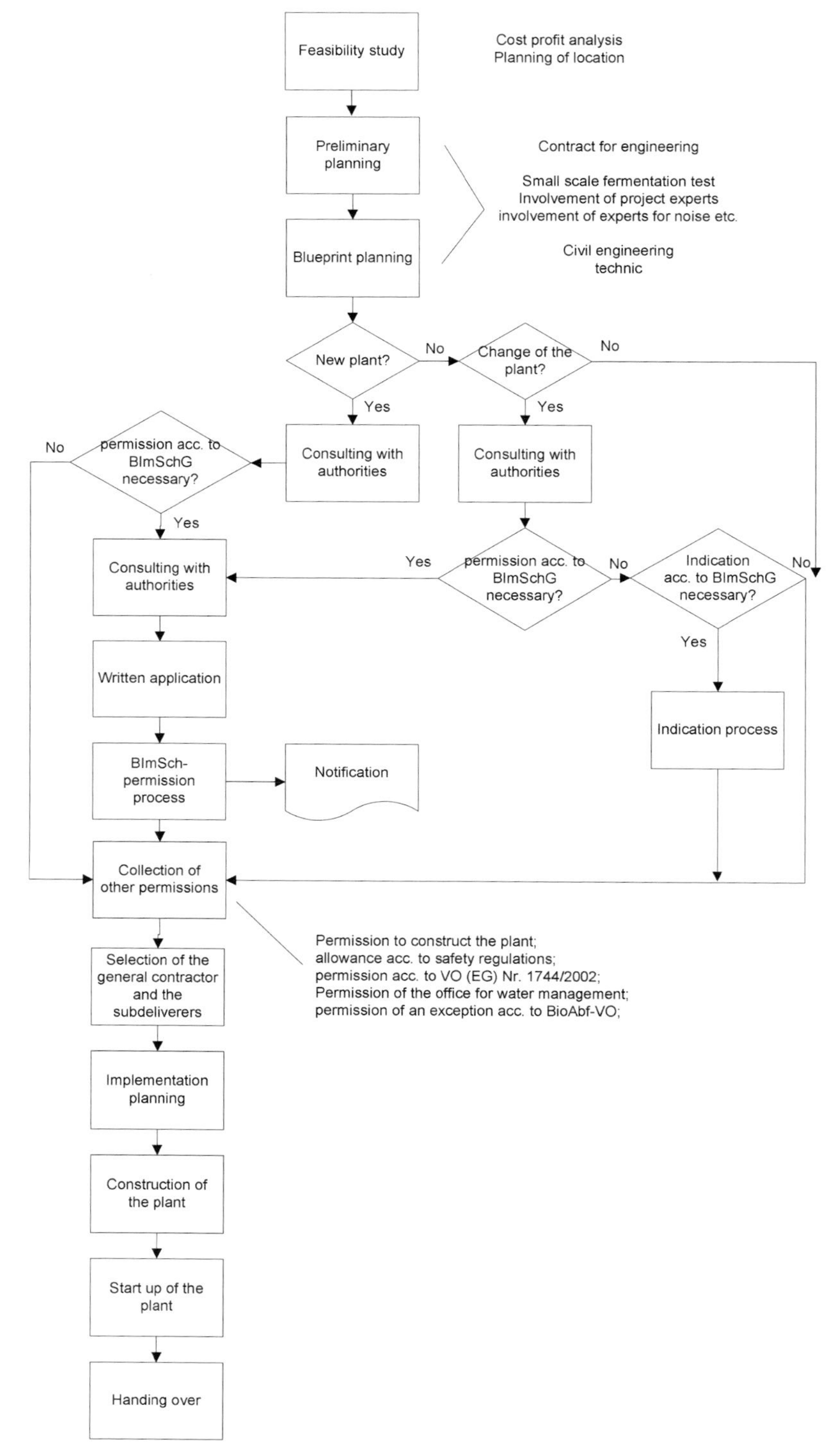

Figure 2.1 Process of planning and setting up a biogas plant.

- short description of the project with the key development goals
- process flowsheet, building, and general layout plans
- time schedule
- costs, planning and financing
- determination of basic data by, e.g., exact computation of the quantity of substrate and yield of biogas, also exact analyses and listing of all raw materials with minimum and maximum quantities required per day
- verifying the acceptability of such matters as soundness of construction, including fire resistance of load-bearing construction units, protection against fire, noise, and heat, structural industrial safety, etc.

During planning, the necessary permission from the authorities have to be obtained. Only small projects do not need such permission. Independently of all permission, all regulations and the state-of-the-art technology have to be considered.

Application for a permit according to laws relating to environmental protection[47)]

An application for permission for the construction of a biogas plant should be broken down into:

General

Purpose of the plant and short description of the process
Location of the plant
General layout plans
Detailed description of the processes and operation of the plant
Description of all technical equipment including costs
All raw materials and products
Measures for the protection of the environment (emissions, noise, vibrations, heat recovery, soil protection, water protection)
Measures for safety-related risk analyses, explosion protection, operation manuals
Measures for the shut down of the plant (plan for recultivation of the places)

Journal for the construction and operation[48)] *and factory regulations*

At the latest, at the start-up of the plant factory, regulations for the smooth operation have to be published at the entrance of the plant with all safety regulations.

47) Cp. DIS 2

48) Cp. BOK 15

For proof of the normal operation of the plant, a maintenance and/or operational journal has to be written indicating in particular the following data:

- origin, quantity, kind, waste code, delivery note, and if necessary identifying analyses of the substrates
- special occurrences, above all operational disturbances, e.g., escape of gases including their causes as well as measures for prevention
- times of operating and outages of components of the entire plant
- all maintenance work, e.g. spark plug changes of the gas engine, oil injection nozzle change at the Diesel engine, any substantial repair work, and all changes of the engine settings
- results of daily performance controls (standby status, normal condition)
- results of the routine emission measurements (NO_X and CO) as well as the orienting emission measurements, which are usually carried out at least annually during the maintenance period. The records should be integrated into the journal, so that they can be monitored by the authorities
- results of regular monitoring of the H_2S content of the biogas and/or the SO_2 content in the engine exhaust gas. These analyses should be carried out by the operators in the performance test of the desulfurization plant in order to maintain the engine by minimization of wear or optimization of the oil change intervals
- results of the regular pH value measurements and organic matter as COD measurements in the bioreactor, in order to maintain optimal gas production conditions, which influences the plant feed rate, feed composition, and the capacity of the plant
- residues of the plant and their disposal with date, kind, place and quantity.

The journal must be submitted to the supervisory authorities regularly for inspection.

2.3 The construction process

For the construction process, companies have to be contracted. In order to avoid problems at the interface between different companies, a general contractor is recommended. Otherwise the following contractual obligations are to be considered:

- Contract with an engineering consultant for the detailed engineering and the supervision of the construction
- Contracts with several subsuppliers for underground and structural engineering, tank and bioreactor construction, foil cover and gasholder, soil circular scraper, conveying engineering, heating, CHP, electrical installations, control engineering, pumping and agitating technology.

The contractors have to guarantee the buildings for five years and all the other technology two years starting from the time of the total acceptance.

The handing over of the plant to the owner can be accompanied by the engineering consultant. The latter is responsible for the final inspection and the final inspection of the documentation.

3
Financing

Biogas plants are financed from own resources, credit, and public promotion. The basis is a feasibility study which convinces investors and creditors of the technical feasibility, economy, environmental compatibility, and general creditworthiness. Therefore the following questions have to be answered:

- Who will be the general contractor and what relevant experiences has he?
- Will the plant work economically, based on a turnover and a profit preview with cost-benefit calculation?
- Is the risk well distributed amongst the project partners?
- Is the substrate supply guaranteed and also the complete logistics?
- Have agreements been concluded with future suppliers?
- Is the technology suitable for smooth operation of the plant?
- Is a suitable place available for the erection of the plant?
- Are qualified personnel to be hired for the operation of the plant?
- Who are the relevant customers for the biogas, the electrical power, or the produced heat?
- Are the requirements for public promotion fulfilled?
- Is there a detailed plan for the financing (own resources, guarantees, and collateral)?

Plants with an electrical power output of less than 75 kW are considered to be not economical if the power is supplied into the public network. Larger plants make possible an income of up to 3000 US$ per hectare – presupposing a good harvest.[49]

The governments of the European Union, states, regional counties, cities, districts, municipalities, and power supply companies promote regenerative power production. The promotion consists in loans at a low interest rate, partial remission of debts, special depreciation rates, and guaranteed prices for the power above

49) Cp. WEB 40

Biogas from Waste and Renewable Resources. An Introduction.
Dieter Deublein and Angelika Steinhauser

ISBN: 978-3-527-31841-4

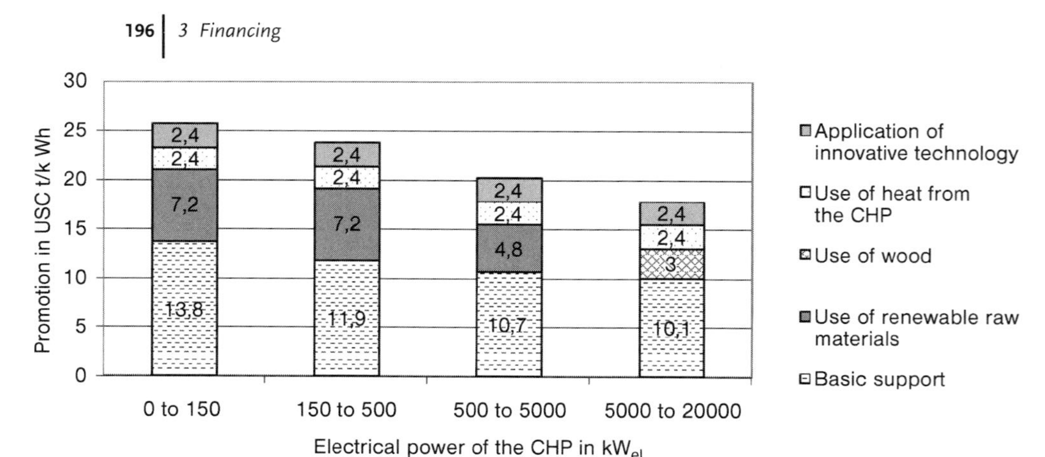

Figure 3.1 *Promotion provided by German government to support biogas production in Germany.*[50]

the normal tariff. The take-up of the promotion has been rapidly increasing since its introduction.[51),52)]

Since 2002[53)] a new law on financing renewable energy has been in force, which is considered to be a good example for law-makers in other countries. This law obliges the electrical energy supplier to pay for electrical power produced from renewable resources a fixed rate. A precondition is that the plant is completely new. The fixed rate decreases by 1.5% per year, so that owners are forced to act as soon as possible. The actual rate is given for the next 20 years after start-up of the plant.

The rates depend on the raw material which is fermented and are indicated in Figure 3.1.

Additionally to these promotions, a credit at a very low interest rate can be granted from a governmental bank. Such credits can cover the total costs of the plant.

50) Cp. LAW 6
51) Cp. BOK 30
52) Cp. WEB 111
53) Cp. LAW 9

Part V
Process engineering

The engineering of anaerobic processes is characterized by various industry-standard procedures and equipment, which are sometimes differently named but actually not different. In an attempt to bring all these processes into an order, the following general procedures can be distinguished.

There are procedures by which the liquid biomass flows through, participates in the entire process, and is discharged afterwards. Such procedures are particularly usual in agriculture. Concentrated biomass is diluted before processing it and cocentrated again after processing.

There are other procedures by which the liquid biomass is processed and the water is separated after processing and recycled to the waste water treatment plant. Such procedures are applied to sewage sludge fermentation and in recent years increasingly for anaerobic purification of waste water from industrial companies.

Other interesting technologies have been established for the fermentation of bio wastes apart from the above procedures. One process is characterized by the separation of solid biomass after hydrolysis and methanation of only the hydrolyzate. In some plants, the water is separated continuously during the methanation of the hydrolyzate. Also, the fermentation of stacked biomass occurs when a process is applied in which the water percolates through the biomass.

Products of the fermentation are solid residues and waste water. Both are further processed separately. The solid residues become aerobically rotted and then represent a valuable compost. The duration of the rotting depends on the anaerobic fermentation process. The waste water is distributed on the fields or in agricultural biogas plants or is recycled to the water treatment plant.

One- and multi-stage plants for all kinds of bio wastes and renewable raw materials are available on the market.[1] The investment and operating costs of the plants depend on the kind of the substrate and the requirements.

1) Cp. JOU 20

Biogas from Waste and Renewable Resources. An Introduction.
Dieter Deublein and Angelika Steinhauser

ISBN: 978-3-527-31841-4

1
Parts of biogas plants

1.1
Tanks and reactors

Substrates contain many deoxygenating substances, which cause fish die-off or groundwater contamination.

In accordance with laws,[2] equipment, tanks, and pipework which contain such substrates must be reliably tight, so that no substrates can penetrate into the groundwater. The tightness of the components, above all the connections, valves, and in particular the mechanisms for leakage recognition, must be easily and reliably controllable.

Corrosion is induced by sulfuric acid, ammonia, and nitric acid, particularly in the gasholder; in all tanks, the area where the water surface meets the wall of the tank or bioreactor is prone to leakage if pH values are low (<0.6).

Also, the central column which carries the dust cover in many vertical bioreactors is particularly liable to corrosion.

Plant and other equipment are often built underground and must then be accessible, so that only a few materials can be used for construction (Table 1.1).

1.1.1
Brick tanks

Tanks can be built from clay bricks (Figure 1.2). When preparing the ground for the tank, the base must be particularly well rammed. For the base plate the following materials are applicable:

- quarry stones with cement mortar filling and screed (Table 1.2)
- brickwork with screed or concrete.

Bricking of a curved bowl is simple. One needs only a center, e.g. from a heap of stone, which is removed afterwards, and a radius stick. In contrast to this, concreting is more difficult because a framework is necessary.

2) Cp. LAW 27

Biogas from Waste and Renewable Resources. An Introduction.
Dieter Deublein and Angelika Steinhauser

ISBN: 978-3-527-31841-4

Table 1.1 Suitability of tank or reactor materials.

Material	*Preparation tank*	*Bioreactor*	*Residue storage tank*	*Gasholder*
Reinforced concrete	++	++	++	––
Concrete blocks, sometimes with coating	++	++	++	––
Sheet steel (welded on-site, parallel fold)	+	+	+	+
Wood	–	–	–	–
GRP (glass fiber-reinforced plastic)	+	+	+	+
Thick plastic	–– (suitable for very small tanks only)			
Plastic canvas cover	+ (suitable for basins)			

++ corrosion resistant and accessible
+ corrosion resistant but not accessible
– perhaps corrosion resistant, not accessible
–– not suitable

Brickwork and mortar

Mortars and bricks should have about the same strength. Bricks of low quality require thicker walls.

Mortar for brickwork consists of sand, water, and binding agents. Cement as binding agent results in stable, waterproof, but brittle mortar. Lime as binding agent results in smooth and elastic mortar. In order to get good waterproof brick-work, a mixture of cement and lime should be used as the binding agent. The sand for bioreactor brick-work must be finely sieved (grain size max. 3 mm). It must be clean and should not contain loam, dust, or organic components. Mortar sand with a high portion of dust or loam needs more cement than clean sand.

The cement plaster must fulfill the local quality requirements and processing rules[3)] and must receive a surface protection. The plaster has to be consolidated by strong, circular rubbing. All edges must be chamfered and all covings must be rounded.

1.1.2
Reinforced concrete tanks

The reinforced concrete must be free of cracks and resistant under the special conditions during fermentation over the entire period of utilization.

With reinforced concrete, the acidic substrate can penetrate to the reinforcement and corrode it if carbon dioxide-containing air also permeates to the

3) Cp. LAW 11

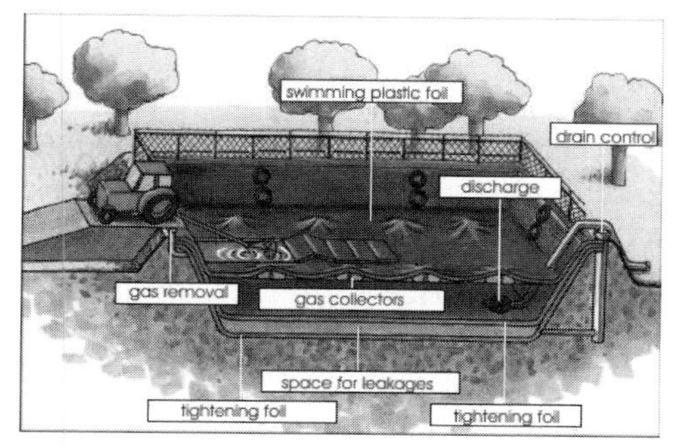

Figure 1.1 *Reinforced concrete in Europe (top left),*
Heating and agitation devices inside a vertical bioreactor (top right),
Construction of a basin (middle left),
Uncovered basin (middle right),
Thoeni Kompogas-bioreactor in a plant in Austria (botton left),
Simple bioreactor in the Philippines (bottom right).

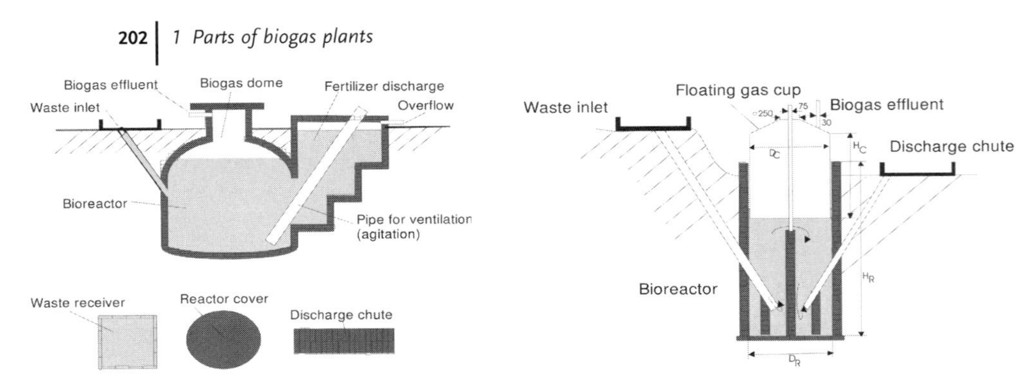

Figure 1.2 Bricked tank: Gasdome (left), Floating cup (right).

Table 1.2 Mortar and plaster mixing ratios.

Mortar (2)	*1 (cement)*	*1 (lime)*	*6 (sand)*
Plaster (1)	1 (cement)		4 (sand)
Plaster (2) (better)	1 (cement)		3 (sand)

reinforcement. Carbon dioxide converts calcium hydroxide in the concrete to calcium carbonate. The pH value in the concrete around the reinforcements decreases to values below 9 and the steel begins to corrode.

In order to prevent such damage, the local laws[4] stipulate a quality-supervised concrete with high resistance to strong chemical attack (concrete B II) when the concrete is exposed to pH values below 4.5 over longer periods. The parts of structure can consist of water-repellent concrete with high frost resistance (class B 25).

The following rules have to be considered for the construction of reinforced concrete tanks:

- Angular gas spaces have to be avoided in principle.
- The transition from the arched slab to the vertical wall must lie in the gas space.
- Break-throughs for inlet and discharge may not lie in the gas space; an exception is the entrance opening.
- Joints and connections of pre-cast segments (e.g., base plate and vertical walls of the tanks) have to be tightened durably and flexibly. A declaration by the manufacturer about the quality of the construction or an appropriate general test certificate is necessary.

The base plate transmits the weight of the tank to the ground. It must withstand bending and pressure differences like unequal loads from the inside (undulation)

4) Cp. LAW 11

and from the outside (underground stones). The tank wall rests with all its weight on the baseplate. It is important that the ground is compact and clean.

For static reasons, a curved plate is most suitable as foundation. A conical foundation can be excavated most simply. For agricultural plants, almost all foundations are even.

Rules for the concreting:

- The concrete has to be applied and consolidated immediately after mixing when mixed on site, or immediately after delivery when ready-mixed concrete is used.
- Water may not be added for maintaining or improving the fluidity: The make-up water increases the w/c value (weight ratio of water/cement) and causes higher porosity and lower strength.
- The concrete must be compressed sufficiently, particularly around the reinforcement.
- The concrete must dry slowly; i.e. after casting it must be covered for about 7 days with a protective plastic foil or sprayed with water in order to prevent evaporation. Otherwise, not enough water will be available for the setting of the concrete, which results in a less pressure-resistant reinforced concrete wall.
- Only when the concrete is correctly hardened it may be loaded.

A surface protection layer must be applied to the concrete.[5]

Surface protection by painting

The internal surfaces (bottom and wall surfaces) of silos[6] and bioreactors of concrete or brickwork with cement plaster corrode under the influence of acidic substrates. Therefore a surface protection must be applied.

Special care has to be taken to attach the protective layer: it is essential that the surface is mechanically stable and dry and withstands large mechanical loads. If this is the case, the surface must be cleaned with a hard broom, with putty, and with a wire brush or a high-pressure water cleaner. Residues of protective oil at new silos, adherent dirt, and loose hardened cement particles are to be removed. Sandblasting is recommended, and etching with dilute acid (e.g., 10% phosphoric acid) is particularly recommended for cement sludge-containing material. Any cracks and cavities in the surface are to be repaired. Only then can the protective coating be applied. It has to be certain that a protective layer develops, which is

- Completely sealed, in order to prevent the penetration of acidic substrates
- Acid resistant
- Weather-proof (UV, thermal radiation from the sun)
- Not liable to wear

5) Cp. BOK 32

6) Cp. JOU 8

Table 1.3 Composition of asphalt concrete for a 35–50 mm thick bottom layer in silos.[7]

Acid-resistant lime-free additive, e.g. granite, particle size <8 mm	*45–50% by weight*
Bitumen	6.5–8
Filler	8–12
Sand	30–40.4

- Does not release particles, which are harmful for animal and humans.

If the concrete is to be painted (thickness up to 0.3 mm) or coated (thickness up to 1 mm), products made from the following raw materials can be used:

- Bitumen
- Dispersions of various plastics, e.g., PVC
- Polyurethane (PUR)
- Epoxy, partly in combination with bitumen
- Asphaltic concrete (Table 1.3), especially for coating the bottom: the plastering of the bitumen layer onto the concrete base has to be carried out with a bitumen bonding emulsion, which is sprayed on. Very important for asphaltic concrete is that at the wall junctions a bitumen sealant, a so-called Tok sealant, is attached, which prevents penetration of liquids into the foundations. Painting of asphaltic concrete must have the properties as specified in Table 1.3.
- Plastic-based mortar (e.g., with epoxy as bonding agent): this has not yet become generally accepted as a wear-resistant layer, but it would be advantageous in many cases, because it can be applied in "do-it-yourself" fashion.

Of the paints mentioned, plastic dispersion and epoxy are the only ones which are solvent-free. When painting with one of the other paints the workplace must be well ventilated, especially when high or low silos are painted internally.

At least two paint layers must be applied. The paint needs a certain time for hardening, until it is accessible or fully loadable. This is to be considered when planning the painting. A surface painted with PU is accessible, e.g., after approximately 8 h, but is only fully loadable after approximately 4 days.

Even with the best painted and coated surfaces, the durability is limited (Table 1.4).

7) Cp. BRO 8

Table 1.4 Durability of correctly applied concrete protective cover on the bottom and the walls of a silo and its costs calculated for a silo with and without roof.[8)]

Concrete protection from	*Durability of a silo in years*				*Material costs per area [US$ m^{-2}]*	*Working time per area [h m^{-2}]*
	Without roof		*With roof*			
	bottom	*wall*	*bottom*	*wall*		
Bitumen	1	1–2	1–2	2	0.40–0.65	0.150
Plastic dispersion	1–2	3	2	3–4	1.25–2.30	0.125
Polyurethane (PUR)	3–4	5	5	6	1.25–3.60	0.075
Epoxide	4	5	5	6	1.90–3.75	0.150
Asphalt concrete	20		20		15.00	

Paint ages faster and is less durable under the influence of UV and thermal radiation in spaces without a roof, in particular when the paint is exposed to the sun as, e.g., in drive-in silos. Therefore bright paints are to be preferred.

A surface painted with PU or epoxy lasts for two years longer if quartz sand is incorporated, but in the bottom area such paint is not satisfactory. Here, thicker coatings (thickness over 1 mm) are unavoidable. These form a wear layer with longer resistance.

The painting of bitumen is highly recommended in all areas of application (wall/bottom, tanks with/without roof) from the point of view of material costs. If the cost of the paintwork is taken into consideration, painting with PU proves to be most economical in all areas of application. Asphaltic concrete is an interesting alternative for bottom coatings from the aspect of costs. With a layer of 40 mm thickness, complex patching up can be avoided.

Surface protection by tanks with double walls

If no quality-supervised special concrete is used, the concrete can be protected by segmented linings of HDPE or PP, which are welded together. GRP coatings with a total thickness of 2–3 mm can protect components which are less corrosion-endangered.

1.1.3
Tanks of normal steel sheet metals with enamel layer or plastic coating

An enamel layer protects the entire steel surface durably. It is glass-like and extremely resistant.

Such tanks are completely prefabricated from steel sheet segments. For enamelling, the segments are prepared in different dipping baths, i.e. cleaned, derusted, etc. Then the enamel powder is blown on in an even layer.[9)] In the kiln (the heart

8) Cp. WEB 35

9) Cp. BRO 15

of an enamel factory), the powder-coated single metal sheet is heated up to 860 °C, so that the enamel powder melts and forms a strong bond with the surface of the metal. The single metal sheets are connected together by means of special screws.

Normal steel tanks from curved and hot-galvanized (80–120 μm layer of galvanization) steel sheets can also be provided with plastic coatings on both sides. Double 2-component-paints based on epoxy resin are applied (2 × 40 μm thick) or a solvent-free PU layer is sprayed on in a hot spraying procedure.

Such tanks are easy, fast, and safe to construct, even in self-assembly with a modular system.

1.1.4 Tanks of stainless steel

Stainless steel tanks usually consist of welded steel sheets of the quality 1.4301, 1.4404, 1.4436, 1.4435 or 1.4571, but they are occasionally built from stainless steel plates screwed onto a steel structure made of hot-galvanized profiles. The plate segments are sealed to each other with methane-gas-tight elastic PU sealing bands.[10] Screws, nuts, etc. have to be made from stainless steel.

The combined material Verinox™ consists outside of a carrier material, steel hot-galvanized on both sides, and inside of a stainless steel lining. The stainless steel sheets from an endless coil are of a high crystalline density due to cold-rolling. The surface so improved guarantees a high corrosion resistance. The carrier material and the inner lining are held apart by an endless acrylic band.[11]

According to the static requirements, such tanks have either curved covers or flat covers with appropriate stiffenings. In the latter case a static proof has to be adduced.

Before beginning welding stainless steel tanks, workshop drawings must be on hand indicating all welding seam specifications. All data for welding must be stipulated in a manual, so that the quality of the welding seam is ensured (welding agent and auxiliary materials, welding methods, etc.). The weldings have to be carried out in such a way that corrosion and in particular intergranular corrosion are prevented; e.g., by avoidance of open gaps, avoidance of interrupted welding seams, welding of fillet welds on both sides, welding root protection, removal of annealing colors, etc.

1.1.5 Ground basin with plastic foil lining

Ground basins with plastic foil lining are the cheapest bioreactors and tanks in biogas plants.

They are usually 3.0 m deep, about 1.5 m deep in the ground and with a 1.5-m high earth wall around the basin, made from the excavated earth. The basins are covered with a gas-tight foil.

10) Cp. WEB 24

11) Cp. WEB 56

Ground basins are made of a lining of 2.0-mm thick, welded polyethylene foil. This foil is absolutely resistant to UV light, frost, liquid manure, and seeping silo juices. It has an extensibility of ca. 700% and can withstand the highest mechanical loads – even growing roots. All welding seams are examined by means of a compression test for 100% tightness.

Besides polyethylene, RMP (red mud plastics), Trevira, Butyl, and similar plastics were found to be suitable.

Nevertheless, even particularly enduring plastics cannot be used for longer than five years and have to be replaced completely after that time.

1.2 Equipment for tempering the substrate

For continuous processing the heating method is important.

The heat requirement results on the one hand from the volume flow of the substrate, its specific heat capacity and the temperature difference between the substrate and the processing temperature. On the other hand, heat requirements result from heat losses at the bioreactor surface.

In Table 1.5, the energy consumption of a mesophilic bioreactor is shown as an example. The smaller the biomass concentration in the substrate the lower is the biogas yield and the higher is the percentage of biogas that has to be used for the heating of the substrate.

Simple plants are equipped with internal heating systems (see Figure 1.3). Deposits and cakings are accepted – these impair the heat transfer and can be removed only mechanically. Heating pipes of stainless steel (1.4571) are to be preferred, because PVC pipes sag and have therefore to be supported more.

Heating coils cast into the bioreactor concrete wall do not work satisfactorily, since the concrete chips off. With steel tanks, the heat pipes can be installed inside and outside, e.g., even at the bottom of the bioreactor. However, then layers of sediment could interfere with the heat exchange.

Table 1.5 Rough estimate of specific energy consumption[12] in a biogas plant related to the energy produced in the plant at 5% DM content (load: 1.6 $kg_{COD}/(m^3.d)$; temperature of the substrate: +15 °C; temperature of the air: −8 °C).

Heating of the substrate	11–15%
Heat losses	7–9%
Agitation and pumping	2–3%
Other components	2–3%
Total	22–30%

12) Cp. BOK 16

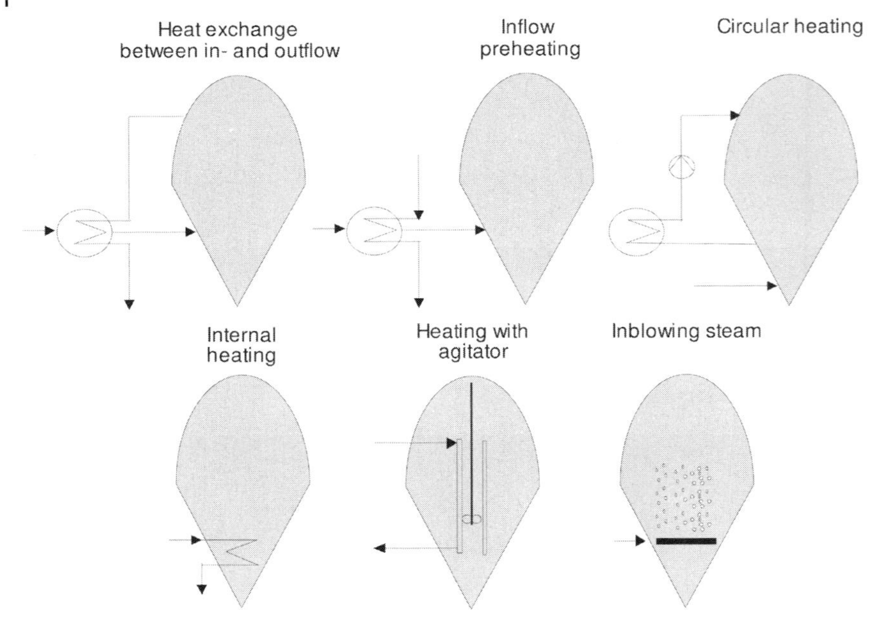

Figure 1.3 Overview of heating systems.

In technically more complex plants and in plants in which hygienization is necessary, external heat exchangers are often installed through which the biomass is circulated by an external pump. External heating systems can be maintained and cleaned more easily. To extend the service life the direction of flow should be reversible.

The following heat exchanger types are used for heating waste water and sewage sludge:

- Double jacket pipe heat exchanger,
- Plate heat exchanger
- Shell and tube heat exchanger
- Spiral heat exchanger.

Usually the heat exchangers run in counter-current flow. The pipes should not be less than 100 mm in diameter if there is no filter or shredder upstream. The flow rate of substrate or sewage sludge is usually $v_W = 0.8–1.5\,m\,s^{-1}$ in the heat exchanger pipes and $v_W = 0.6–1.5\,m\,s^{-1}$ if water is pumped through. The larger the flow rate the better is the heat transfer. The heat transfer coefficient through the heat exchanger pipe walls is $k = 100–300\,W/(m^2.K)$. With steam-heated exchangers, the heat transfer coefficient rises to approximately $k = 800\,W/(m^2.K)$.

At all inlets and outlets of the heat exchanger a thermometer has to be installed.

The heat exchanger is mainly heated with warm water, hot water, or low-pressure steam. For hot-water heating, the heat from the biogas CHP is used. Hot-water heating can be easily regulated and gives good heat transfer coefficients in the heat exchangers.

The natural circulation of warm water is forced by differences in the density due to the temperature gradient between supply and return water pipes. Warm-water heating with natural circulation is sometimes used in small bioreactors. It is characterized by large cross-sections of the pipes.

Low-pressure steam heating at a pressure less than 0.5 bar, where steam is injected directly via nozzles into the sludge, is no longer state-of-the-art.

If the heat supplied by the CHP is not sufficient, then additional heat can be generated by a heat pump, e.g., from the waste water, from groundwater, or from the air. A heat pump works economically only for covering the peak demand, when it is driven by power produced from biogas, either electrical power or the gas itself. But mostly the relatively high investment is not made because of the short use.

1.3 Thermal insulation

Depending on whether the tank is underground, above ground, or in a building, it must be more or less thermally insulated in order to avoid heat losses and/or to offer contact protection when the reactor is run in a thermophilic process.

As insulating material, expanded plastic slabs of polyurethane are used within the lower zone of the wall. They are equipped with moisture barriers in order to prevent the penetration of water. In the upper zone of the tank wall, expanded polystyrene slabs or mineral wool mats are often installed, or alternatively plastic foam is attached. As blinds and as protection from humidity, the thermal insulation is covered with riveted metal sheets.

1.4 Piping system

The piping system of biogas plants should be installed above ground to allow the easy detection of corrosion and leakages caused by any acidifying agents in the plant.[13]

Such an above-ground system requires good insulation of frost-exposed pipes and armatures with additional trace heating.

Any subterranean pipes should be installed deep enough so that no damage can occur because of above-ground weight. All the pipes should be frost-protected, and special precautions must be taken to avoid any potential leakage.

13) Cp. JOU 17

The inlet and delivery pipes of all tanks should have at least one slide valve. Two gate valves and an additional compensator are mandatorily required only for such pipes that pass through the bioreactor wall below the sludge surface. The valves should be installed directly one after the other.

Regular checks of the whole system are required to guarantee that the pipes are all intact. Therefore, sight glasses need to be installed which should be rinsable and illuminatable.

In case of emergency, all the pipes should be of adequate dimensions and should be installed with a slight inclination to allow the complete system to be emptied and deaerated. The gate valves should always be on top. For the gas pipes, steam or foam traps need to be installed, sufficiently dimensioned to accommodate the maximum gas volume and a velocity of $v_F = 10\,m\,s^{-1}$ in the discharge pipe and $v_F = 5\,m\,s^{-1}$ in the intake pipe.

The material of partly filled delivery pipes or pipes that carry biogas and exhaust air should ideally be HDPE, PP, or PE. The connections must be welded. It is common practice that the company in charge of the installation proves and signs off the tightness of the whole piping system for normal daily conditions and for emergency cases where the pressure is at its maximum.

All the external plastic pipes need to be resistant to UV light. Pipes that carry the substrate are wrapped with aluminum foil for special protection against UV radiation.

1.5
Pump system

Pumps (Table 1.6) are in general needed in order to transport substrate to and from the equipment in the plant.

Centrifugal pumps are to be found in 50% of all biogas plants. In 25% of all plants, positive-displacement pumps are installed to handle high solid concentrations. Ca. 16% of all biogas plants are operated without pumps, depending on geographical conditions. Other pumps like mono pumps or gear pumps, etc., are seldom used. Cutting pumps are used preferably in the pretreatment tank in particular for long-fiber materials in liquid manure such as straw, fodder remainders, grass cuttings, etc. They have hardened cutting edges at the impeller and a stationary cutting blade at the housing.

Pumps must be easily accessible, because

- they have to be controlled and maintained regularly.
- the moving parts of the pumps are wearing parts, which are subject to special stress and have to be replaced from time to time, especially in biogas plants.
- blockages of the pump occur despite various precautionary measures and prudent planning, and these have to be eliminated promptly.

Table 1.6 Comparison of pumps.

Type	*Centrifugal pump*	*Positive displacement pump*
Design	Submerged pump Submerged motor pump	Eccentric screw pump Rotary piston pump
Features	High throughput Relatively low pressure head Not self-priming in general Simple and robust Suitable up to 12% DM in the substrate	Dosing Self-priming Constant pressure Suitable for substrates of high viscosity Accident-sensitive and therefore to be combined with a macerator
Pressure and throughput	Maximum pressure head: 0.4–2.5 bar Power consumption: 3–15 kW Throughput: 2–6 $m^3 min^{-1}$.	*Eccentric screw pumps* Self-priming up to 0.85 bar Maximum pressure head: 24 bar Throughput: less than centrifugal pumps *Rotary piston pumps* Maximum pressure head: 2–10 bar Power consumption: 7.5–55 kW Throughput: 0.5–4 m^3/min

1.6 Measurement, control, and automation technology

The process control equipment is used for the supervision and regulation of the operation of the plant and for the limitation of damage.[14]

In cases of emergency, e.g., breakdown of the electrical power supply, the biogas plant must be automatically transferred to safe operating conditions by the process instrumentation. Necessary electrically driven devices must be supplied with emergency power.

1.6.1 Mechanisms for monitoring and regulation

The operation of many simple agricultural biogas plants is dependent on the substrate accumulation and/or the gas consumption of the gas engine. They do not have regulation and the bioreactor is fed once or twice a day. The gas production is estimated from the running time of the CHP. If it runs inefficiently, substrate is fed in order to increase gas production. The only measuring instruments are a thermometer and a manometer at the CHP.

14) Cp. LAW 12

What is going on inside the bioreactor remains unconsidered in such biogas plants. Whether the microorganisms are satisfied or surfeited, are hungry or suffer in any way, is not known. Unexplainable operational disturbances occur with various consequences:

- acidification of the biology
- over-foaming of the plant
- widely varying gas quality
- odor emissions, possibly resulting in annoyance of the neighborhood
- hold up of the CHP and loss of income.

When in contrast the plant runs in a stable and troublefree fashion, then it is usually not working at full capacity and its full potential is not tapped.

By automatic control techniques and recording of significant parameters it is possible

- to supervise the plant parameters in real time and to recognize and correct aberrations immediately.
- to run the plant closer to its optimum and thereby to save resources and costs
- to make recordings for the journal at least partly automatically.

Often the measurements are disturbed by improper installation of the measuring instruments. Measuring sensors installed in the substrate tend, e.g., to be overgrown or to be influenced by settling particles or by accumulation of gases. Measurements in the biogas space or gas stream are impaired by water condensation, foam, or precipitation of sulfur. If the gas flow rate gage is not installed in the straight pipe some distance behind a bend, it shows wrong readings. Before selecting devices and installing measuring sensors, the following have to be considered:

- Measuring sensors may not be attached in neutral zones or close to the bottom of the tank.
- Gas must be able to escape out of protective sleeves around measuring sensors.
- It must be easily possible to clean and calibrate measuring sensors regularly. They should always be installed with a bypass fitted so that they can be mounted and demounted without problems.
- Measuring instruments must supply reliable readings; the indicated value of measuring devices for flow rates should not be influenced by gas bubbles in the liquid.
- Gas analyzers need gas drying or other procedure for condensate separation upstream. In some cases, hydrogen sulfide has to be removed.

For monitoring the processes in biogas plants, the following measurements are recommended:

- substrate mass flow
- concentration of dry matter and organic dry matter (oDM or TOC)
- BOD_5 value and COD value
- degree of decomposition and inhibition
- temperature
- pH value and/or redox potential (both correlate)
- acid value and nitrogen content
- concentration of nutrients
- ignition value of the residue
- sludge volume index
- biogas quantity and quality (CH_4 content, CO_2 content, H_2S content).
- heavy metals in the residue

Some of these are described in more detail in the following.

1.6.1.1 Dry matter concentration in the substrate

The dry matter concentration in the substrate is an indication of the gas yield. It is usually given as a percentage and is calculated according to the following formula:

$$DM = \frac{mass\ of\ dry\ matter}{total\ mass} \cdot [\%]$$

The Weender analysis and the van Soest analysis or alternatively the NIR (narrow infrared spectroscopy) give values for the

- ADF (Acid Detergent Fiber)
- NDF (Neutral Detergent Fiber)
- ADL (Acid Detergent Lignin)

1.6.1.2 Organic dry matter content and/or total organic carbon (TOC)

A very easy and rapid method for the evaluation of substrate regarding its degradability is the determination of the concentration of organically bound carbon (TOC). Firstly, inorganic carbon (mainly CO_2) is removed from the substrate. Then, the substrate with its content of organic matter, is burned in an electrically heated furnace. The quantity of CO_2 formed is measured.

1.6.1.3 Biochemical oxygen demand (BOD)

The biochemical oxygen demand (BOD) is the quantity of oxygen which is consumed by microorganisms in order to degrade organic materials in the substrate at 20 °C. The degree of decomposition depends strongly on the temperature.[15] Two phases of the decomposition have to be distinguished. In the first phase, carbon

15) Cp. JOU 36

compounds (sugar, fats, etc.) are degraded predominantly. The BOD measured in this phase is often called the carbon BOD. The first phase lasts for more or less 15 days. In the second phase, nitrogen compounds (protein materials, amino acids, ammonia) are degraded. This procedure is often called nitrification, because nitrate is formed as the final product.

When measuring the BOD value, some inhibitors are added (e.g., allyl thio urea ATH) in order to prevent the nitrification and thus to be able to determine only the carbon BOD. Another agreement is that the measurement of the BOD is broken off after 5 days. The degraqdation is not completed within that time. The BOD value determined in such a way is called the BOD_5. It is given in $mg\,L^{-1}$ oxygen used. It is an important characteristic value for the load of a substrate with biologically degradable organic materials.

For the determination of the BOD_5 the manometric and the dilution procedure are preferably used. With both measuring methods, the sampling and the sample preparation are of the greatest importance.

Manometric BOD_5 determination

The manometric determination of BOD_5 is based on a reduction of the air pressure correlating with the oxygen consumption. The pressure decrease can be measured with a manometer. A sample of the substrate is mixed by agitating, so that the substrate is saturated with oxygen from the air. For the degradation of the organic materials by microorganisms, oxygen is converted to carbon dioxide. The carbon dioxide is absorbed above the sample surface by sodium or potassium hydroxide. The measuring instrument develops a negative pressure, which can be read off daily.

Dilution method (simplified procedure)

If a bottle is completely filled with an oxygen-saturated sample, locked hermetically, and held at 20 °C, then the oxygen concentration in the bottle decreases depending upon the BOD_5 of the sample. By measuring the oxygen concentration in the bottle at the beginning and after 5 days the oxygen demand of the sample can be determined. The difference in the oxygen concentration at the beginning and after 5 days corresponds to the BOD_5.

For degradation, only about $9\,mg\,O_2\,L^{-1}$ (concentration of saturated oxygen at 20 °C) are available for the microorganism. Since the degradation of the organic compounds is slowed down below $1–2\,mg\,O_2\,L^{-1}$, the final oxygen concentration in the bottle must be higher than $2\,mg\,O_2\,L^{-1}$. Consequently the BOD_5 of the sample must be lower than $7\,mg\,O_2\,L^{-1}$. Since this is mostly not the case and the BOD_5 of samples is mostly higher than $7\,mg\,O_2\,L^{-1}$ the sample must be adequately diluted.

1.6.1.4 **Chemical oxygen demand (COD)**

The chemical oxygen demand in the residue in relation to the organic matter (COD in the substrate) is a measure of the degree of decomposition.

For the determination of the COD value, nearly all organic carbon compounds are degraded to CO_2 and H_2O by a strong oxidizing agent (potassium dichromate) and the oxygen consumption is measured. Two hours at a temperature of 150 °C are sufficient for the complete oxidation and the determination of the COD value.

According to the standards, an exactly measured volume of the previously well-mixed sample of substrate is heated in the presence of concentrated sulfuric acid, potassium dichromate, silver salts, and mercury salts. Part of the potassium dichromate is consumed in chemical reactions with the organic compounds. The remaining potassium dichromate can be determined either chemically by titration or photometrically. Subsequently, the COD value is calculated.

In sewage treatment plants, the COD value is usually determined with cuvette tests because of the complex and dangerous procedure. On-line measurements are becoming more and more popular, and the control of the entire biogas plant is based on these on-line measurements. A procedure which has become widely accepted for COD value determinations is on-line photometric UV measurement.

1.6.1.5 Degree of decomposition

The degree of decomposition is the number of disintegrated cells expressed as a fraction of the entire number of cells. The degree of decomposition can be measured either by the COD value or by the oxygen demand. It is to be noted that the value determined by the oxygen demand (A_S) is always larger by the factor 2–4 than the value determined by the COD value (A_{COD}).

Degree of decomposition (A_S) determined by the oxygen demand The microorganisms in substrates are aerobic or facultative anaerobic. Therefore they need oxygen for their metabolic procedures. The activity of substrates can be determined by the measurement of the oxygen uptake rate. When the microorganisms are destroyed or inactivated through decomposition, they are not longer able to take up oxygen. Therefore the ratio of the oxygen uptake rate (OUR) at present to that of the untreated substrate (OUR_0) is a measure of the degree of decomposition.

$$A_s = 1 - \frac{OUR}{OUR_0}$$

The sample should be diluted in such a way that the oxygen consumption rate is within the range of 0.5–2 mg/(L.min), because otherwise the results of the measurements could be negatively influenced.

Degree of decomposition (A_{COD}) determined by the COD value Organic material is set free by disintegration of the sludge, which increases the chemical oxygen demand in the substrate. The degree of decomposition (A_{COD}) corresponds to the COD value expressed as a fraction of the maximum possible COD value COD_{max}, which can be determined, e.g., by degradation with caustic soda. The COD value COD_0, which is to be measured in the projection of the untreated sample, has to be subtracted from the resulting value.

$$A_{COD} = \frac{COD - COD_0}{COD_{\max} - COD_0}$$

1.6.1.6 Acid value

The acetic acid values of the feed and discharge of the reactor for hydrolysis are important measured values for the evaluation of the fermentation.[16] Normally, heavy and mostly unexplainable fluctuations can be observed (Figure 1.4), but a constant acid value is precondition for a smooth operation of the biogas plant.

Therefore it is recommended to determine the acid value in the laboratory regularly during the entire period of operation of the plant.

The acid value can be influenced in the hydrolysis by variation of the residence time, the temperature, and the pH value, e.g., by adding caustic soda solution. In many cases the acidification can be controlled by recirculation of residue from the bioreactor.

Each change in the load (start-up, restarting operation after a break for filling) must be carried out in consideration of the degree of decomposition of the substrate.[17] If new substrates are used with unknown behavior of degradation, the acids (acetic, propionic-, butyric, and valeric acid) in the substrate should be determined by means of gas chromatography. Overacidification can be prevented in this way.

1.6.1.7 Determination of nutrients (nitrogen and phosphorus compounds)

The degradation of the substrate depends on the availability of different nutrients, e.g., nitrogen and phosphorus. Both are to be found in the substrate in the form of different compounds.

NH_4-N	Ammonium nitrogen
NO_2-N	Nitrite nitrogen
NO_3-N	Nitrate nitrogen
TKN	Kjeldahl nitrogen

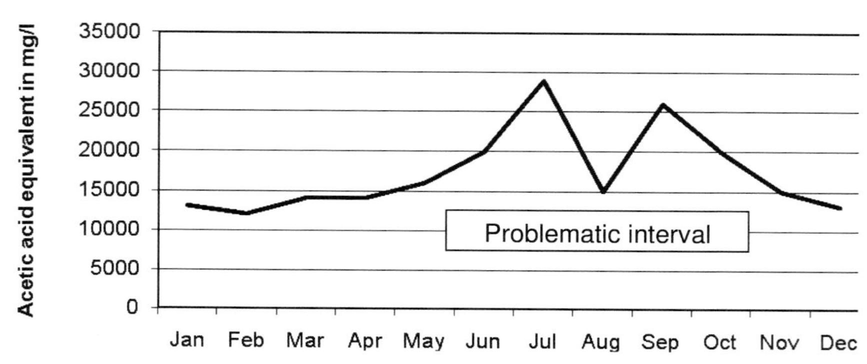

Figure 1.4 Operation at the upper limit of the acetic acid concentration.[18]

16) Cp. BOK 36
17) Cp. BOK 75
18) Cp. WEB 14

$N_{tot.}$ Total nitrogen
PO_4-P Phosphate phosphorus (ortho-Phosphate phosphorus)
P_{tot} Total phosphorus

Often phosphorus is lacking and must be added as nutrient. If the ammonium concentration in the substrate is measured, possible inhibition of the activity of the microorganism can be asummed. Test kits for rapid determinations with optical comparisons are very useful for such measurements.

1.6.1.8 Sludge (volume) index (I_{SV})

The sludge volume index indicates the kind of the microorganisms participating dominantly in the process of degradation. If the value is higher than 40–150 mL g^{-1}, then bulking sludge can be assumed, caused by a high fraction of filamentous microorganisms. Measures to prevent bulking sludge have to be applied.

The sludge volume index indicates the volume that 1 g dry matter takes up after a time of settlement of 30 min. It is a measure of the thickening ability of the sludge. The sludge index in mL g^{-1} can be calculated as the sludge volume in mL L^{-1} divided by the dry matter in g L^{-1} of the same sample.

$$I_{SV} = \frac{sludge\ volume}{sludge\ dry\ matter} \cdot \left[\frac{ml}{g}\right]$$

1.6.1.9 Ignition loss

The ignition loss is an indication for the content of organic materials. The better the sludge is degraded the lower is the ignition loss. Usual values are in the range 40–70%.

For the determination of the ignition loss dried sludge is heated in a porcelain crucible at least for 2 h at 550 °C. The ignition loss is then calculated by dividing the difference between the weight of the original sample and the weight of the burned sample by the weight of the original sample.

1.6.1.10 Biogas yield and quality

For the indication of the gas flow rate, vortex shedding devices, bellows-type gas flow meters with condensate discharge (plastic version), and pitot tubes with measuring orifice are approved to be suitable. Turbine flow meters are less applicable.

In sewage treatment plants explosion-proofed impeller metering devices[19] are installed.

1.6.2 Equipment to secure the operatability

Large damage in the operation of biogas plants is avoided by the adherence to the general safety regulations (see Part IV). In addition, attention must be paid to the following.

19) Cp. WEB 48

1.6.2.1 Foaming

Foaming can cause shutdown and must be prevented, e.g.,

- by a pressure release safety device
- by sufficient space for storage
- by adding antifoaming agents controlled by a foam detector
- by controlling of the substrate composition
- by foam traps
- by gas pipes which are generously dimensioned in their cross sections.

1.6.2.2 Blockage

The emergency overflow of the reactor can clog due to precipitates and coatings of magnesium ammonium phosphate (MAP) developed in the bioreactor. This can be prevented through

- addition of antiagents against coatings
- installation of plastic pipes
- lowering of the pH value
- installation of an overflow on top of the bioreactor, equipped with an airjet, which blows time-controlled air into the ascending overflow pipe.

It must be ensured that the maximum liquid level in the bioreactor and after-fermenter is not exceeded, e.g., by a frost-protected ascending pipe (overflow) for the residue from the bioreactor to the liquid manure tank.

Since the relationship between substrate and biogas quality is very important for the economical success of the plant, all experience with the plant should be documented in details in a dairy. Into this manual at least all load limits as well as the exact specifications of the feed and discharge material flow and operational funds should be mentioned. All actions in the case of disturbances should be documented in the journal.

1.6.3 Safety devices for humans and the environment

There are many devices for safety, prescribed by the authorities. Some are listed below.

1.6.3.1 Safety device before the gas flare

Biogas plants are often equipped with a gas flare, so that in case of emergency CH_4 can be burned and does not escape into the atmosphere. However, this enhances the risk that the flame from the gas flare strikes back into the plant and sets fire to the whole plant. Therefore the safety devices shown in Figure 1.5 are prescribed in the gas pipe before the gas flare.

1.6.3.2 Overpressure and negative pressure safety device (Figure 1.6)

With a U-tube, the over- and negative pressure are limited simply by the length of the water gauge.

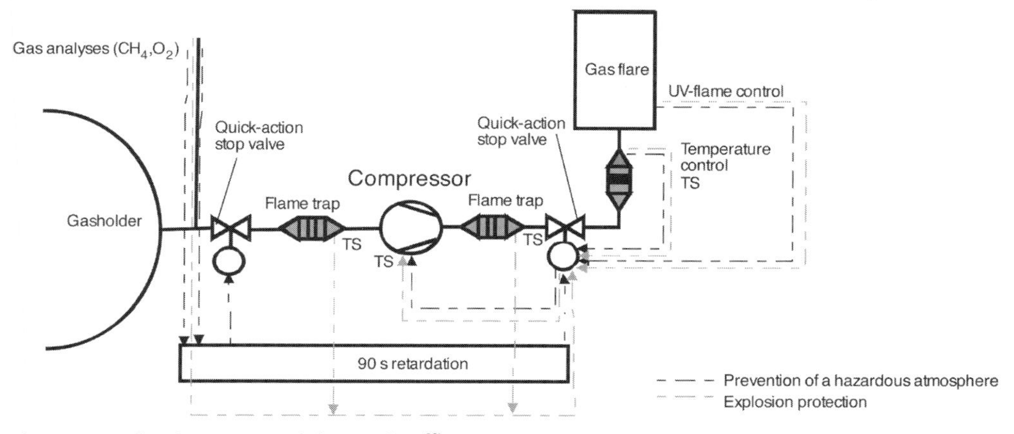

Figure 1.5 Safety devices around the gas flare.[20]

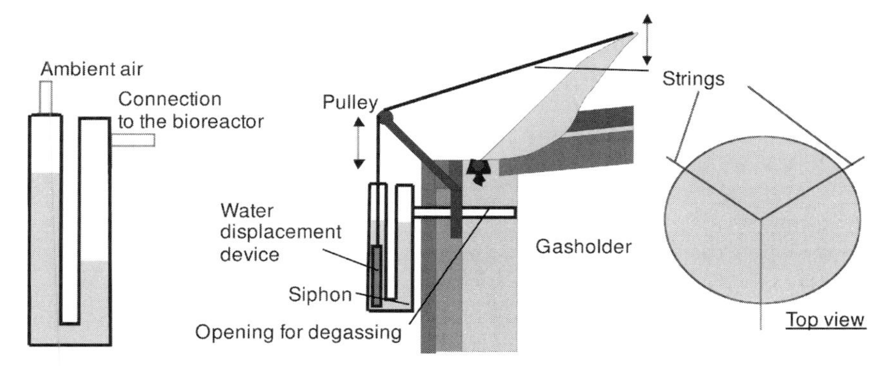

Figure 1.6 Safety device for high and low pressure in an agricultural plant: U-tube (left), Bioguard™ (right).

With the Bioguard™, a displacement body hangs on a string, which is fixed on top of the gas hood. The displacement body more or less dips into a siphon and thereby changes the level of liquid in the siphon. With too high pressure, the gas hood is lifted and with it the displacement body. The sealing fluid drops, so that biogas can leak out of the bioreactor. By the pressure loss, the water level, the gas hood, and the displacement body are lowered, and the gas system is locked again by the sealing fluid. In the case of negative pressure, the sealing fluid in the siphon is raised until air can flow under the siphon. The minimum negative pressure is limited to 4 mbar by a drilled hole 2 cm above the siphon upper edge.

In sewage sludge and bio waste fermentation plants the safety devices are more complicated (Figure 1.7). The shown over/negative pressure safety device closes the biogas space in the bioreactor with a diaphragm. The diaphragm is covered with a water layer. The space above the water layer still belongs to the bioreactor space and is therefore filled with biogas. At negative pressure the diaphragm is

20) Cp. LAW 2

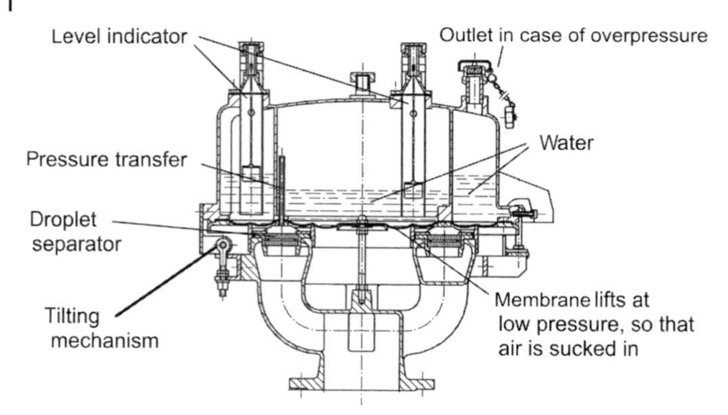

Figure 1.7 Membrane valve to regulate high and low pressure in a biogas plant.

raised and ambient air can flow into the bioreactor from below the diaphragm. At positive pressure the water is pressed into an outside ring chamber, so that gas can bubble out of the bioreactor beneath the ring chamber wall.

1.7 Exhaust air cleaning

At various places in the biogas plant exhaust air emerges in even larger quantities.[21]

The exhaust air pipeline from the hydrolysis reactor can be connected to the biogas pipe or, e.g., to the central air discharge pipeline of the entire plant. When the exhaust air is added to the biogas, the gas quality deteriorates and the risk of explosions rises. pH value fluctuations during the hydrolysis result in sudden CO_2 and H_2 production, which can sometimes cause a shutdown of the gas engine.

Each uncontrolled escape of exhaust gas into the environment represents a source of odor and also of danger.

Therefore the exhaust air must be cleaned. Bioscrubbers, chemoscrubbers, and biofilters are approved for this purpose.

Bioscrubbers[22] work with circulating water and are preferred when the concentration of odorous substances is high and the concentration of H_2S is less than $20\,mg\,m^{-3}$. For high H_2S concentrations a chemoscrubber is recommended, which operates with a caustic soda solution at a pH value of abt. pH = 10. But for deodorization a biofilter has to be installed downstream. Biofilters[23] are suitable for H_2S concentrations in the exhaust air up to $20\,mg\,m^{-3}$ H_2S.

In order to avoid general odor problems, e.g., from long piplines for the return of the water from the bioreactor to the sewage treatment plant, it is recommended to connect downstream to the biogas plant an aerobic treatment of the waste water with nitrification and denitrification procedures.

21) Cp. JOU 31
22) Cp. LAW 28
23) Cp. LAW 29

2
Area for the delivery and equipment for storage of the delivered biomass

Already during transport the fermentation process starts: biomass is particularly converted by bacteria into lactic acid, acetic acid and butyric acid. After approximately 4–6 weeks storing the maize silage releases even biogas. Since the acids mentioned are smell intensive, special precautions have to be taken during transportation, storing, feeding and exhaust air treatment. Driveways and areas around the biogas plant equipment should be well solidified.[24] The solidified surfaces must be cleaned regularly, without dispersing dust. Possibly a vehicle sluice is necessary, in which vehicles are cleaned.

In residential areas, smell nuisances have to be anticipated. There, only closed vehicles can be used for biomass transportation.

It is best if the vehicles are emptied in the same building in which the biomass is stored.

Since the odor-intensive preacidification starts during the storage, all conveyors from the storage to the biogas plant, even short ones, should be enclosed.

The exhaust air from the well-aerated storage can be used as combustion air in the CHP or can be fed into the central exhaust air pipe.

24) Cp. WEB 42

Biogas from Waste and Renewable Resources. An Introduction.
Dieter Deublein and Angelika Steinhauser

ISBN: 978-3-527-31841-4

3
Process technology for the upstream processing (Figure 3.1)

There are several reasons to prepare substrates and co-substrates before the fermentation in the biogas plant, e.g.:

- The substrate possesses a too high concentration of dry matter. For wet fermentation a concentration of 10–12% DM is ideal, where the fraction of organic dry matter should be just as high.
- The substrate must be sanitized due to legal restrictions
- Fibrous materials,e.g., green cuttings and straw must be comminuted in order to avoid process disturbances in the bioreactor.

Table 3.1 Pretreatment of different substrates (+: essential process; –: not essential process; 0: under certain circumstances recommendable process).

process	*Agricultural plant*	*Sewage plant*	*Unstructured wastes*	*Structured wastes*	*Residual wastes with a biological fraction*
Sorting and removing of disturbing substances	–	–	+	+	+
Removing of metalls	–	–	0	0	+
Homogenisation and comminution	0	–	+	+	+
Adjustment of the water content	–	+	+	–	0
Drying	–	+	+	+	+
Hygienization	–	0	0	0	+
Stabilisation	0	+	+	–	0
Heating	+	+	+	+	+
Adjustment of the pH-value	0	0	0	0	0

Biogas from Waste and Renewable Resources. An Introduction.
Dieter Deublein and Angelika Steinhauser

ISBN: 978-3-527-31841-4

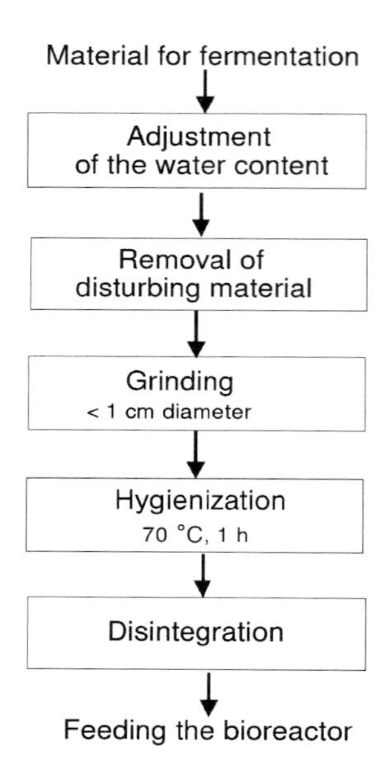

Figure 3.1 Front-end treatment of the organic residues before the processing in the bioreactor.

Bio wastes (market waste, leftovers, wastes from households) are to be fermented under dry more than 15% DM or liquefied (less than 15% DM) conditions. Wet fermentation is much more frequently used than dry fermentation. Sometimes special procedures such as percolation or squeezing-off procedures are used for the separation of a fermentable liquid (pressing water, extraction water) or a compostable residue.

3.1
Adjustment of the water content

Wet and dry fermentation are equal alternatives (Table 3.2) especially for agricultural and bio waste fermentation plants.

The water content of the substrate for wet or dry fermentation is adjusted by feeding pure or slightly contaminated water.

3.2
Removal of disturbing/harmful substances

Disturbing and/or harmful materials in the substrates can cause disturbances in the liquid flow and possibly remain as noxious matters in the residue.

Particularly with biogas plants for bio wastes sorting and removal of fractions of the biomass is inevitable.

In the dry pre-treatment and fermentation process, large particles can be separated by hand, metals, e.g., with a magnet, and sand, etc., with filters.

Table 3.2 Comparison of wet and dry fermentation.[25]

Wet fermentation ***Water content in the substrate >85%***	***Dry fermentation*** ***Water content in the substrate <85%***
Advantages	
• Considerably wider spectrum of substrates applicable – many multistage plants work after this process • Better suitable for stronger water-containing, pasty substrates • Easier mixing by mechanical agitators or compressed gas • Easier adjustment of the solid-liquid ratio • More easily controllable pH value, DM-concentration, NH_3-content, and content of volatile fatty acids • Substrate easier to transport and therefore more hygienic • Better exchange of material and heat • Less salt enrichment in the residue because salts remain dissolved in the water • Safe gas liberation out of the substrate	• Compact construction of the complete plant • Low maintenance: robust technique with only a few rotating parts – low wear, but modern automatic control technique required • Better suitable for structural substrates (green) • Low process energy consumption (below 15% of the produced energy) • No foaming • Odor avoidable • Less disturbing material • Sometimes higher gas yield and gas quality (abt. 80% methane content, abt. 20 ppm H_2S), so desulfurization may not be necessary
Disadvantages	
• High process energy consumption: 30–45% of the produced energy • Demixing, e.g., formation of a scum (retention of gas in the substrate or sediment layer (difficult discharge). • Foaming can be problematical • Odor can be problematical	• Additional costs for biomass conveyors • Energy-consuming, pretreatment necessary. • Mixing often difficult • Sludge recycling as inoculum necessary • Occasionally enrichment of inhibitors; e.g. NH_3 or volatile short-chain organic fatty acids • Supply of nutrients and removal of gases (CO_2 / CH_4) can be problematical • Manual selection of disturbing material questionable from point of view of hygiene – special devices to maintain hygiene may be required

In the wet process, the bio wastes are mixed with water in a garbage dissolver and rinsed by a filter with 8 mm punched holes. Fermentable organic fractions are separated from the residue. They remain in diluted form or as defibrated solids in suspension. In the next stage, floating particles (plastics, textiles, wood etc.) are removed by a hydraulically operated rake, while heavy materials (minerals,

25) Cp. WEB 46

non-ferrous metals, glass, bone etc.) are separated by a heavy material air-lock at the bottom of the garbage dissolver. Related to the total amount of dry matter there is about 5% by weight iron, 10% by weight heavy materials and 25% by weight screenings. The floating particles are drained in a press and the water is recirculated as process water. The wet process works with a ratio of 4–5 parts process water per part of bio waste.

The suspension from the garbage dissolver is pumpable and mixable and contains 8–10% dry matter. It is buffered before it is fed continuously into the biogas plant. During buffering, degradation of the substrate starts. The pH value drops to pH = 6.25–7. Ca. 4% of the substrate is degraded in this early stage.

3.3 Comminution

Large lumps of excrement, agricultural residual substances, renewable raw materials, and bio wastes must usually be pre-comminuted. Solids with haulms, e.g., straw with haulms over 10 cm long would otherwise block the pumps or would form a gas-impermeable scum in the bioreactor.

Above all slow-running multiple screw mills (up to 100 U/Min.) are used, which are known from the composting technology. Tearing devices at the feed conveyor and/or at the metering screws into the bioreactor worked satisfactorily for comminution. They have relatively small maintenance costs (approx. 3% of the Investment costs per year) with a good degree of comminution (approx. 85% of all material <200 mm). Comminuting machines such as crop choppers are well known for the comminution of bio wastes for composting.

3.4 Hygienization[26)]

Hygienically suspect substrates must be sanitized before or after the treatment in a biogas plant. There are processes for thermal and chemical hygienization.

Thermal hygienization requires a temperature of at least 70 °C and 1 h residence time, constant mixing during the entire process, and a grain size below 10 mm.

For wastes from a rendering plant, 20 h residence time at 55 °C or a chemical hygienization is prescribed.

For chemical hygienization, a 40% calcium hydroxide solution is mixed with water in a ratio of 40–60 L m^{-3}. The impact time must be at least 4 days.

Since the hygienization is technically laborious and cost-intensive, not all substrates are sanitized, but only the really suspect ones and those, for which a hygienization is prescribed by law.

If hygienization is used, a proof in form of a temperature/time diagram must be given.

According to European laws,[27)] residues of biogas plants must be harmless on leaving the plant from the point of view of an epidemic or the phytohygienic aspect.

26) Cp. BOK 9

27) Cp. LAW 30

The effectiveness of the hygienization of the respective bio waste fermentation plant is to be evidenced to the authorities by means of various verification procedures.

- By a single verification procedure of the hygienization with a so-called direct inspection
- By regular controls of the adherence to the temperatures to be applied by so-called indirect inspections
- By regular controls of the hygienic safety of the treated bio wastes by so-called final product examinations.

3.4.1 Direct inspection

For direct inspection, small quantities of bio waste are taken, are infected in high concentration with special, thermoresistant test organisms and exposed somewhere in the biogas plant in diffusion germ carriers (volume: ca. 20 mL). After the processing in the plant, the diffusion germ carriers are taken off and examined to determine whether the test organisms are sufficiently inactivated. In the case of a successful inactivation of the test organisms, it can be assumed and accepted that the pathogens appearing in low concentrations in the bio wastes will also be sufficiently inactivated when they are similarly treated.

The direct inspection (Table 3.3) is accomplished in new installations as a start-up examination within the first 12 months and repeated when applying new procedures or substantially new techniques within 12 months after start-up.

For direct inspections, four test organisms are applicable:

- By an inspection using the bacterium *Salmonella senftenberg* W775, it is determined whether human and veterinarian hygienically relevant epidemic exciters are sufficiently inactivated.
- By an inspection using fungus (*Plasmodiophora brassiceae*), a virus (tobacco mosaic viruses), and tomato seeds, it is determined whether the phytopathogenic organisms are inactivated.

3.4.1.1 Salmonella

Salmonella belong to the family of the *Enterobacteriaceae*. It is very resistant to heat and is classified as human-pathogenic as it causes characteristic diseases such as typhoid fever, bacterial dysentery, and plague.

Today, more than 2000 salmonella variants (Serovare) are well known which are classified in 2 different sub-groups:

Typhoid Salmonella

This species of infectious pathogen primarily occurs in human excrement. The germ is transferred directly by a smear infection or indirectly by means of food or particularly potable water. It has a small infection dose of 100–1000 cells. Without

Table 3.3 Scope of direct testing.[28]

Parameter	*Prescriptions*
Testing period	New plants: within 12 months after start-up. After important changes of the plant: within 12 month after the changes.
Number of test series	2 test series: With open plants 1 test in winter.
Organisms to test hygiene of epidemics.	1 test organism (*Salmonella senftenberg* W775, H_2S-negative)
Organisms to test hygiene of phytochemicals	3 test organisms (*Plasmodiophora brassicae*, tobacco mosaic virus, tomato seed)
Objectives for testing hygiene of epidemics	Composting: evidence of salmonella in 50 g test substrate negative Fermentation: evidence of salmonella in 1 mL test substrate negative
Objectives for testing hygiene of phytochemicals	*Plasmodiophora brassicae*: effect index <0.5 per test area Tobacco mosaic virus <8 lesions per plant per test area Seed of tomato: <2% germinable seeds per area
Number of samples:	
Test of hygiene of epidemics	Throughput <3000 $Mg\,a^{-1}$: 12 test series Throughput >3000 $Mg\,a^{-1}$: 24 test series
Test of hygiene of phytochemicals	Throughput <3000 $Mg\,a^{-1}$: 3 test series of 6 samples Throughput >3000 $Mg\,a^{-1}$: 3 test series of 12 samples

therapy, about 15% of the infected people die, while with adequate therapy the percentage is about 2%.

Enteric Salmonella

This type of bacterium is found in the intestinal tract of humans and animals. Meat is the primary source for infection for humans. Animals are usually infected by feed, but quite often without showing any symptoms. Even for humans the infection dose is relatively high (about 10^6 germs), but then quickly results in diarrhoea and vomiting. It is very rare that this infection leads to death.

It is compulsory to notify any infections with salmonella to avoid its spreading in the environment. Therefore residues from biogas plants are not allowed to contain any salmonellae. 74% of untreated bio wastes contain salmonellae – up to 1.6×10^4 CFU g^{-1} DM. So in most cases the bio wastes have to be sanitized.

For the direct inspection of the process, specimens with high concentrations of *Salmonella Senftenberg* W775 are employed. This species exists in an H_2S-positive and an H_2S-negative form, but for inspection purposes only the H_2S-negative variant is allowed. Dealings with Salmonellae in general are subject to approval by authorities first. Special care has to be taken during the direct inspection of the

28) Cp. LAW 30

process, because relatively high concentrations of *Salmonella senftenberg* W775 are applied. The suspensions are prepared with 2×10^7–2×10^8 CFU/g DM of salmonellae. The test is positive when no augmentable salmonella is present anylonger in one milliliter of the suspension.

Despite its strong resistance to heat, *Salmonella senftenberg* is usually quicky inactivated at sufficiently high temperatures in thermophilic biogas plants and during the 1-h hygienization treatment at 70 °C.

3.4.1.2 Plasmodiophora brassicae

Plasmodiophora brassicae, an obligate endoparasite, is classified as a fungus and belongs to the group of the plasmodiophoromycota. It causes swellings at the main and side roots of cabbage and cruciferous plants, blocking the uptake of nutrients and water. Permanent endospores are formed which remain infectious in the soil for many years.

The heat resistance of the fungi varies widely depending on the origin. In most cases (3/4 tests) the fungus is inactivated at temperatures in the range 50–55 °C, allowing one to consider this range as a relatively safe guideline. The results of such a quick temperature check help to assess whether the temperature in the hygienization stage is set sufficiently high.

3.4.1.3 Tobacco mosaic virus

The tobacco mosaic virus (TMV) causes a widely spread infectous disease of plants. About 25 different biological plant families with more than 60 species are affected, especially solanaceous herbs like tobacco, tomato, and pepper. The virus is extremely resistant. Even after a residence time of more than 21 days at 70 °C, the fungi remain virulent. Only after 35 days at temperatures of 55 °C and 60 °C does the infectiousness of the virus decrease to less than 3%.

This test allows optimization of the residence time of the substrate in the hygienization stage at temperatures higher than 70 °C.

3.4.1.4 Tomato seeds

Untreated bio waste still contains numerous germinable seeds and parts of plants. The seeds of tomatoes are used as an indicator, since these seeds have high moisture content and are relatively resistant against high temperatures.

Only at temperatures of >60 °C are tomato seeds quickly inactivated. Hence, thermophilic fermentation and pasteurization at 70 °C for 1 h should be applied.

The results of the test with tomato seeds indicate how effectively weed seeds are inactivated during the process.

3.4.2 Indirect process inspection

Untreated bio waste provides enough nutrients for pathogens to grow (high water content, high amount of easily decomposable organic substances, etc). Treated bio

Table 3.4 Scope of indirect testing.[29]

Parameter	*Prescriptions*
Testing period	Continuous time period
Process of testing	Continuous temperature control with automatic recording Minimum 1 measurement per working day at 3 representative zones within the process flow which are relevant for the thermal inactivation of pathogens. Documentation of all data (e.g., temperature, feeding intervals of the ferment). Archiving of data for minimum 5 years
Objectives for fermentation plants	Thermophilic processes: >55 °C, >24 h; hydraulic residence time >20 days. Mesophilic processes: pretreatment or aftertreatment (at >70 °C, >1 h) or aftertreatment of the residual bio waste by aerobic rot.

waste, however, hardly contains any nutrients and offers only poor conditions for demanding pathogens.

The objective is to keep the level of nutrients in the bio waste as low as possible.

This can be achieved by strictly following guidelines which regulate the parameters (temperature/residence time) to be set during the hygienization process (Table 3.4).

This specific process control is to be applied for all the big and small particles to meet the requirements. Therefore it is mandatory to prove that the "actual and effective residence time" for the thermal inactivation organisms is long enough and in line with the "instructions given as guidelines".

3.4.3 Control of the finished goods

The finished goods need to be regularly controlled to ensure that the treated bio waste is hygienically safe, meeting the requirements as stated in (Table 3.5).

Additionally, the concentrations of the following different types of bacteria are to be determined by analyzing the bio waste to ensure that they are below the respective guidance values:

Total content of bacteria at 37 °C: $<5 \times 10^8\,\mathrm{CFU\,g^{-1}}$
Faecal coliform bacteria: $<5 \times 10^3\,\mathrm{CFU\,g^{-1}}$

Residues have additionally to be analyzed for
Enterococcus: $<5 \times 10^3\,\mathrm{CFU\,g^{-1}}$.

The degree of reduction indicates the efficiency of the treatment, especially of the hygienization process.

29) Cp. LAW 30

Table 3.5 Scope of final testing.[30]

Parameter	*Prescriptions*
Test series	Several test series required per year: – every half year (with a throughput <3000 Mg a^{-1}). – every quarter (with a throughput >3000 Mg a^{-1}).
Target in hygienic testing to avoid epidemics	Evidence of salmonella in 50 g product should be negative.
Target in hygienic testing against phytochemicals	Less than 2 germinatable seeds and plant parts per liter substrate
Number of samples	Samples are mixed (ca. 3 kg) from more than five smaller samples. Samples are taken from different charges of the product. 0–3000 Mg a^{-1} throughput: 6 samples/year. 3000–4000 Mg a^{-1} throughput: 7 samples/year. 4000–5000 Mg a^{-1} throughput: 8 samples/year. 5000–6000 Mg a^{-1} throughput: 9 samples/year. 6000–6500 Mg a^{-1} throughput: 10 samples/year. 6500–9500 Mg a^{-1} throughput: 13 samples/year 9500–12 500 Mg a^{-1} throughput: 14 samples/year. (for each added 3000 Mg a^{-1} throughput: 1 additional sample/year.

A failure in the final test shows a lack of hygiene in the production area. The bio waste may have been re-contaminated by contaminated deliveries, polluted processing water, or cross-contamination by untreated bio waste.

The responsible institution is to be notified about the negative results of the quality control, which are recorded together with the resulting measures taken to address the issues.

If the direct process inspection does not show any negative results, the final quality control in accordance with the guidelines will be redundant. If direct process control is impossible, an extended final test is required to ensure the product meets all the qualitative requirements.

3.5 Disintegration

Disintegration means the breakdown of organic cells, e.g., from microorganisms, to give products that will ferment more easily. The degree of disintegration depends on the characteristics of the substrate, the application of energy, and the technique. Most of the tests[31),32),33)] on disintegration were carried out with sewage sludge.

30) Cp. LAW 30
31) Cp. BOK 6
32) Cp. JOU 1
33) Cp. WEB 52

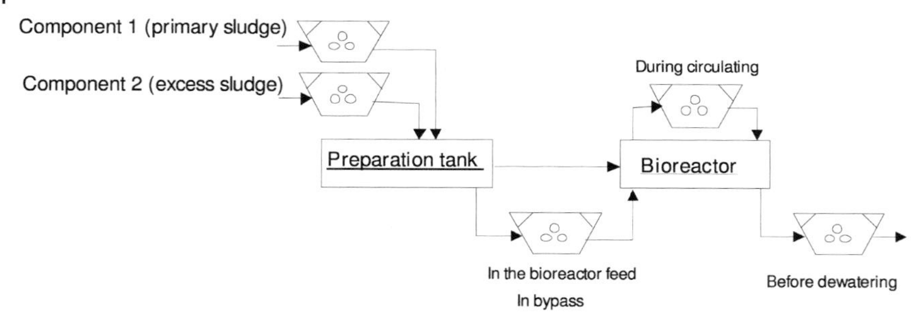

Figure 3.2 Recommended stages for the disintegration process.

Primary sludge and excess sludge can be decomposed at different speeds and to different degrees because of the difference in the content of intact cells. With minor application of energy, the flaky structure of the cells is dissolved, while the application of high energy allows the cells to be completely decomposed and provides the cell components for further degradation.

It is recommended to apply disintegration to accelerate the relatively slow biochemical process of biological decomposition.[34] Only 3–4 days (instead of 20 days) may sometimes be sufficient to yield maximum saccharification.

The disintegration can be performed at different stages of the process (Figure 3.2).

The following technologies can be used for degradation purposes.[35]

- Mechanical processes
  - High pressure homogenizer
  - Agitator ball mill
  - High-performance pulsed reactor
  - Baffle jet equipment
  - Lyzing centrifuge
- Acoustic processes
  - Ultrasonic cavitation
  - Sonochemical reactions
- Chemical processes
  - Acid-/base-treatment
  - Ozone-oxidation
  - Detergents, salts
- Thermal processes
  - Pyrolysis
  - Thermal cell disruption
- Physical processes
  - Freeze and thaw
  - Alternating pressures

34) Cp. BOK 18

35) Cp. BOK 51

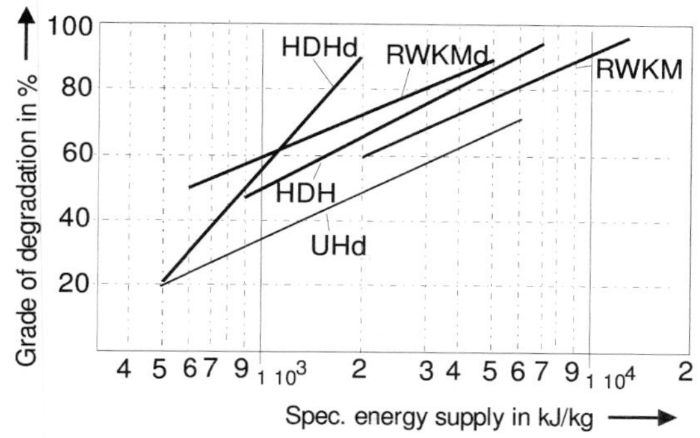

Figure 3.3 Comparison of different processes for the mechanical disintegration of sewage sludge (UHd = Supersonic with pre-thickening to abt. 40 g/kg; HDH d = High pressure ho-mogenization with pre-thickening to abt. 40 g/kg; HDH = Homogenization at abt. 10 g/kg; RWKM d = Agitation with ball mill with pre-thickening at abt. 40 g/kg; RWKM = Agitation with bal mill at abt. 10 g/kg.

- Biological processes
  - Enzymatic lysis
  - Phage
  - Electrical processes
  - Electric shock
- Thermo-biological processes
  - Composting
  - Fermentation-tempering-fermentation.

All listed processes have very specific advantages and disadvantges and require trials on the laboratory and pilot scale prior to a large scale application.

The relationship between the degree of degradation and the specific energy supply is shown in Figure 3.3.

In general, thermally disintegrated sludge has a much lower filtration resistance than sludge disintegrated in any other way.

Ultrasonic disintegration leads to a 17–45% improvement in the sedimentation behavior of sludges. Similarly, mechanical disintegration in an agitator ball mill can give about 37% improvement. 40–60% improvement is achieved by using a high-pressure homogenizer, and about 68% by using a lyzing centrifuge.

In order to integrate the disintegration process in a sewage water treatment plant, special equipment is needed, and additional aids such as flocculants are used to clean the sewage sludge from trash and to thicken it. Special equipment is also needed to treat and use the additional biogas and the wastewater, which is directed back into the sewage water treatment plant.

In total, the installation of an integrated disintegration process requires an investment of 10% of the total costs of a sewage water treatment plant and operational and maintenance costs of about 15%.

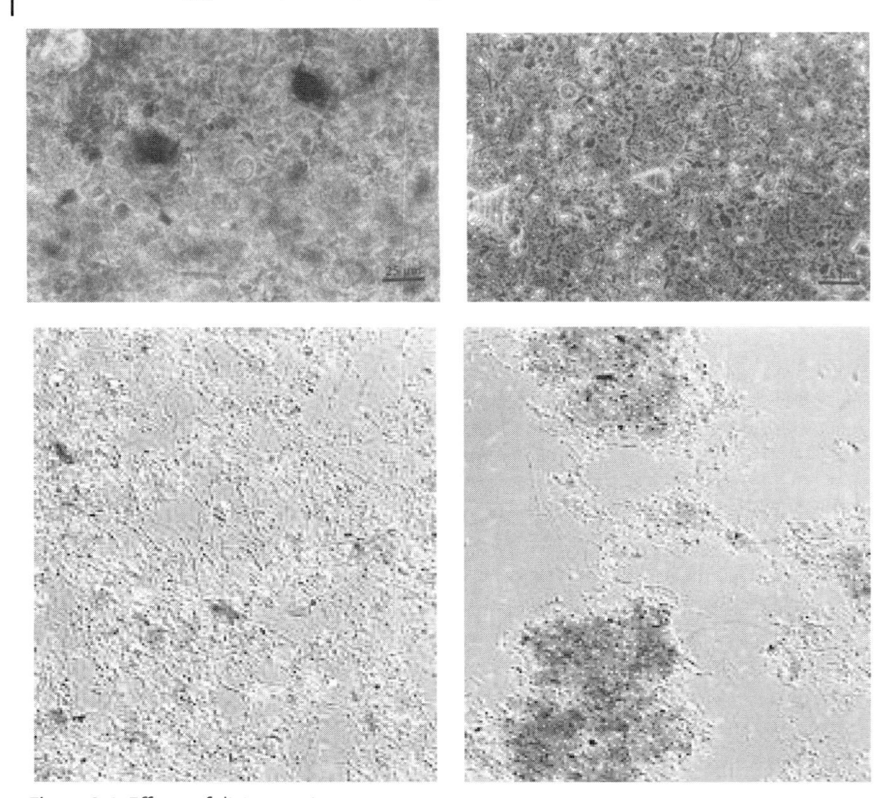

Figure 3.4 Effects of disintegration
Sewage sludge: not disintegrated (above left) and ozonised (above right), the length of the bar equals 25 μm.
Scum: before thermal disintegration (below left) and after disintegration (128 °C; 70 min) (below right).

The energy consumption of about 3 kWh per population equivalent per year makes up about 10% of the total energy consumption of a sewage water treatment plant, which needs 30 kWh per population equivalent per year. But disintegration leads to a 20% increase in the degree of decomposition, which annually yields about 1 kWh per population equivalent per year of additional power generated from the biogas. This is about 30% of the energy consumption of the disintegration system.

Even with a negative energy balance associated with the disintegration system, its installation may still be economic. Significant savings may be achievable because of lower disposal costs for the sludge due to a reduction in the quantity of accumulated sludge.

In the following section, some processes are explained more in detail (See Figure 3.4).

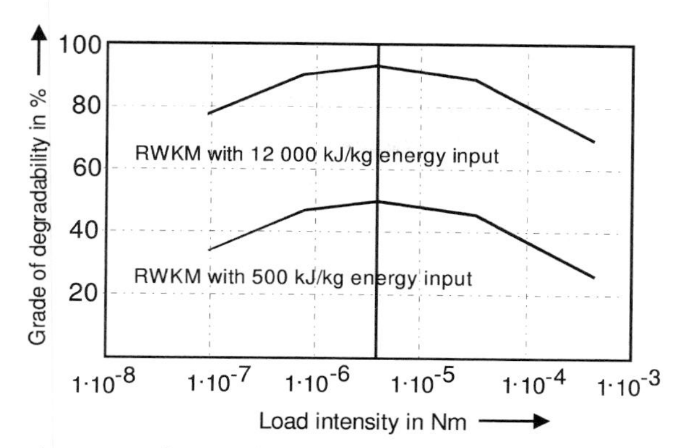

Figure 3.5 Influence of the load intensity on the grade of degradation at constant specific energy (RWKM = agitated ball mill).

3.5.1 Mechanical processes

Most favorable is the mechanical disintegration system after pre-thickening the sludge. Agitator ball mills and homogenizers were found to be best suited.

In a system applying an agitator ball mill, the disintegration intensity B is dependent on the diameter of the grinding ball d_{MK}, its density ρ_{MK}, and the rotational velocity of the agitator sysem v_u.

$$B = d_{MK}{}^3 \cdot v_u{}^2 \cdot \rho_{MK}$$

The disintegration intensity indicates the maximum energy of loading when a grinding ball hits. It can be optimized (Figure 3.5).

3.5.2 Ultrasonic process[36)]

Ultrasonic energy generates a high alternating voltage which causes cavitation beyond the human audio range. This leads to the formation of small bubbles of gas during the cell disruption process caused by an internal negative pressure. The bubbles collapse immediately after their appearance. The resulting shock pressure (several 100 bar) and temperature peaks (several 1000 °C) lead to the destruction of the cell membrane. If the substrate experiences an ultrasonic field of 31 Hz, the disintegration of the cell membrane takes place within 100 s only. The longer the ultrasonic treatment, the smaller is the quantity of accumulated mud bacteria (normal size around 100 μm). The ultrasonic disintegration is mainly applied in a bypass system where 30% of the sewage sludge is treated.

36) Cp. JOU 23

In sewage sludge bioreactors, the digestive period is halved, the gas yield is increased by 40%, and the organic dry matter is reduced by about 40%.

Modern ultrasonic emitters deliver frequencies in the range between 20 kHz and 21.6 MHz, amplitudes up to 250 μm, and variations of energy density in the range 12–4000 W. The low to middle range of frequencies up to 100 kHz is preferred, because fewer disintegrated cells and more radicals are formed at high frequencies (100–1000 kHz). Abd. 40 kHz were found to be best in many applications.

3.5.3
Chemical processes

The addition of acids or bases leads to the saponification of the cell components, which hence become soluble and are much easier to decompose during the subsequent fermentation process. About 55% of the dry matter can be decomposed, especially when using NaOH to saponify. Other chemicals like ammonia and quicklime are quite often used to increase the pH. Similar results can be achieved by decreasing the pH value to 2 using sulfuric or hydrochloric acid at higher temperatures of 140–160 °C.

A well-known process is the APTMP process (alkaline-peroxide-thermo-mechanical pulping) which is often applied in the paper industry. It combines thermo-mechanical and chemical process steps which are well suited to the decomposition of wood, especially softwood of broad-leafed trees with a low content of lignin, e.g., poplar. Hardly any waste water is produced during the process, and substances contained in it are biologically degradable. Compared to the acid CTMP process (chemo-thermo-mechanical pulping) the APTMP requires a lower specific energy input. The lignocellulose-containing biomass is first milled into wood chips and then impregnated, helping to soften the structure of the fibers and decrease the specific milling energy. The impregnation depends on the quality of the wood and the amount of chemicals. Usually, a one-level or an up to three-stage process is applied in which the filtrate of the second or third screw extrusion press is re-circulated into the preceding impregnation stage and supplemented with fresh chemicals. Trash (heavy and lightweight materials) are separated with a sluice for heavy materials or in a hydrodynamic elutriator (hydrocyclone). Each stage requires a specific input of mechanical energy of about 50–60 $kWh\,Mg^{-1}$. In the end, a refiner is needed for the mechanical defibration of the substrate before the biomass is thermally disintegrated. Overall, less energy is applied for milling hardwood than softwood as a pretreatment. But it still accounts for 15% of the electrical energy yield of the biogas.

With the application of ozone, the structure of the molecules present in the mud changes (Figure 3.6) due to the following effects:

1. Ozone undergoes a chemical reaction with the double-bonding of unsaturated fatty acids in the cell membranes. The cells become permeable and cytoplasm is discharged.
2. Water-insoluble macro molecules can be decomposed into smaller water-soluble molecules of lower molar mass.

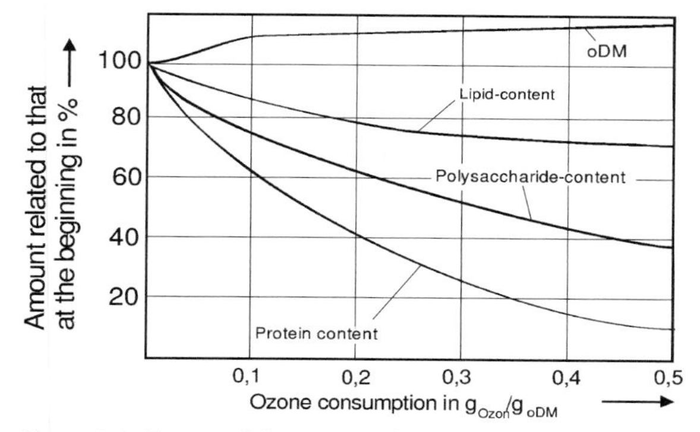

Figure 3.6 Change of the sewage sludge matrix due to partial oxidation by ozone.

Water-insoluble proteins are 90% decomposed,
polysaccharides 63% and lipids 30%.

About 0.05–0.5 g of ozone is commonly used for 1 gram of organic dry matter. In general, a higher input of ozone leads to more organic carbohydrates being released. Sometimes this may even increase from 10% to about 40%.

When using ozone, the disintegration of digested sludge in sewage water treatment plants works best using the following processes:

- If the digestion towers are in line, the ozonization can take place between the first and the second digestion tower.
- In smaller plants which have one digestion tower only, the ozonization can occur in the bypass. 4 to 20% of sludge can be removed from the digester, be ozonized, and then be transferred back.
- It seems to be more advantageous to only ozonize the biomass which has been separated externally and is recirculated.

Disadvantages:

- The methanization may become delayed by more than 10 days (lag phase of about 9 days), since the system first needs to re-adapt itself to the substrate, which, after the ozonization, contains a higher concentration of substances which are toxic toward methane-producing microorganisms, e.g., formaldehyde.
- The concentration of free carbon increases with the amount of ozone consumed. This carbon cannot be degraded with anaerobic digestive processes but just gets partly decomposed (about 27%) in an aerobic environment. Therfore it is recommended not to feed too much ozone in order to minimize the recontamination of water with organic substances.

Even hydrogen peroxide (H_2O_2) acts in a disintegrative way, releasing highly reactive hydroxyl radicals when used at a concentration of $800\,mL\,m^{-3}$.

3.5.4
Thermal processes

The yield of biogas is 30% higher if the substrate is thermally disintegrated before it enters the bioreactor. This process occurs at pressures above 10 bar and temperatures above 150 °C. The fermentation only takes half the time of other processes and sanitization is automatically included.

The higher the temperature and residence time, the better is the degradation. The optimum temperature is about 135–220 °C with a residence time of 2 h.[37] Parameters exceeding these settings often yield a higher COD/BOD_5 ratio in the filtrate of the secondary sludge. But this indicates the formation of substances which are difficult to degrade. At temperatures above 230 °C, thermal degradation processes of substances containing lignocellulose may be more activated, depending on the substrate. This leads to yield loss, mainly caused by phenols, furfuraldehydes, and hydroxybutyric acid.[38],[39]

The re-loading of the sewage water plant with dissolved matters, indicated by a high COD-value, high N_{tot} and PO_4-P content, is only marginally increased when applying the thermal disintegration process.

But all the germs and worm eggs become completely inhibited.

The process only runs economically with regenerative heat exchange between the supply and the downpipe of the thermo reactor (Figure 3.7).

The sewage sludge is first pre-heated in the first shell and tube heat exchanger, then heated up to the disintegration temperature, then held in the disintegrator and then cooled down by heat exchange with the pre-heater, so that the temperature is slightly above the temperature in the digester. After thermal disintegration it is subjected to the circulation in the closed fermentation circuit.

The process hence requires only slightly more heat than a conventional fermentation. The regenerative heat exchange is achieved with a water circulation through the shells of the first and third heat exchangers. The final heating process in the second heat exchanger occurs with thermal oil under slight pressure.

Alternatively, quite expensive components can be used, e.g., spiral heat exchangers, which are quite robust and less sensitive to clogging.

3.5.5
Biological processes

Biological processes are less expensive but up-and-coming.

Very early in the process, during the process of ensilage or the production of hay, disintegration takes place[40] by the lactic acids which decompose complex

37) Cp. BRO 9
38) Cp. BOK 14
39) Cp. BOK 21
40) Cp. WEB 112

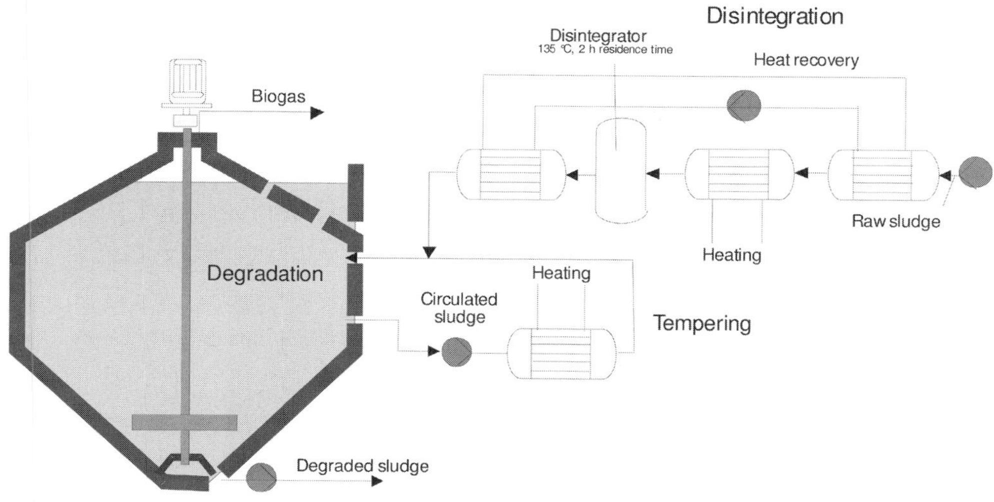

Figure 3.7 Flow diagram of a thermal disintegration process with indirect heat exchange and water circulation.

components of certain substrates. Therefore, clover should be harvested and processed before blooming, while maize is best cut at the end of the maturation.

The disintegration of primary sludge with enzymes has been quite successful, especially using cellulase, protease, or carbohydrases at a pH of 4.5–6.5 and a residence time of minimum 12 h, better several days.[41] Excess sludge is much more difficult to disintegrate biologically.

The yield of methane can be significantly increased by adding a small amount of NaOH to the substrate and applying a 4-day long biological autolysis at 60 °C before the methanation.

3.6 Feeding

After the completion of adequate preparation steps, the substrate is ready to be fed into the biogas reactor. The bioreactor is usually loaded once or twice a day. In agricultural wet fermentation plants, where special preparation procedures are not needed, a preparation tank, which serves also as buffer tank, is placed ahead of the bioreactor. It is equipped with a mixer or helical agitators. This is very important, especially when using silage as co-ferment. The inlet and the downpipe should be placed about 10 cm above the bottom plate, since clods, stones, and sand may sediment. The preparation tank is usually covered. The displacement air from the filling stage is purified or fed to the combined heat and power plant.

During the filling of the preparation tank or the bioreactor, any swirls and/or release of aerosols should be avoided. Therefore, in agricultural plants, just for

41) Cp. JOU 28

Figure 3.8 Equipment for upstream processing
Closed belt conveyor (above left),
Feeding the bioreactor (above right)
Feeder (vielfraβ®) for agricultural plant (middle left),
Pump chute (middle right),
Hygienization installation (below).

filling, the cover is removed for a short period. The accessing pits are less practical, because they are prone to clogging, can only be filled batch by batch, and are not fully tight, so that air may enter into the reactor.

Ideally, the inlet should be below the liquid level. The following processes are now implemented:

- Dungshaker-like units
- Feed blending carriages
- Skimmer suspension chain bottom
- Screw conveyor
- Hydraulic units
- Punching press
- Automatic pulp preparation vessel and metering tank.

The injection regenerative raw materials and solid dung is easy with an open steel funnel of 2 m^3 volume and a screw conveyor with an automatic interval control.[42] Such a feeder transports maximum 1 $m^3 min^{-1}$ at an energy consumption of 0.5 $kWh\,m^{-3}$.

In order to load the dry fermentation plant, storage tanks should be equipped with chain conveyors in the inside, so that no dry bio waste is overlaid at any place in the storage tank.

In some plants the substrate is pre-heated in an external heat exchanger up to the temperature to which it is exposed when transferred into the bioreactor.

42) Cp. WEB 6

4
Fermentation technology

In general, different types and technologies of bioreactor plants can be found in the industry (Figure 4.1).

In each of the different processes it is important to have enough buffer tanks to avoid a sudden load from, e.g., a short interruption of the operations, and to ensure a constant load of the methanation.

4.1
Batchwise and continuous processes without separators

For agricultural plants, it is important to have simple and good value equipment. The farmer should be able to lend a hand himself with problems that may arise in the plant. Therefore, single-stage plants have been installed most successfully in agriculture. Such plants would require more measurement and monitoring for optimal running, and the conditions for biochemical reactions still need to be more investigated. But these technological disadvantages seem to have been accepted.

Even for the anaerobic purification of industrial waste water, such simple plants, called ASBR-reactors (Anaerobic Sequencing Batch Reactors), are used. All the different stages of the waste water treatment (filling, biochemical reaction, sedimentation, decanting) happen sequentially in one and the same tank. Because of the low load, a larger reactor volume is required than in plants with a continuous process. The whole process is susceptible to toxic substances.

In the batch process (Figure 4.2), the digester is completely filled all at once. The substrate degrades without anything being added or discharged until the end of the residence time. This leads to a temporal variation in the production and the composition of the gas. The production of gas starts increasing until it reaches its maximum at about half the residence time. It then starts slowly decreasing, impacting negatively on the gas motor, which cannot be evenly fed any longer and starts running at suboptimal conditions.[43] At the end of the process the fermenter is emptied into the storage tank, and only small amounts remain to inoculate the next load. Fresh substrate is mixed with the remaining fermented substrate to continue using the microorganisms.

43) Cp. WEB 113

Biogas from Waste and Renewable Resources. An Introduction.
Dieter Deublein and Angelika Steinhauser

ISBN: 978-3-527-31841-4

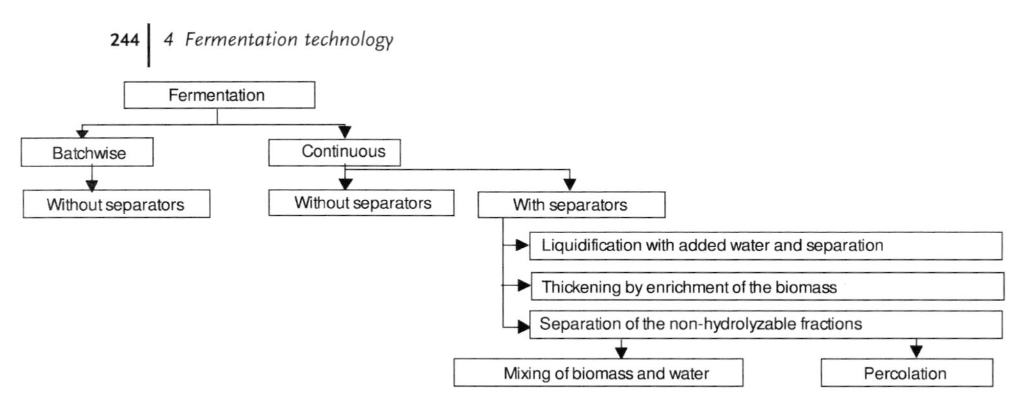

Figure 4.1 Overview of different processes to produce biogas.

All the negatives can be overcome by installing changeover tanks, but at higher cost.

The residue storage tank should be covered because of emissions of methane and odorous ammonia. In the flow-through process, the residue storage tank can be used as the methane reactor, making it a two-stage process.

At the beginning of the 1990s, biogas plants in Europe were most frequently built as flow-through plants with horizontal cylindrical steel tanks (plug flow reactor).

Flexible storage flow-through systems (agitated vertical tanks) are preferred for larger plants because of their larger volume. About 50% of all the agricultural biogas plants in Germany are of the storage flow-through system.

Just 4% of all biogas plants work, with one tank as bioreactor and residue storage tank. The system was not very successful in industry because of the discontinous gas production and the heat losses, because even the volume, which in other plants serves as a storage tank, has to be kept at mesophilic temperature.

4.1.1
Systems engineering

Strongly scum-forming substrates like rumen content or cut grass are fed into the bioreactor using mixing macerators. The scums are removed with slowly agitating windlass mixers. Bioreactors should be preferred with larger liquid surface, which develop a respectively thin scum, and have a particularly large downpipe system.

Figure 4.3 shows a simple one-phase plant as is often found in European agriculture (Table 4.1). It consists of a building which serves as a storage space for the co-substrates and a preparation tank which is loaded below the liquid level. The air in the building is sucked off and serves as combustion air in the combined heat and power plant. The building is large enough so that trucks can drive in to deliver the co-substrates. The liquid manure is directly fed in from the stables.

In the preparation tank, the mixture of biomass is homogenized before it is pumped into the totally gas-tight closed bioreactor, where it remains until it is sufficiently fermented. From time to time the substrate is mixed.

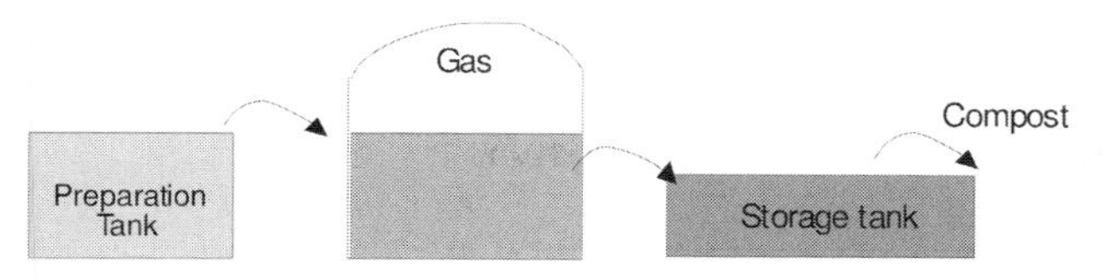

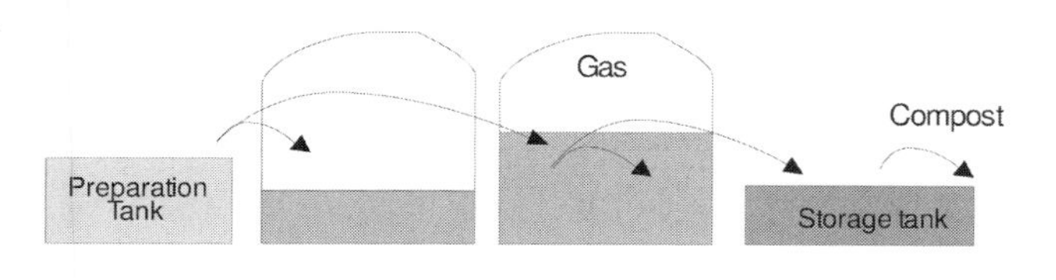

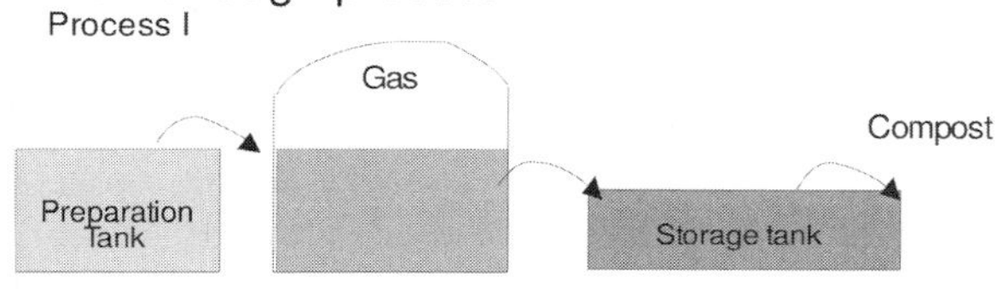

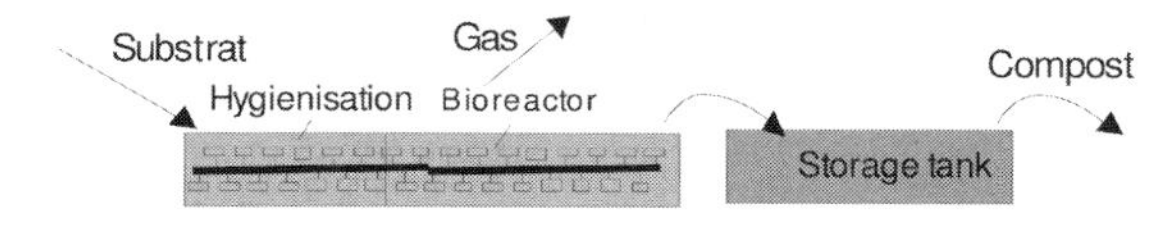

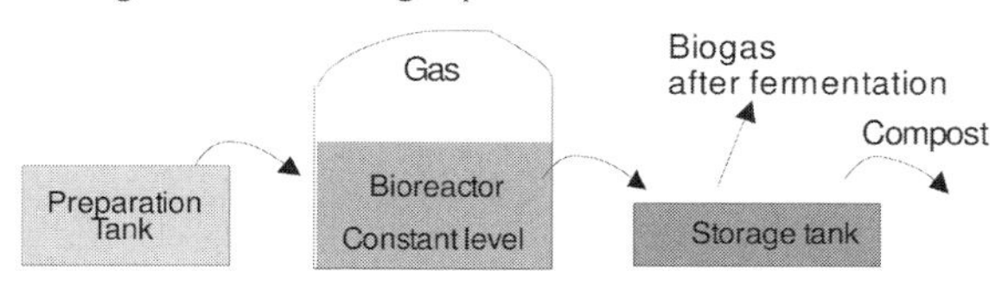

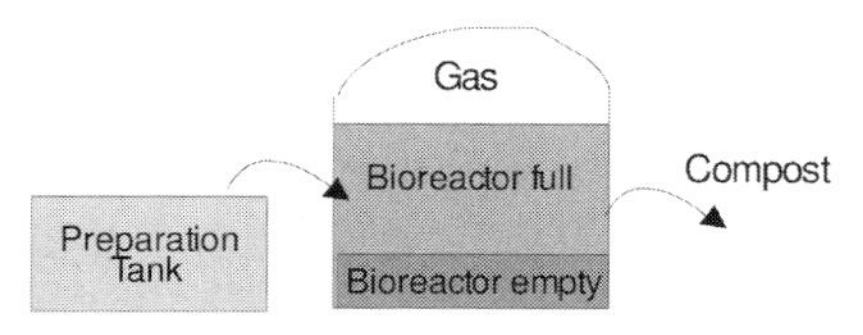

Figure 4.2 Operation modes of agricultural biogas plants.

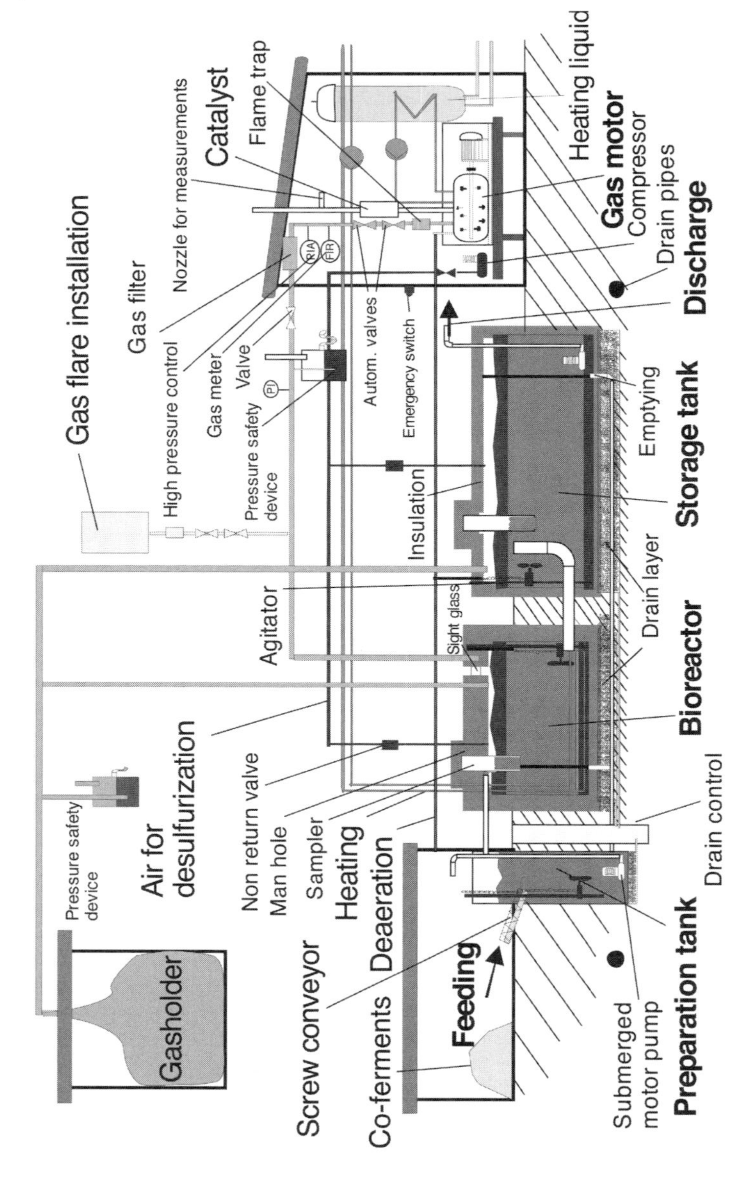

Figure 4.3 Simple European agricultural biogas plant to add to habitations.

Table 4.1 Technical features of agricultural biogas plants.

GVE	*Unit*	*50*	*60*	*65*	*70*	*75*	*85*
Substrate	M^3/d	2.2 Liquid manure	3.5 Liquid manure	3 Liquid manure	6 Liquid manure. dung	4.5 Liquid manure. dung	4.5 Liquid manure. dung
Co-ferment	m^3/d	0.55 SPR	1.1 FAF	5 FAF. BAF	0.1 GAF. AAF	0.3 KAP	0.35 FAF
Bioreactor volume	M^3	80	n.a.	n.a.	n.a.	n.a.	n.a.
Temp.	°C		37	36	26		30
Power consumption	%	8	n.a.	n.a.	n.a.	n.a.	n.a.
Heat consumption	%	50	n.a.	n.a.	n.a.	n.a.	n.a.
Gas production	m^3/d	179	57	150	22	57	42
Elec. engine capacity	kW_{el}	20	15	31	n.a.	25	13
Th. engine capacity	kW_{th}	44	25	51	n.a.	42	20

GVE	*100*	*100*	*110*	*110*	*200*	*300*	*500*	*1000*
Substrate	4.4 Liquid manure	4.4 Liquid manure	7.5 Liquid manure	4.5 Liquid manure	12 Liquid manure	18 Liquid manure	30 Liquid manure	60 Liquid manure
Co-ferment	0	1.10 SPR	1 FAF	0.65 FAF	0	0	0	0
Bioreactor volume	250	350	n.a.	n.a.	300	450	750	2 × 750
Temp.			39	36				
Power consumption	8	8	n.a.	n.a.	n.a.	n.a.	n.a.	n.a.
Heat consumption	50	50	n.a.	n.a.	n.a.	n.a.	n.a.	n.a.
Gas production	140	359	44	39	250	375	625	1'250
Elec. engine capacity	20	50	2 × 15	30	19	29	48	97
Th. engine capacity	44	110	2 × 25	50	89	127	112	156

FAF = grease-removal tank flotate; BAF = bio wastes; GAF = vegetable wastes; AAF = milky wastes; KAP = potatoe mash; SPR = left over

The residue is transferred into a tightly closed storage tank for the afterfermentation. Again the residue is sometimes stirred up. At the end of the process the storage tank is emptied and the residue is ideally applied in agriculture.

The biogas from the bioreactor and the residue storage tank is sucked off into a foil gasholder.

The bioreactor and the residue storage tank can be emptied into the preparation tank from where the substrate and residue can be removed with pumps and tube pipes.

The CHP is installed in a building located at a safe distance from the biogas plant. It is fed with biogas and if necessary with ignition oil. The current produced is directly fed into the public electricity network. The waste heat from the CHP is transferred into a heat storage facility and used to maintain the temperature in the bioreactor. The surplus heat can be used outside the biogas plant.

The desulfurization happens biologically by injecting air for the desulfurizing microorganisms.

4.1.2
Reactor technique

Most of the agricultural biogas plants consist of either an over- or underground cylindrical bioreactor with a vertical fast-rotating agitator (Figure 4.3). The cylinders themselves are locally made of concrete or are constructed of prefabricated elements made of concrete. Some cylinders may consist of enameled steel slabs, wound iron sheet plate, or high grade steel plates. In-ground tanks with movable ceiling, large height and small tank diameter are preferred. Bioreactors protruding out of the ground are mostly protected by tear-proof foil supported by compressed air which is generated by a small fan.

Small biogas plants and those with tendency to develop a sinking or floating-layer, e.g., fed with liquid chicken manure, often consist of a lying steel tank with horizontal agitator shaft. Such bioreactors are manufactured industrially and have to be transported by truck, limiting their size to a maximum volume of $280\,m^3$. Horizontal (lying) cylindrical bioreactors of concrete can be manufactured locally and can thus be larger. They serve mostly as pre-fermenters, which store the substrate for 15–20 days. Herein the substrate slowly flows through the digester in order to avoid horizontal mixing.

4.1.2.1 Reactor size

Common volume loads in agricultural bioreactors are shown in Figure 4.4. It can be seen that the volume load usually does not exceed $B_R = 4\,kg\ oDM/d.m^3$ of the

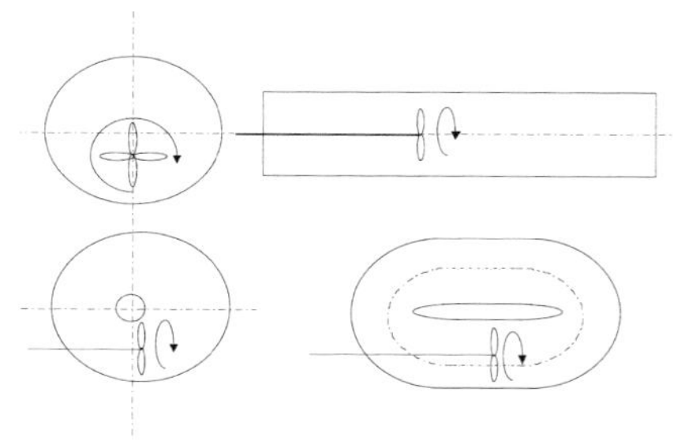

Figure 4.4 To view of different shapes of bioreactors: vertical cylinder (top left); Horizontal cylinder (top right); Torus form (bottom left); Long torus form (bottom right).

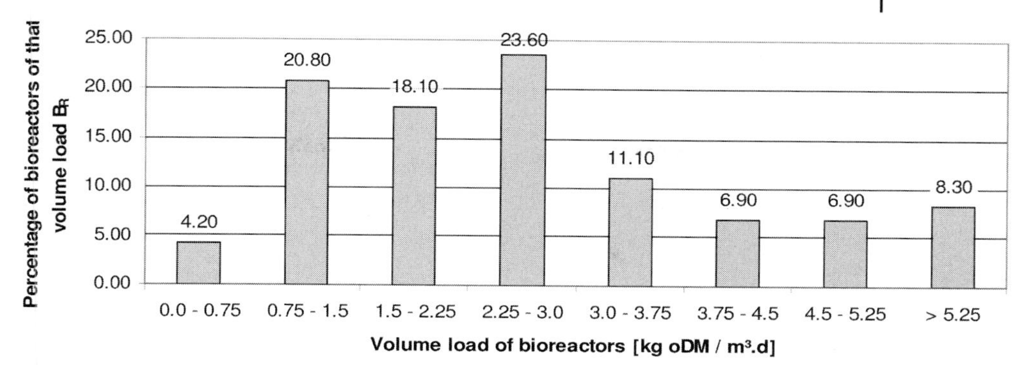

Figure 4.5 Volume load of bioreactors.

Table 4.2 Common bioreactor sizes.

Bioreactor size [m³]	*Percentage of bioreactors of this volume load*	*Diameter [m]*	*Height [m]*
<200	12	Mostly horizontal cylinders	
200–400	51	10	5
400–600	24	12	6
600–800	9	14	6
>1000	5	fitted	6

fermenter volume.[44] But in the charging area the volume load can be up to B_R = 10 kg oDM/d.m³, because during fermentation water is released and the volume load therefore reduced.

From the volume load and the amount of available substrate, the bioreactor sizes and dimensions can be calculated (Figure 4.5).

The bio-reactors are commonly 4 to 6 m high and filled up to 40 cm below the brim. When using larger sizes, the mass can no longer be mixed properly, leading to problems with the heat distribution in the tank.

It is recommended only to work with larger bioreactors (up to V_{BR} = 5000 m³, diameter D_{BR} = 15 m and height H_{BR} = 20 m) when using the "Danish Technology". Here the agitator is positioned centrally at the top. Furthermore, the substrate is pumped. There is an external heat exchanger for tempering.

The residence time and consequently reactor size can be reduced with increasing fermentation temperature (Table 4.3). But it has to be taken in consideration that the gas quality (methane concentration) decreases with increasing fermentation temperature.

44) Cp. WEB 114

Table 4.3 Temperature ranges.

	Temperature range	*Residence time*		*Percentage of all plants*	*Advantages*
Psychrophilic (cold fermentation)	15–30 °C	abt. 60 d		3%	
Mesophilic (warm fermentation)	30–50 °C	17–22 d	l. man.fr. poultry	92%	Stable process
		22–28 d	l. man.fr. pigs		Low heat consumption
		28–38 d	l. man. fr. cattle		Good biogas quality
		35–45 d	excrement		
Thermophilic (hot fermentation)	>50 °C	abt. 15 d		5%	High degradabilty and speed of degradation Hygienization possible

4.1.2.2 **Reactor Designs**

Particularly in developing countries, bioreactors are built of clay bricks. They mostly have round shapes and are of not more than 10 m^3 capacity.

In developed countries, reinforced concrete reactors are more widely used.

These are built rather flat, to reduce the pressure stress. Some have a circular foundation for the ring wall.

Enameled and plastic-coated steel reactors are used more rarely because of their sensitivity to impact.

Even ground basins are seen very rarely. These are built directly at their final location from plastic foil. Foil is put into the basin's base as an underlayer. Then, a polyethylene safety foil of thickness 0.2–0.8 mm, depending on the application, is laid onto the underlayer. The foil is usually laid out in an overlapping manner. In water protection zones, the foil-courses are welded together.

A leak detection mat (1000 g m^{-2}), a polypropylene foil with good draining properties, is then spread out onto the whole surface. The drainage liquid is passed into a drainage pit with a diameter of 300 mm. The final polyethylene foil constituting the basin is then spread out.

For collecting the biogas, the surface of the substrate is covered (sealed) with a gas-tight foil with openings and pipelines to the gasholder.

According to the guidelines for the building of liquid manure tanks (see Chapter 4), the inlet and outlet have to be positioned outside the border of the basin. Thus, no substrate can flow out of the basin and no gate valves are required. The inlet begins at one side of the basin with a connection for the liquid manure hose or tank and ends at the basin's bottom fixed to a tire filled with concrete.

The homogenization operation of the biomass must not damage the bottom foil. Therefore, either a shaft agitator with tractor actuation, a fixed submerged motor agitator, or bridges with submerged agitators have to be used.

Shaft agitators with tractor actuation should at least be 7.00 m long. In some cases, the agitator is in a separate trolley. The agitator's drawbar is fixed to the edge of the basin. The agitator can be freely moved in the basin through the drawbar.

It is also possible to circulate the bio-mass through the inlet and outlet. For this configuration, the pump is situated outside the basin.

A so called "balloon" installation consists of a plastic sack. The gas accumulates in the upper part of the balloon. Inlets and outlets are directly fixed to the outer skin of the balloon, which is inflated only slightly. The balloon is in fact not very elastic. Through the movement of the balloon's skin, the sludge is gently mixed, promoting the digestion process.

Balloon installations are cheap, and easy to transport, clean, and maintain. They are recommended for locations where no damaging of the outer skin will occur and constant high temperatures are present.

In a Finnish biogas plant in Vaasa, natural caves are used as bioreactors. They protect the fermentation process against cold snaps, especially during the cold Scandinavian winter.

4.1.2.3 Covering of the bioreactor

Lately, coverings of textile mass are used which are hung up on a central pillar and twisted at the edges (Figure 4.6).[45]

Such coverings are more gas-tight and more corrosion-resistant than concrete. They can, for instance, consist of EDPM rubber[46]. This material stands out because of excellent UV and ozone stability. Additionally, it is very elastic and particularly durable, and as such can be built in without wrinkles.

A disadvantage compared to concrete is that the microorganisms that are required for the desulfurization are not able to settle down on these coverings. Therefore, additional "settling-down" areas in the reactor have to be installed.

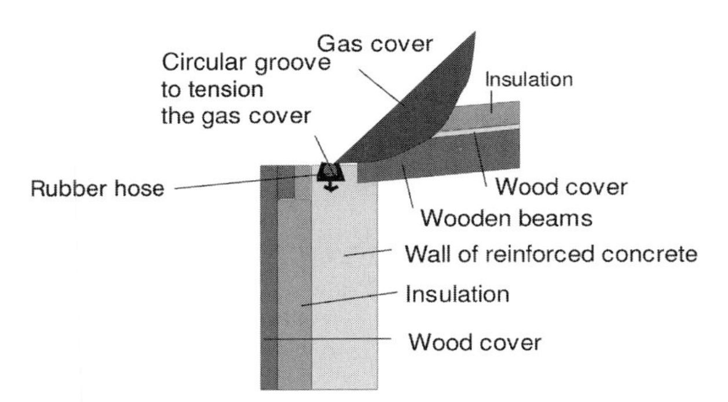

Figure 4.6 Components of a gas-tight cover for a bioreactor.

45) Cp. WEB 22

46) Cp. WEB 4

All connections between the covering and fermenter have to be gas-tight. For this purpose, the foil can, for instance, be fixed inside the bioreactor below the liquid level.

If a tight connection is achieved by a rubber hose in a pipe, one has to make sure that the pressure in the hose is always sufficient (pressure control by alarm gauge and daily inspection of the manometer to be included in monitoring procedure) and that the rubber material of the hose retains its elasticity.

The floating dome system consists of plastic foil in the shape of a socket open at the bottom and dipping into the substrate. The floating dome system can also be used as a gasholder. Floating dome systems did not became popular for the following reasons:

- High costs of the floating socket
- Difficult disposal of the scum
- Many rust-prone parts, leading to a shorter lifetimes (up to 15 years, or in tropical regions 5 years for the socket) and regular maintenance costs because of paintwork.

The most suitable materials for the socket are fiberglass and high-density polyethylene. These materials are rather expensive.

The fixed-dome type consists of a reactor with rigid covering and a balancing basin for the displaced substrate. The gas is stored in the upper part of the bioreactor. When the gas generation begins, the substrate is forced into the balancing basin. The gas pressure increases with the amount of stored gas. In these circumstances, the volume of the bioreactor should not be larger than $20\,m^3$.

The installations have to be gas-tight without any porosity or cracks. Fixed-dome installations are cost-effective and have a long lifetime.

4.1.2.4 **Access door and inlet**[47)]

Access doors must have a size of at least 600 × 800 mm. All access doors and manholes, especially those for maintenance and repair tasks, have to be equipped with aeration devices.

The filling operation has to be done as described in chapter 5.3.4. Inlet holes have to be positioned in such a manner that nobody can fall into them:

- Filling funnel with a height above 1.30 m in combination with a cover
- Filling funnel without covering and a height above 1.80 m
- Fix installed grid with a distance between the bars of 5–20 cm
- Self-closing flap when using a vertical inlet
- Inlet flushing channels, whose vertical openings are covered.

The openings for filling should always be orientated downwind, so that gases are blown away from the control panel.

47) Cp. LAW 16

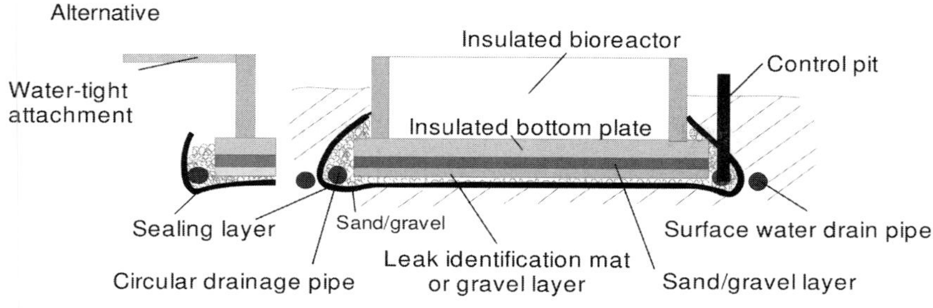

Figure 4.7 Drainage of a bioreactor.

Filling funnels have to be equipped with a place for operation guiding of the rinsing hose for rinsing the filling funnel to be done safely.

4.1.2.5 **Drainage layer below the bioreactor**

Single-wall kettles for liquid manure and silage-seeping liquids have to be equipped with leak-detecting drainage (Figure 4.7) in order to conform to the administration guidelines of various European states.

Between the lower edge of the construction and the sealing layer a 10–20 cm thick drain layer of gravel (grain size at least 4–8 mm) is to be provided, unless it needs to be thicker for freeze protection. It must have a slight inclination of at least 1% to the drainpipes and/or to the control pit. For tanks with flat bottoms, circular drainage is sufficient. For tanks with a volume of more than 1000 m^3 a plain drain layer is required or alternatively a coarser gravel layer (at least 8–16 mm) Leak identification mats may only be used in combination with a sealing layer of water-tight plastic foil. The control pit has to be set, so that samples can easily be taken.

The drainage layer is built as follows:

- Bottom excavation.
- Laying out of the sealing layer foil (0.8 mm).
- Insertion of a sand and gravel layer (150 mm) and the ring drainage (pipe diameters larger than 100 mm). The gravel layer can be replaced by a leak identification mat as an alternative.
- Laying out of the leak identification mat (PP, 1000 $g\,m^{-2}$) with a connection to the ring-drainage.
- Application of the sand filtering layer.
- Forming of the tank with a bottom plate having a diameter 1.2 m larger than the vessel diameter.
- Turning up the sealing layer foil to the tank wall.
- Insertion of the control pit. The control pit, with a diameter larger than 20 cm, has to be equipped with a sump for sample taking. For a tank diameter larger than 10 m, two control pits are required.
- Laying of the surface water drain pipe and filling of the trench.

- Installation of the surface-drainage for discharging of the rain water on the surface and draining of the water on the roof.

4.1.2.6 **Heat insulation**

Since the fermentation temperature is higher than the outside temperature, bioreactors usually require heat insolation. However, if the heat of the engines can be used for maintaining the fermentation temperature, insulation is not necessary.

The heat insulation of bioreactors has at least to be normally flammable.[48] In the areas within 1 m of gas exits, the heat insulation has to be at least "difficultly" flammable.

Foamed polyurethane or polystyrene sheets are usually put into the lower wet bottom area.

For the upper tank surface, mineral wool mats, mineral fiber mats or foamed plastics are alternatively preferred.

4.1.2.7 **Agitators** (Figure 4.8)

Mixing by filling in of new substrate, by thermal convection flow, and by raising of gas bubbles is mostly not sufficient for agricultural biogas plants. Only small plants can be operated without agitators.

The mixing operation is done mechanically by devices moving in the bioreactor, hydraulically by pumps located outside of the tank, or pneumatically either by pumping in biogas or by using the autonomously generated gas-pressure for pump work, potentially with two-component nozzles (Table 4.4).[49]

Excessive mixing increases power consumption, deteriorates the energy balance, and disturbs the microorganisms, thus ultimately decreasing the gas output. Consequently, various different mixing strategies have been developed:

- Slow-moving agitators with arms and paddles reaching the whole of the fermentation space, with low energy consumption and operating continuously.
- Medium speed agitators running intermittently or continuously with modest energy consumption.
- High speed, intensely operating agitators with high energy consumption, which are switched on repeatedly but only for a short time, e.g., every three hours for 15 min during a day.

Screw agitators with submerged motor adjustable vertically, horizontally, and with a changeable tilt angle are mainly used.

The required agitator intervals and times have to be determined individually for each biogas installation. Shortly after installation, mixing should be performed more frequently and for longer times as a precaution.

48) Cp. LAW 5

49) Cp. PAT 2

Figure 4.8 Agitators.
Axial agitators: steplessly adjustable in its height (top left) Casting for lifting the agitator out of the reactor (top right); Adjustable shaft mixer driving and agitator of 1.2 m diameter (middle left); Rührgigant FR for biogas-fermenter (middle right); Paddelgigant™ (bottom).

Table 4.4 Application frequency as a percentage of all plants with different agitators in Europe.

Type	*Application frequency*	*Remarks*
Submerged motor propeller agitators	44%	These agitators are vertically installed in vertical bioreactors, where temperatures are below 70 °C. The motor (2.5–25 kW) is liquid-tight encapsulated. The propeller agitator with 2 or 3 blades is directly fixed to the motor shaft. In vertical bioreactors with a much bigger diameter than the height, they are installed horizontally through the wall.
Axial agitators, shaft mixer	16%	Axial agitators are installed to produce a torus-like vortex with an upward stream along the wall. Through this liquid stream it is possible to destroy permanently the floating cover by sucking it to the bottom in the middle of the reactor. The speed of rotation of the electric motor must be low. The shaft mixier with its motor at the upper end of the shaft can be vertically and horizontally positioned in the reactor. Therefore the opening in the cover of the bioreactor is flexible. Additionally to the agitator itself, cutting devices can be installed to cut fibrous material.
Paddle agitators	10%	Paddle agitators are best suited to generate horizontal plug flow in a horizontal cylinder. Across the flow direction they balance differences in concentrations and support the heat exchange. Agitators[50] with four inclined paddles is designed to be used in vertical bioreactors with substrates of high concentration of fibrous material like grass silage. The long paddles hinder the formation of a scum. They rotate slowly and therefore save energy.
Pneumatically forced rotation	ca. 12%	A part of the biogas is cleaned and then circulated by a compressor into the reactor, so that gas bubbles ascend in the substrate, and it is agitated.
Other agitators	18%	Some agitators, called “Grindel agitators”, have a vertical shaft and are installed eccentrically. They form a vertical vortex, which is always broken at the reactor walls. Grindel agitators are not suitable to destroy scum and a sinking layer.

As a rule, a brief but intense mixing is better than a prolonged but weak one. In practice, a mixing frequency of 1–10 times a day and mixing duration of 5–60 min has proven to be useful.[51]

All electrical devices in a continuously run bioreactor have to be explosion proof, suitable for zone 2. In discontinuously run bioreactors the requirements explosion proof for zone 1 have to be met.

50) Cp. WEB 5

51) Cp. BOK 33

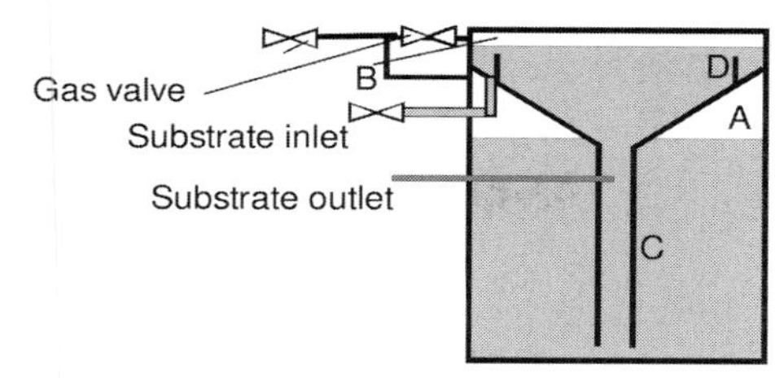

Figure 4.9 Bioreactor without rotating agitator.

Submerged motor agitators or submerged motor pumps have to at least conform to protection class IP 68 according to the European standard EN 60529 and can only be run when submerged. An appropriate standard operating procedure is required.

Installations with artificial leveling do not require a rotating agitator (Figure 4.9). The bioreactor[52),53)] mainly consists of a cylindrical container with a central, cylindro-conical tube C and overflow D.

Initially the bioreactor is filled up to the upper brim of the overflows. The gas valves are closed at first. As a consequence of the gas development, the gas accumulates in compartment A below the conical insert and the liquid is pushed up through the cylindrical insert C.

A part of the liquid flows into the overflow ring D. The gas in compartment B is compressed. Once the fermentation time is completed, fresh substrate is pumped into the cylindrical container C. The same quantity of fermented substrate is discharged from the overflow ring.

After the refilling operation, the gas in compartment A is withdrawn into compartment B through valves, causing the liquid level to fall and the newly inserted liquid to be pushed into the bottom part of the bioreactor. Thus, an intensive mixing of the reactor compartment takes place and the floating layer is destroyed.

4.1.2.8 **Heating**

The following heating systems can be considered (see Section 1.2.).

- Wall heater consisting of stainless or normal steel tubing fitted along the inner walls. This type of heater is frequently installed.
- Floor heater consisting of tubing laid spirally on the floor. This type of heater is suitable for converted liquid manure storage, but is difficult to clean, so that it cannot be recommended.
- Submerged heater. The heating power decreases as the heater gets dirtier, so that the heater occasionally has to be removed for cleaning.

52) Cp. BRO 5

53) Cp. WEB 115

- Heat exchanger. This type can only be used in conjunction with a substrate pumping system. They work well, but can get dirty over time.
- Agitator heater. The heating is done through the hollow agitator shaft. This system cannot be recommended because the heating area is too small.

4.1.3 Efficiency

The efficiency of agricultural biogas installations depends on the installations:
- The capacity or size of the plant
- Fraction of autonomously generated power
- Installation standard and automatization level.

The relationship between installation performance and available biomass can roughly be estimated based the figures given in Table 4.5.

Both investment and operational expenses (labor, maintenance, repairs, insurance) (Figure 4.10) for agricultural biogas plants, including total costs for bioreactor, mixing, residue storage tank, gasholder and gas preparation, and the components for the conversion of the gas into energy, vary significantly depending on the automation level, especially for smaller installations.

For installations with a co-fermentation of liquid manure and, for instance, bio waste, the expenses for the preparation (e.g., break-up and sanitizing) have also to be considered in addition to the cost of an overall larger installation.

Table 4.5 Typical figures for the design and operation of an agricultural biogas plant.

Biogas yield	
1 animal unit (see Part II)	200–500 (800) m^3 biogas/a
Ensilaged maize from 1 ha	8000–10 000 m^3 biogas/a, biogas amount from 20 animal units
1 ha meadow grass	6000–8000 m^3 biogas/a
1 Mg liquid manure	20–40 m^3 biogas
1 Mg ensilaged maize	170–200 m^3 biogas
1 Mg meadow grass	80–120 m^3 biogas
Bioreactor size	
100 animal units	Horizontal cylinder reactor: 150 m^3 volume, vertical cylinder: 250 m^3 volume
1 ha ensilaged maize	ca. 10–20 m^3 volume
Bioreactor volume load	2–4 (max.7) kg oTS / m^3.d at 38 °C
Energy yield	
1 animal units	0.15–0.20 kW installed power
Ensilaged maize from 1 ha	3–4 kW installed power
12–15 m^3/d biogas	1 kW installed electrical power

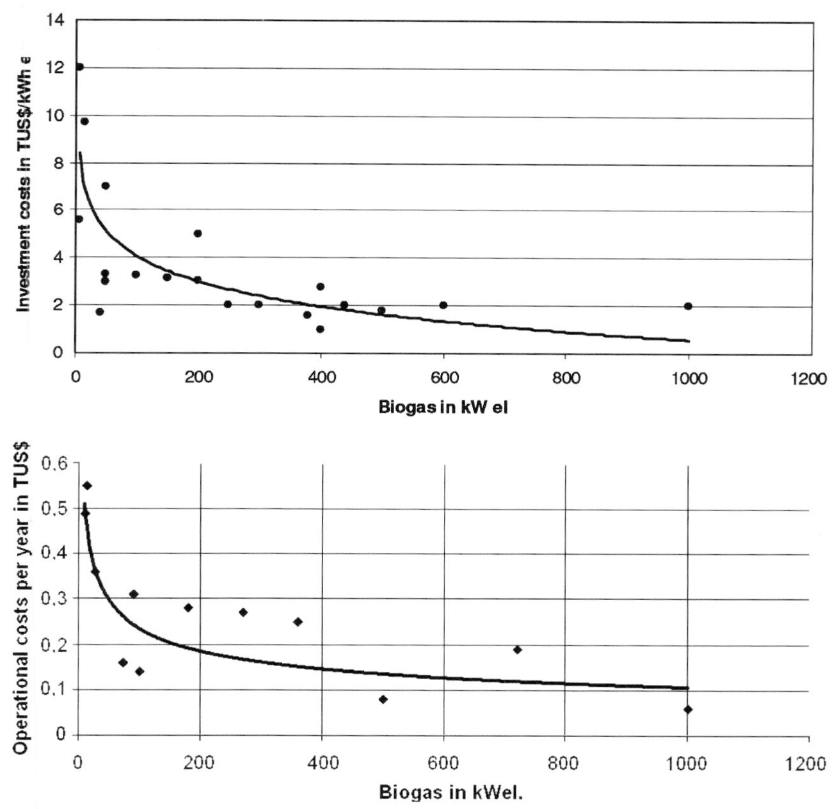

Figure 4.10 Investment costs[54),55)] (top), operation costs per year[56)] (bottom) for agricultural plants; some of the plants are fed with co-ferments.

Nevertheless, the investment costs for every kWh produced are often up to 25% lower than without co-fermentation.

The typical figures given in Table 4.6 can help in performing a rough estimate of the initial planning of an installation.

4.2 Existing installations by different suppliers

The installations described above are often found in agriculture. They are built autonomously by the farmers who purchase single-installation components.

54) Cp. BOK 10

55) Cp. WEB 116

56) Cp. BOK 10 and own values

Table 4.6 Typical figures for costs of agricultural biogas plants.

Cost basis	*Typical figure*
Invest. costs for CHP per kW_{el}	500–1500 US$
Invest. costs per $1\,m^3$ reactor volume	300–500 US$
Invest. costs per 1 animal unit	450–700 US$ (self construction)
	650–1 800 US$ (industrial construction)
Invest. costs per 1 kW installed power	2400 US$ (large plants >300 kW)
	6000 US$ (small plants)
Invest. costs per $1\,m^3/h$ biogas	4000–7000 US$ (plant alone)
	5500–9000 US$ (including silo for maize silage)

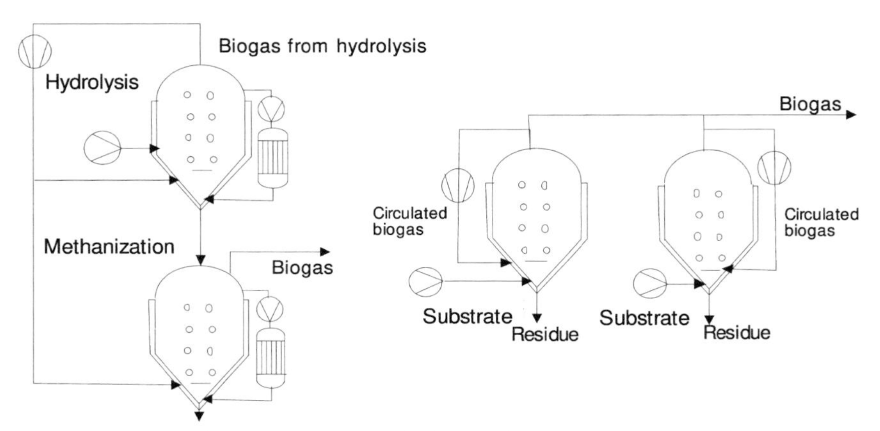

Figure 4.11 Biogas plants in Vaasa (Finland) (left), and Bottrop (Germany) (right).

The installations described below are based on the same processing principle. They were developed for the generation of biogas from waste.

4.2.1 WABIO-Vaasa process (Figure 4.11)

The WABIO-Vaasa process was invented by Ecotechnology JVV OY of Finland. The process is a single-stage mesophilic. The first installation for $10\,000\,Mg\,a^{-1}$ was built in 1991 in Vaasa in Finland. The second installation for domestic waste with a capacity of 6500 Mg waste per year was erected in 1993 in Bottrop, Germany.

A characteristic of the process is the fact that the mixing is done by injected biogas from hydrolysis or recycled biogas.

The fermentation residue is dehydrated to 40% DM by a screw press or decanter and then aerobically rotted.

For this type of installation, it soon became clear that the bio- or hydrolysis gas led to so many impurities that the compressor soon became dirty. A consequently

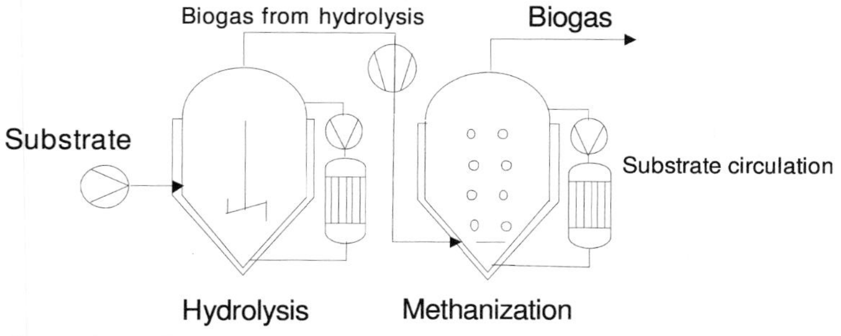

Figure 4.12 DUT process.

necessary gas cleansing is too expensive. As such, the WABIO process based on the Finish development has not been followed up further in recent years.

4.2.2 **DUT process**[57] (Figure 4.12)

The DUT Proces is named after the company which developed it. The process runs with a DM content of 20% and a 25-day residence time of the solids in the anaerobic process (hydrolysis + methanation).

A characteristic of the process is that the hydrolysis is performed at max. 25–28 °C in order to minimize the methane development. The hydrolysis gas, which at low hydrolysis temperatures mainly consists of CO_2, is injected into the methane reactor to support mixing.

The suspension in the hydrolysis and thermophilic methane reactor is circulated continuously through an external heat exchanger.

In this process, the hydrolysis gas has to be cleansed in order to avoid excessive dirtying of the compressor. Implementation of this process in a larger installation has not yet been reported.

4.2.3 **WABIO process** (Figure 4.13)

The WABIO process[58] works differently from the WABIO-Vaasa process. The substrate is moved without agitator or gas recirculation. The gas is withdrawn continuously in the middle of the ceiling. The reactor is divided into two ring chambers. The substrate is continuously fed into the outer ring chamber. In the outer ring chamber's gas space, a gas cushion develops which pushes the substrate into the inner ring chamber, where it can flow off at the top after fermentation, with few interruptions.

To start the mixing operation, the valve between outer and inner gas chamber is opened, so that the biogas having accumulated there escapes and the substrate

57) Cp. BRO 4

58) Cp. WEB 115

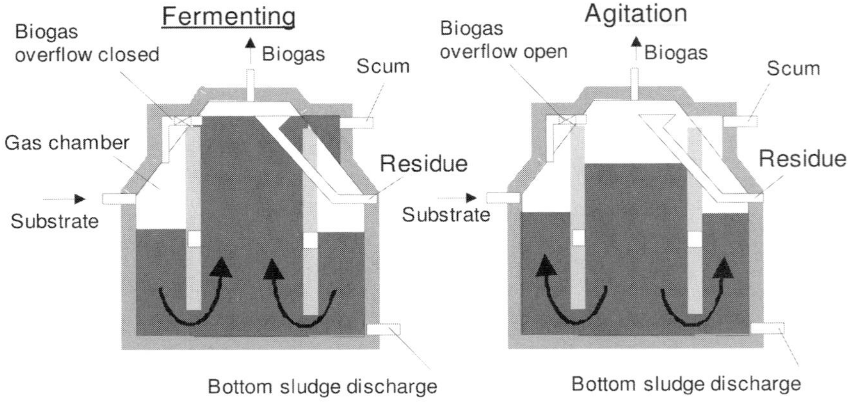

Figure 4.13 WABIO process.

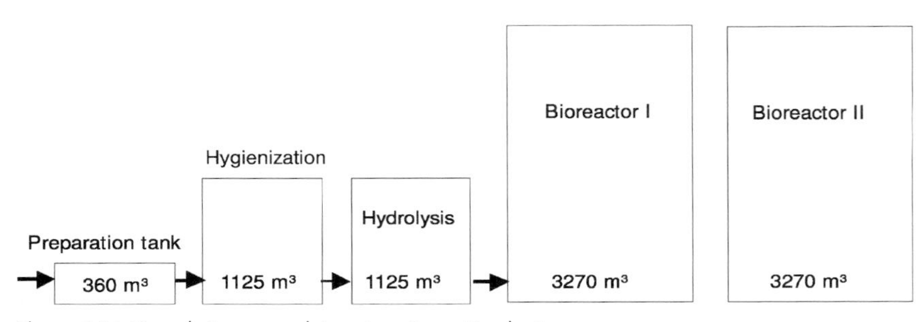

Figure 4.14 Vessel size exemplatory in a Farmatic-plant.

flows from the inner to the outer chamber. Afterwards, the valve is closed and the gas cushion develops again.

The process stood the test on an industrial scale several times. However, because of the probably insufficient mixing effect, it is mainly suitable for special bio waste, as for instance food waste. The process attracts attention because of its low energy consumption, no additional power for mixing being required.

4.2.4 Farmatic™ biotech energy installation

The Farmatic process is a simple, single-stage mesophilic process including an upstream sanitization for the substances to be sanitized. The installations, including larger ones, are individually planned and assembled accordingly to customers' requirements.

It is interesting to note the vessel's proportions in an installation with two reactors running in parallel (Figure 4.14).

These installations are based on technology tested over many years and proved to be reliable for various applications.

4.2.5
Bigadan™ process (formerly Krüger process) (Figure 4.15)

The Bigadan process is a one-level mesophilic process with sanitization. It includes an interesting and complex heat recovery system.

The installations were manufactured in the mid 1990s in Germany, with converting capacities between 40 000 and128 000 $Mg\,a^{-1}$ liquid manure with admixing of bio waste.

The heat recovery system is certainly interesting from a technical point of view, but in biogas installations enough heat is normally available anyway; therefore, the complicated and difficult to clean heat exchanger might not be required.

4.2.6
Valorga™ process (Figure 4.16)

The Valorga process was specifically developed for the treatment of communal bio waste, food waste, and sewage sludge. The process is based on "dry" anaerobic fermentation with a DM content of 25–35%.

The fermentation process – starting with hydrolysis and up to methanation – takes place under anaerobic conditions in only one vertical cylindrical bioreactor.

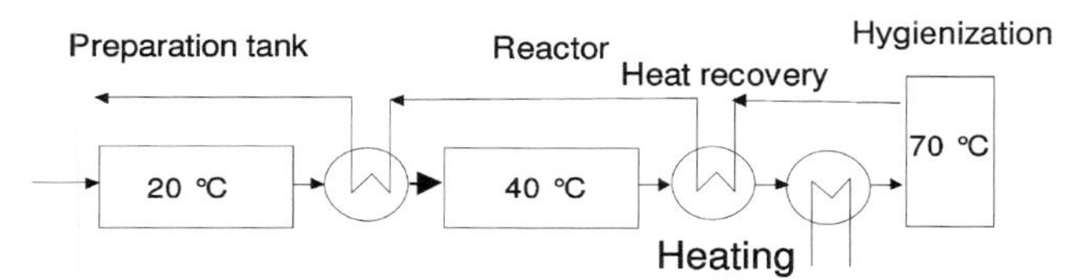

Figure 4.15 Bigadan process.

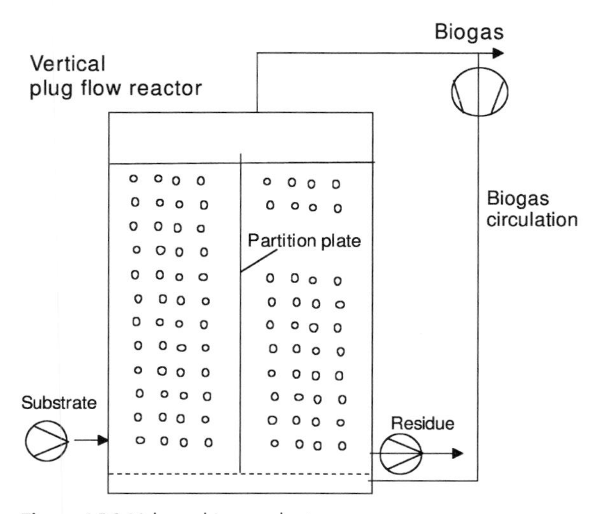

Figure 4.16 Valorga-biogas plant.

Because of the vertical arrangement of the bioreactor, the fermentation residues can be discharged by gravity without the aid of mechanical handling devices. The energy consumption of the process reaches 25–50 kWh Mg^{-1} power and 20–30 KWh Mg^{-1} heat.

A middle wall subdivides the bioreactor in order to achieve "plug-like", horizontal flow. The openings for the feeding in and discharging of the substrate are located mid-wall on opposite sides. Thus, the mixing of the biomass over the whole reactor volume is promoted. The Valorga process does not require any mechanical devices within the bioreactor. The special reactor geometry together with the vertical mixing system guarantees a defined residence time (2–4 weeks) and a complete sanitization of the substrate without bypass flow in the reactor and a gas yield of 80–180 $Nm^3 Mg^{-1}$.

Even this installation uses biogas for the mixing, requiring elaborate gas cleansing to avoid pollution of the compressor.

4.3
Installation with substrate dilution and subsequent water separation

Biomass, as for example bio waste which cannot be pumped and therefore not mixed during fermentation, is diluted by the addition of fresh water.

In principle, in these installations the correct water content of minimum 75% is set up in a mixing vessel prior to fermentation (Figure 4.17). After fermentation, the water is separated again, because often the fermentation residues cannot be used in the immediate proximity.

These installations are defined as dry fermentation installations, even though the fermentation is carried out wet at high water content.

4.3.1
Equipment

For the mixing of biomass and water, cylindro-conical standing vessels with very pointed bottoms are used. The bottoms end at the inlet of large volume piston pumps, which are also known from the pumping of concrete.

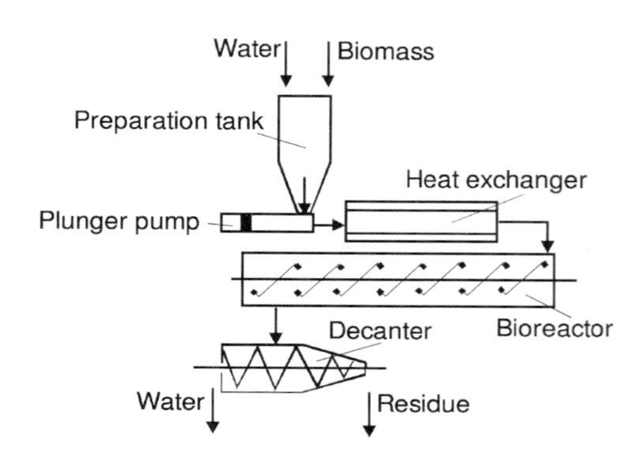

Figure 4.17 Dry fermentation process with liquidification.

A jacketed tube, through which the moistened biomass is slowly pushed with a residence time of approximately 1 day, serves as a heat exchanger. It is heated with hot water. The subsequent fermentation takes place in a horizontal bioreactor with a flow-through process. The reactor consists either of steel or of concrete coated on the inside with a thin steel panel lining material. After fermentation, dehydration is performed on belt-type presses, filter presses, or decanters (Figure 4.18).

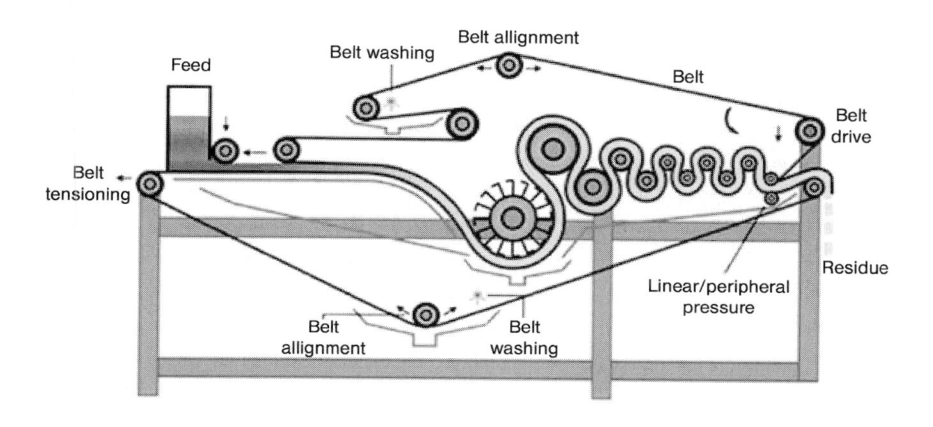

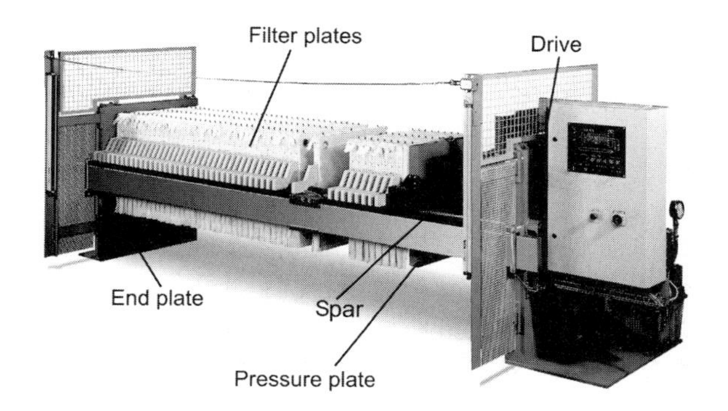

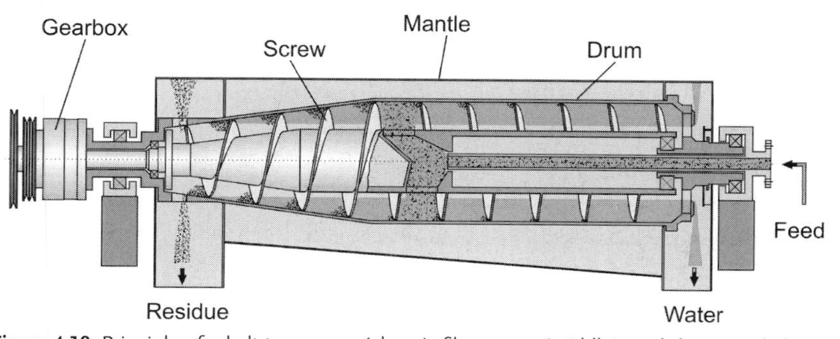

Figure 4.18 Principle of a belt-type press (above), filter press (middle), and decanter (below).

Table 4.7 DM concentration in % depending on the residual and the dehydrating process.

Features of the sludge with PE or Me^{3+} as flocculant	*Decanter*	*Belt-type press*	*Chamber filter press*	
			Without lime addition	With lime addition
Easy for dehydrating	>30	>30	>40	>45
Not so easy for dehydrating	22–30	22–30	22–40	30–45
Difficult for dehydrating	<22	<22	–	30–35

The DM concentrations that can be achieved in the residue by dehydrating equipment are shown in Table 4.7.

Belt-type Press

A belt-type press consists of two endless belts guided over several rolls. The belts are pressed against each other over a certain length. Both belts are permeable to water, but not to fermentation residues.

The fermentation residues are fed between the two belts where the water is pressed out.

Filter press

Filtering presses can be classified as frame filter presses, compartment filter presses, and membrane filter presses.

A filter press in general contains up to 150 square filter frames with a side length of up to 4 m, which are movably mounted on 2 to 4 shafts. The required filter frame surface depends on the filtration velocity, which ranges between 0.05 and 1 m^3/m^2.h. Filtering materials (cloths or layers) are clamped between all filter frames. The filtering materials tighten the frames against each other only marginally; consequently liquid permanently flows out of the frames and is then collected by a bottom tub and redirected into the filtration system again.

The frame bundle is bordered at the face sides by a fixed face plate and a movable pressure plate including the closing and switching mechanism. The frames are pushed together by the closing mechanism with a clamping force at least 10% greater than the forces developing during filtration.

Initially, the filter press is firmly closed. Sometimes the closing force is obtained with the aid of hand spindles. These permit closing pressures up to 600 kPa. Other small presses (800 × 800 mm frames) are brought to a closing pressure of 1500 kPa by small hydraulic pumps. Larger presses are equipped with hydraulic pumps with an automatic closing pressure controlled at 2500 kPa.

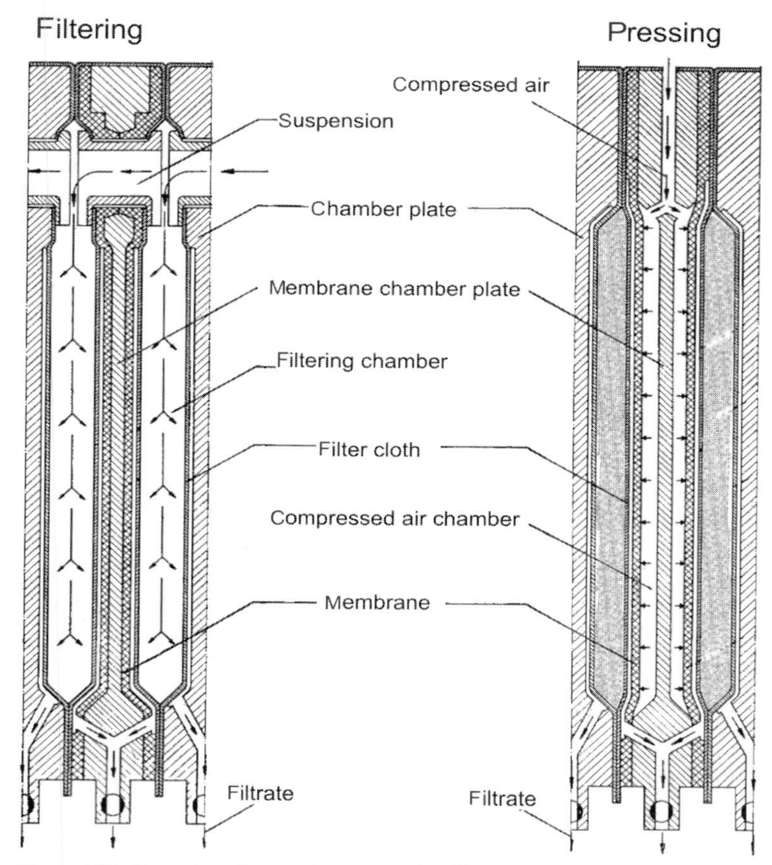

Figure 4.19 Operation of a membrane chamber filter press (cross section).

Next, the suspension is pumped in through the face plate. The filtration materials retain solids, while the filtrate flows through the filter agents into the frame, where it can drain. Once the press is filled with solids, the in-feed of substrate is stopped and the remaining solids are washed out of the frames if necessary. Then the frames are opened to allow the solids trapped between the frames to fall out. The filtering materials are cleaned and the cycle can start again.

In contrast to frame filter presses, a compartment filter press (Figure 4.19) is made up of compartment plates, which are end-to-end filter plates. The space for solids entry is provided by a two-sided indentation of the plates. During filling, the filtering cloth is put into the indentation and supported by knobs on the plate surface.

The membrane chamber press possesses alternatively chamber and membrane plates. For the membrane chamber plates, on the membrane are situated knobs behind which pressurized air can be injected. The membrane is then pressed into the chamber and the solids are squeezed additionally.

Decanters

Decanters are horizontal cylindro-conical or conical centrifuges with a drum diameter up to 2 m and a ratio of length to diameter of $L_D/D_D = 1.3–5$. A characteristic is the conveying screw in the centrifugal space which rotates at a slightly different speed than that of the drum.

The suspension is usually fed in at the center. At the end of the operation, the liquid phase is withdrawn from the larger diameter through a tube or overflow, whereas the solid phase is withdrawn at the opposite side with a conveying screw.

To increase the performance, flocculation agents such as polyethylene granules or trivalent metal ions are added to residues that are difficult to dehumidify.

4.3.2
Implemented installations of different manufacturers

For the usually single-stage plug flow process (Figure 4.20),[59] the bio waste is transported through a horizontal plug flow fermenter with rectangular cross section at a total solids content of 15–45% DM in the substrate. The transport is performed by a pressure gradient or by transverse, slow-running mixing paddles. Sometimes an intensive fermentation at 40–45 °C for hydrolysis and specific acidification is performed preceding the anaerobic fermentation with a residence time of 21 days and a temperature at 55 °C.

In a very similar installation, the process runs in a cylindrical instead of a rectangular plug flow bioreactor.

This process was developed for the anaerobic fermentation of bio waste from garden, kitchen and others.

The preparation of the biomass consists of a coarse milling and successive manual sorting of foreign bodies and metal separation followed by fine milling. The milling is carried out by slow-moving multiple screw mills with immediate sieving of fine particles to a particle size smaller than 40 mm.

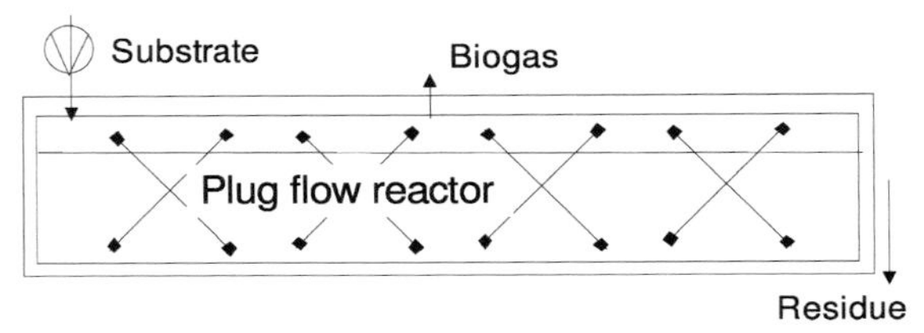

Figure 4.20 BRV bioreactor.

59) Cp. WEB 55

The biomass is intermediately stored in a ventilated bunker of a volume sufficient for 5 days operation. The storage is done by an internal distributor in such a manner that the product with the longest storage time is always forwarded. Before reaching the bioreactor, the bio waste is moistened to a DM concentration of 28% in a twin shaft mixer. The water used is a mixture of sewage and fresh water, as the aim is to withdraw as little sewage as possible from the installation.

The reactor is loaded by a hydraulic reciprocating pump. Directly behind the pump, fermented product is fed in intermittently for inoculation. A double heat exchanger, heating up biomass to the reaction temperature within 12 h before it is conveyed into the reactor, is arranged behind the pump. Lost heat from the CHP is used for this operation.

Next, the 14-day fermentation is carried out as a single-stage thermophilic dry fermentation at 55–60 °C in a horizontal, cylindrical bioreactor with a volume of 200–300 m^3 and a diameter of 5 m, made from reinforced concrete lined with steel sheets on the inside. The substrate homogenization is achieved at constant flow by a paddle agitator. The agitator's shaft runs centrally through the whole fermenter and its bearings are outside the fermenter. The paddle agitator is switched on and off depending on the consistency of the bio waste.

Before post-fermentation, the material is conveyed into belt-type presses by reciprocating pumps and dehydrated there. Afterwards, the solids, still containing approximately 50% water, are post-fermented for approximately 10 days in a compost heap.

4.4 Installation with biomass accumulation

The accumulation of biomass during the fermentation process is useful if the suspension fed in is a thin fluid. The reactor can then be kept small. For reactors working with pelletized biomass sludge, as for instance packed-bed and fluidized-bed reactors, a solids separation occurs if the solids concentration equals to 10% of the inlet COD value, or 500 mg L^{-1} solids concentration.

Both internal and external water separation is known, depending on whether the separation unit is located inside or outside the reactor.

Installations with biomass accumulation are most commonly used for the anaerobic purification of industrial sewage and for the fermentation of sewage sludge.

4.4.1 Sewage sludge digestion tower installation

Sewage sludge fermentation is also known as digestion. It occurs in the so-called digestion towers. Their volume (size) depends on the number of residents of the cities where the sewage comes from.

For settlements from 10 000 to 15 000 residents, sewage water treatment plants with single-stage digestion, where the substrate is mixed a few times per day, are

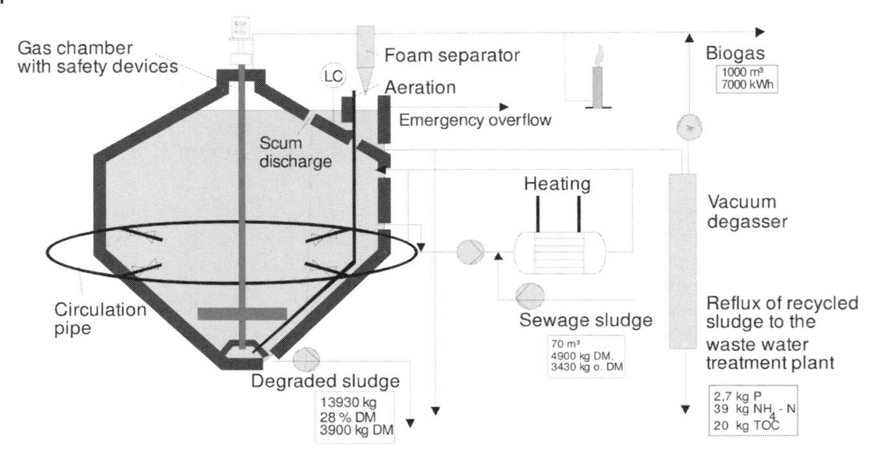

Figure 4.21 Single-stage sewage sludge fermentation plant with internal phase separation.

mainly used in Europe. The plants run at mesophilic temperature.[60] Up to 85% of all European plants are of this type. Figure 4.21 shows such a single-stage sewage sludge digestion plant.

For cities with a larger number of residents (between 100 000 to 1 000 000), two-stage plants are preferred. For instance, at a level of 100 000 residents, often two sewage sludge digestion towers, each of volume 2250 m^3 (diameter ca. 15 m), are built. A schematic drawing of a modern two-stage plant is shown in Figure 4.22. The plant is run with different temperatures in the two towers.

The turbid water on top of the towers is still organically contaminated and has a BOD_5 value of 2000 mg L^{-1} and an NH_3 content of 400 mg L^{-1}. It has to be returned to the sewage water treatment plant.

4.4.1.1 Equipment

In sewage sludge digestion towers, attachments for phase separation and for avoiding post-gassing are necessary in addition to the digestion tower itself with its devices for mixing and heating of the sludge. The most important ones are described below.

Digestion tower (Figure 4.24) The digestion tower should be built in such a way that the following aims are achieved:

- Thin walls for cost reasons
- Small surface to save on insulation
- Easy outgassing of biogas from the substrate
- Intensive mixing for a regular distribution of nutrients and metabolism products

60) Cp. JOU 7

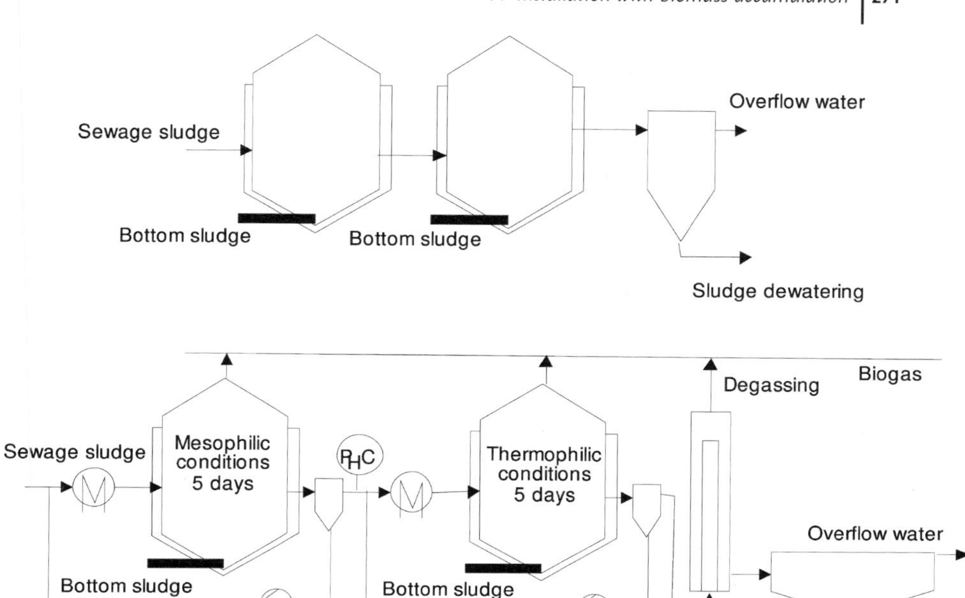

Figure 4.22 Process with multistage fermentation.

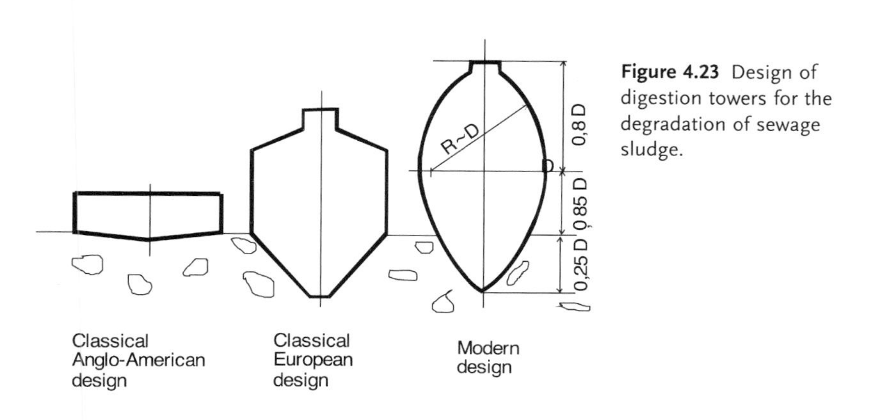

Figure 4.23 Design of digestion towers for the degradation of sewage sludge.

- Avoiding the development of layers, especially of a scum
- Avoiding uncontrolled accumulations.

The empirically developed designs that have become globally accepted are illustrated in Figure 4.23.

A degradation dump – a so-called sludge-bed reactor – is found in the smallest plants.

Figure 4.24 Digestion towers. Bioreactors for the degradation of sewage sludge-
Classical form (top), modern form (middle), Climbing formwork-Digestion tank in Munich (Germany)-Digestion tank in Pa-Li (Taiwan).

For plants with a volume up to 2500 m^3, digestion towers of classical continental form have achieved acceptance. That means that the vessels have a cone shaped bottom, a cylindrical mid part and a frustum as ceiling. The digestion towers are mostly made of steel sheets[61)] and do not require corrosion protection where they are in contact with liquids because of the anaerobic degradation process. Only the gas space is painted with a corrosion protection layer. The outside is heat-insulated and covered with a weather protection layer.

Larger plants with a digestion tower volume of 2000–15000 m^3 show a parabolic or similar shape, also called "a digestion egg".

These often consist of reinforced concrete, which is air- and gas-tight and has good heat insulation and corrosion resistance. Approximately 1/10 to 1/3 of the digestion tower's height is located below ground. The heat insulation above the ground, with a k-value of 0.25–0.35 $W/m^2.K$, is achieved with mineral wool and foamed plates. Below the ground, no insulation is usually applied. The insulation is encased with aluminium sheet strips on the outside. A gap is located between the aluminium sheets and the insulation material in order to avoid it becoming wet and to allow it to dry out.

For construction, the digestion towers are striked.[62)] The conical framework made of steel and pressure rings is mounted on, e.g., movable stairs, from inside the tower. Modern degradation towers are erected with a "climbing" framework. In this case, no framework scaffolding is constructed, but the framework is put on the already completed tower's wall and practically climbs up along it.

Nevertheless, there are also many large digestion towers of the classical form, which are made from steel sheets. This construction method possesses the following advantages:

- The base is more stable, which is advantageous especially for unstable ground.
- If the digester with the bottom conical area is put into the ground, it is supported by a vertical frame. Between the concrete base, the vertical frame, and the container bottom, an accessible space is left where fittings and tubing for heating, sludge inlet and outlet, and sludge mixing devices can be accommodated.
- Comparatively low weight of the base.
- The assembly is hardly affected by weather conditions.
- Shorter building time.
- Additional cost savings are possible by pre-assembly and manual welding of the segment sheets.

At least the following devices have to be mounted onto the gas cover of the digestion tower.

- Gas withdrawal dome with manometer and vacuum meter
- Over- and underpressure safeguard with in-line water trap

61) Cp. WEB 50

62) Cp. WEB 54

- Sight glass with inside and outside wipers and protective cover
- Swiveling spraying nozzles with ordinary water connection (hose coupler)
- Scum removal
- Manhole
- Explosion prevention
- Gas filter
- Foam trap
- Condensate trap.

The scum removal can be constructed either as a rectangular scum gate or a cylindrical scum slider. The allowed minimal opening amounts to 400 × 400 mm for doors and 400 mm for the slider diameter. The closure must to be opened quickly, e.g., through a lever mechanism. The drain must to be opened to such an extent that the bulky floating part of the sludge can be removed easily.

For the internal supervision of the digestion tower, a manhole is required. The minimal diameter of the opening has to be 600 mm. The closing cover must be fitted so that it can be swivelled.

The degradation proceeds asymptotically, and is thus stopped at the technical degradation limit. This limit is reached when 90% of the degradation gas quantity developed at 15 °C is generated. At this point, approximately half of the organic substance fed into the process is degraded.

The digestion tower's volume depends on the sewage sludge quantity, the solids concentration in the sludge, and the residence time. Per resident (IN) and per day, 80 g DM accumulates. The digestion tower volume depends on how much the biomass can be concentrated before digestion and on the residence time.

- If the primary and excess sludge from a sewage treatment plant can be dried to 4%, 2 L/(IN.d) substrate is obtained.
- If a solids concentration of 7% can be achieved, for instance by centrifuges, the sludge volume is 1.425 L/(IN.d).
- Sludge with a DM concentration of 10% can be processed without problems. The sludge volume is then 0.8 L/(IN.d).

Depending on process engineering and machines, the following residence times are normally chosen:

- Fermentation time of 120 days if using unheated bioreactors like ground basins.
- Fermentation time of 60 days if using simple digesters like Imhoff tank.
- Fermentation time of 30 days if applying mesophilic fermentation.
- Fermentation time of 10 days if applying thermophilic degradation. Thermophilic fermentation is used rather rarely, even though the sewage gas yield can be increased by 25%, and the sludge's sanitization and stabilization can be improved if the fermentation lasts longer than 10 days.

If a modern digestion tower is dimensioned with a volume load of B_R = 2–4 kg DM/(m^3.d), a volume of 20–40 L/IN results. German regulations[63] require a minimum volume of 55 L/IN.

For rain water, an additional 25% of the calculated digestion tower volume is necessary.

Trash and inhibiting substances[64] can partially lead to a massive dysfunction of a digestion tower and require significantly longer residence times.

Because of seasonally changing climatic conditions, the fermentation temperature can, for instance, vary between 33 °C and 38 °C in the mesophilic range, leading to considerable variations in the course of the fermentation process and sewage gas development.

In general, the degree of degradation increases with the residence time, and the volume load decreases accordingly (Figure 4.25).

The mix ratio between primary and excess sludge influences the degree of degradation.[65] The fraction of easily decomposable primary sludge leads to rapidly increasing degree of degradation at the beginning of the fermentation, whereas the cell membrane of the microorganisms in the excess sludge gives good protection against the attack of extracellular hydrolyzing enzymes. Thus, the increase in the degree of degradation slows down with the time of fermentation. After a residence time of 20 days, the degradation is largely completed.

For older single-step plants, the tower is loaded two or three times a day with, in total, approx 1/20 of its volume and mixed strongly. In the meantime, supernatant is slowly drained off. By this running system, an ecosystem can develop best which is suitable for the degradation in all four phases, allowing the digestion to continue nearly continuously during feeding.

In modern plants, the sludge is circulated continuously in order to keep an optimal temperature distribution, to destroy the scum, and to avoid a harmful

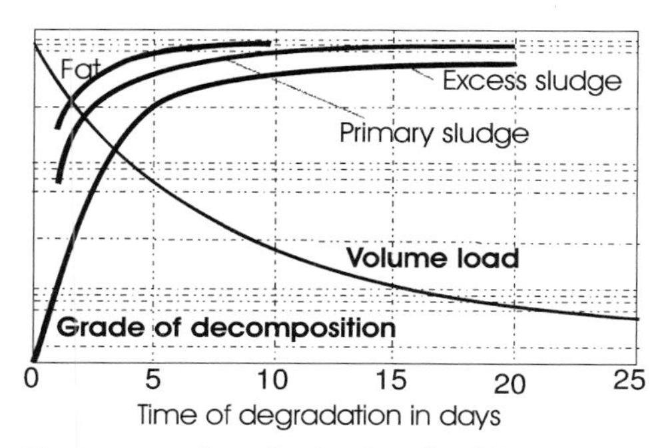

Figure 4.25 Volume load and grade of decomposition for different sewage sludges depending on the time of degradation at mesophilic conditions.

63) Cp. LAW 17
64) Cp. WEB 73
65) Cp. DIS 5

concentration of decomposition products and a sedimentation of the biomass. Then, a residence time of 18–22 days can be enough.

Digestion towers larger than 10000 m^3 are unusual.

Comparison of the characteristic values (Table 4.8 and Table 4.9) with values gained from actual plants (Table 4.10) shows that modern digestion tower plants in Europe are normally built with a surplus factor of 1.5 for the residence time and of a factor of 3–4 for the volume load.

Table 4.8 Characteristic values for digestion tower plants.

Parameter	*Unit*	*Average value*	*Median*
Time of decomposition	d	abt. 27.5–29.0	abt. 25
Volume load	kg oDM/(m^3.d)	1.10–1.22	0.95
DM in the feed	%	4.05–4.81	3.7–4.6

Table 4.9 Characteristic values for digestion tanks.

Size	*EGW*	*<50000*	*50000–100000*	*>100000*
Volume load	kg oDM/(m^3.d)	<2	2.0–3.5	3.5–5.0
Minimum time of decomposition	D	>20	15–20	>15
Recommended time of decomposition[66]	D	22	20	18

Table 4.10 Dimensions of digestion towers.

P.E.	*Time of decomposition*	*Temp.*	*Design*	*Volume*	*Height*	*Diameter*	*Place*
n.a.	n.a.	35 °C	Modern	9400 m^3	40 m	25 m	Pa Li (Taiwan)[67]
28500	20 days	32–35 °C	Classic	1300 m^3	ca. 16 m	13.0 m	Mauthausen (Austria)[68]
2 Mio	n.a.		Modern	4 × 14500 m^3	32 m above ground floor	n.a.	München
n.a.	n.a.	34 °C	Modern	6000 m^3	39.20 m 28.50 m above ground floor	22.90 m	Berlin[69]

66) Guideline of German authorities
67) Cp. WEB 80
68) Cp. WEB 59
69) Cp. WEB 61

Table 4.11 Energy consumption of decomposition.

Purpose	*Explanation*	*Value*
Sludge circulation	$Wh/(m^3_{FR} \cdot d)$	50–100
Heating	$L_{oil}/m^3_{sewage\ sludge}$ depending on the size of the plant	0.17–2
Personnel	Percentage of the total personnel requirement of a sewage water treatment plant	10
Scum removal	Percentage of the sewage sludge	0.1–1

Table 4.12 Size of the plant and power consumption.

Plant size in P.E.	*Average power consumption in kW*	*Average specific power consumption in kW/1000 IN*
<50 000	255	7.0
>50 000 to <100 000	460	6.1
>100 000 to <200 000	745	5.3
>200 000 to <400 000	1480	5.4
>400 000	2655	3.0

The overall power consumption for the sludge digestion breaks down as shown in Table 4.11.

Mixers

The mixing of the reactor contents has an essential effect on the performance of the digestion tower.

- Screw conveyor mixer
- Circulation by externally mounted pumps
- Gas injection.

The mixing is carried out only by charging fresh sewage sludge and discharging water, by temperature differences, by compensation of density differences, or by naturally ascending fermentation gas. But this is mostly insufficient.

By a conical narrowing at the top or by the digestion egg shape, the scum area is minimized, and thus the admixing of the scum is facilitated.

The energy consumption of larger digestion towers is 4.5 kW/m^3.d for the gas injection, 3.5 kW/m^3.d for the screw conveyor mixer, and 2.2 kW/m^3 for the circulating pumps.

The total electricity consumption of the digestion tower and sewage plant fluctuates widely (Table 4.12). In general, it decreases with increasing plant size. The average specific electricity consumption of complete conventional mechanical/biological sewage water purification plants is 18.00 kWh/(E.a). The sludge digestion requires 7–11% of this.

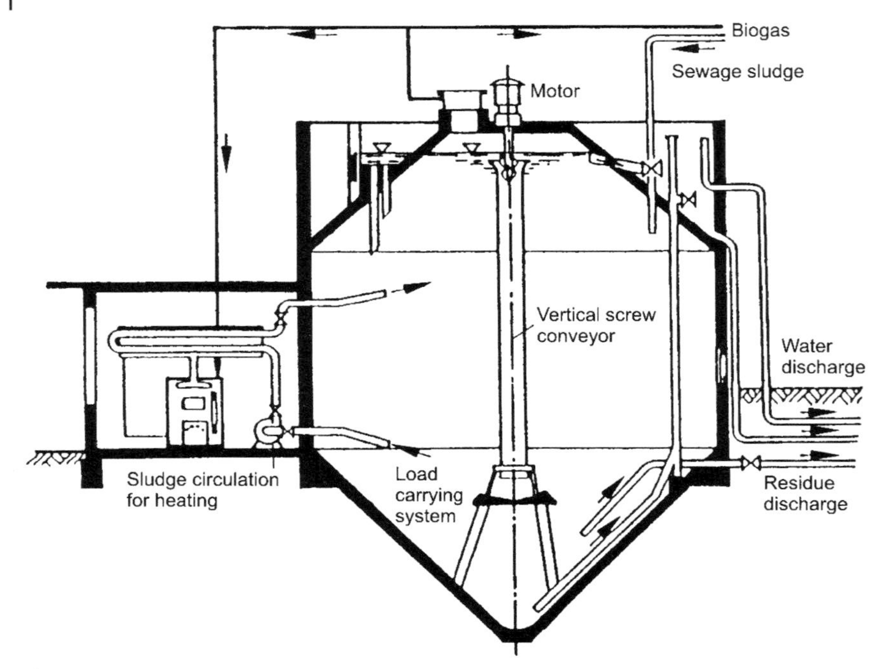

Figure 4.26 Screw conveyor mixer and external heating of the sludge.

Screw conveyor mixer

For a screw conveyor mixer, a steep conical shape is advantageous (Figure 4.26).

The screw conveyor mixer consists of a supporting structure at the bottom, a rising pipe, a mixing rotor shaped as a special propeller with reversing disk, and an explosion-proof motor.

A screw with reversible direction of rotation on the upper part allows the conveying of the sludge from bottom to top and conversely, so that the scum layer is destroyed and residues in the bottom area are avoided. The mixer shaft runs on anti-friction bearings which are regreased by an automatic self-driven grease lubrication system.

Circulation Pumps

The degradation tower's contents should be circulated 2–3 times a day for towers with up to $6500\,m^3$ content, for bigger towers 5 times per day. Based on this data, the circulation pump should be designed for a throughput of ca. $0.12\,m^3 h^{-1}$ per m^3 digestion tower volume. Its power consumption amounts to $2.5–8\,W\,m^{-3}$ fed substrate.

Gentle pumps which are easy to clean should be used, for instance, low-revving ducted-impeller pumps, which are adequate for conveying outgassing sludge. The pump's rotation speed must not be higher than 1000 rpm in order to avoid cavitation. Shafts with stuffing-box seals are to be equipped with a seal water coupling possibility.

Eccentric screw pumps are to be protected by adequate means against dry running and overpressure.

Gas injection into the digestion tower

For gas injection, a cylindrical bioreactor form with a flat bottom cone is preferred.

The required gas quantity is given by the gas velocity in an empty reactor. The following gas velocities in an empty reactor should be chosen:

$w_G = 0.3\,Nm^3/m^2$, if no precipitations occur and the volume load is low

$w_G = 0.6\,Nm^3/m^2$ for volume loads up to $5\,g\,L^{-1}$

$w_G = 1\,Nm^3/m^2$ for high volume loads, particularly with high content of inorganics.

Gas injection plants work with biogas at medium pressure. For gas injection, at least one compressor for each digestion tower and at least one redundant compressor have to be planned. For gas injection, a supply mains coming directly from the gasholder with dedicated own gravel pot has to be installed. The required control equipment includes: starting devices, overpressure valve, over and underpressure safety device, temperature control, and droplet separator.

Particular attention has to be given to the selection of the gas compressor. It has to be suitable for continuous operation with wet, aggressive gas, thus made of steel casting (1.4408, 1.4581) or forged or rolled steel. The maximum pressure head has to be at least 1 bar higher than the static injection pressure. Only mechanical seals should be used. The output of the compressor has to be at least $20\,m^3\,h^{-1}$ per injection lance in the bioreactor.

The piping for the injection lances has to be constructed in such a manner that the lances can be removed during operation without any gas escaping.

Internal phase separation for biomass concentration[70)]

The internal phase separation (Figure 4.27) occurs through sedimentation of the sludge or through immobilization or mechanical separation, e.g., filtration, compression, or decanting. Mechanical separation devices are rarely used for internal phase separation, but are external quite common for external separation.

For sedimentation, the sludge whirled up in the digesting tower is left to settle in a separate, not intermixed inner digester partition. For immobilization, stones etc. are provided, onto which microorganisms can grow, so that they cannot be carried away by the substrate flow. They possibly grow even better in the immobilized state than dispersed. The new microorganisms generated during the fermentation process and the sewage sludge residues which got stuck have to be removed from time to time.

The phase separation is strongly impeded by post-gassing.

70) Cp. WEB 21

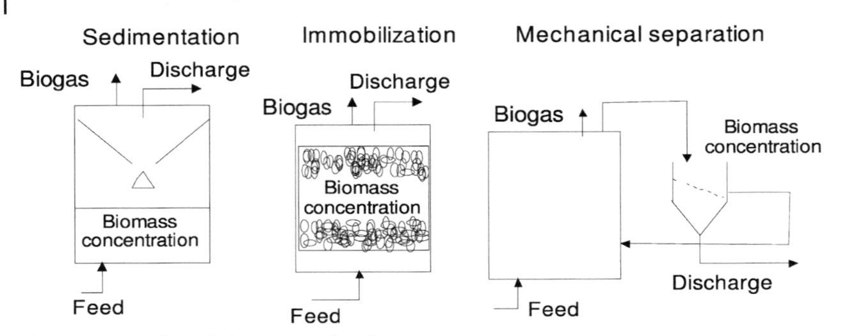

Figure 4.27 Technical alternatives for phase separation.

Facilities to prevent post-gassing[71] Gas bubbles adhering to the sludge flocculates have to be separated in order to

- facilitate the sedimentation of the sludge
- avoid scum development in the sedimentation tank
- minimize foam generation
- possibly recover residual gas
- avoid bad odor problems during the secondary treatment.

The following processes are available:

Ventilation

Ventilation is a simple, straightforward process technology, which is robust against changing flow rates and total solids concentration. In a ventilation tank with a time of passage of 8–10 min, the air requirement is approx 0.75 m^3 air/m^3 intermixed reactor outflow. Therefore, an increased energy requirement arises for ventilation and biogas losses.

The active anaerobic microorganisms are inactivated when they flow through the ventilation tank, but they regenerate in the subsequent anaerobic sedimentation tank.

An intensely smelling gas, explosive when coming into contact with air, can develop at this stage, making a treatment of the discharged air and safety devices for explosion protection necessary.

Vacuum degassing

During the vacuum degassing operation, in addition to the separating adhering gas bubbles, biogas is withdrawn from the water. By pressure release, a part of the gas which is retained in solution at atmospheric pressure is released (Table 4.13).

In order to be able to extract more than approx 8% of the gas dissolved by vacuum degassing, experience shows that a vacuum lower than 0.5 bar has to be

71) Cp. BOK 28

Table 4.13 Vapor pressures of different biogas components.

Vacuum	*Pressure (abs.)*	*Methane (CH_4)*		*Carbon dioxide (CO_2)*		*Hydrogen sulfide (H_2S)*		*Undissolved gases (total)*
		Dissolved	*Undissolved*	*Dissolved*	*Undissolved*	*Dissolved*	*Undissolved*	
Mbar	*mbar*	*L m^{-3}*	*L m^{-3}*	*L m^{-3}*	*L m^{-3}*	*L m^{-3}*	*L m^{-3}*	*L m^{-3}*
0	1000	13.00	–	120.00	–	100.00		–
100	900	12.88	0.11	119.95	0.04	99.98	0.01	0.16
200	800	12.53	0.46	119.81	0.18	99.99	0.05	0.69
300	700	11.48	1.51	119.33	0.66	99.81	0.18	2.35
400	600	9.54	3.45	118.20	1.79	99.49	0.50	5.74
500	500	7.16	5.83	116.03	3.96	98.87	1.12	10.91
600	400	4.75	8.24	111.86	8.13	97.63	2.36	18.73
700	300	2.57	10.24	102.54	17.45	94.63	5.36	33.23
800	200	1.01	11.18	80.19	39.80	85.80	14.19	65.97
900	100	0.29	12.70	42.69	77.30	62.36	37.63	127.63

applied over a time of 2–3 min. For film or thin-film vacuum degassing, the contact time can be fairly short (ca. 2–5 s). Flow-through and contact vacuum degassing have proven themselves useful with time of passage of 15–30 min and an partial vacuum of 15–30 mbar.

The metabolism of the microorganism is not affected by vacuum degassing. The partial reduction of post-gassing by vacuum degassing can be explained by the fact that a fraction of the freshly generated gas dissolves until saturation is achieved.

Compared to ventilation, vacuum de-gassing results in considerably better values of organic dry matter (as COD) in the outflow of the intermediate purification.

Cooling

Especially in plants where enough cooling energy is available, the post-gassing can be reduced by cooling of the influx.

The metabolism of microorganims in the sedimentation tank is reduced by fast cooling, and at the same time the solubility of the gas increases considerably: a temperature reduction of 6 °C (36 °C to 30 °C) results, for instance, in an additional solution potential of approx $70\,L\,m^{-3}$.

If vacuum degassing and cooling are combined, the cooling operation should be performed after vacuum degassing. However, the anaerobic return sludge has to be heated afterwards.

Agitators

Because of the post-gassing, very fine gas bubbles develop which stick to the sludge particles. These bubbles can be removed within 2–3 h by a slowly rotating agitator and the total solid concentration thus doubled to $DM = 24\,g\,L^{-1}$.

Addition of flocculation agents and flocculation-aiding agents

The natural biological flocculation results from the interaction of extracellular polymers of high molecular weight depositing onto the surface of organism flocculates and thus influencing the flocculates development. These polymers include exopolymeric substances (EPS), as for instance exopolysaccharides and polyamino acids, which exhibit a high concentration of negative surface charges.[72)] In addition to natural flocculation agents, weak cationic synthetic polymers can be used.

Organic flocculation agents are bio-degradable. Concerning inorganic flocculation/precipitation agents, these lead to the development of poorly soluble hydroxide sludges, which actually sediment exceptionally well, but accumulate in the reactor and could stop mass transport. The use of weak to strong cationic flocculation agents is only useful if a low organic sludge load ($B_R = 0.3\,kg_{COD}/m^3.d$) is ensured.

The optimal polymer dose for the conditioning of the sludge is achieved at the point where there are no electrostatically repulsive forces between the sewage

72) Cp. BOK 45

Table 4.14 Comparison of technologies for the minimization of sludge-gassing.

Process	*Effect*
Aeration with intensive air contact	Stripping off biogas attached to the sludge floc but also short time of inhibition of the activity of the microorganisms (gas production)
Aeration with air contact at a minimum	Stripping of the biogas attached to the sludge flocs
Aeration under vacuum without or with little air contact	Sucking off the attached biogas bubbles and gasification of dissolved biogas in the substrate at low pressure
Cooling without air contact	Inhibition of the metabolism of the anaerobic microorganisms and increase of the biogas solubility in the substrate
Slowly rotating agitator without or with little air contact	Separation of the biogas bubbles at minimum stress on the substrate with the consequence of easier natural flocculation
Addition of flocculants with little air contact	Forming of big compact flocs, which release biogas bubbles more easily, and reduction of the proportion of floating single flocs (bacteria).

sludge components. This depends on the surface area of the microorganisms, the pH of the solution, the type of mixing operation, and the type and dosage of flocculation agent. Anaerobically activated sludge, for example, tends to flocculate less than aerobically activated sludge. As the pH value of the reactor outflow (pH = 6.9–7.2) is mostly not in the optimal range for the flocculation agent (pH = 4–6), the flocculation agent has to be overdosed.

However, the formation of bulking sludge at more intensive organic sludge loads (B_R = 0.5–0.6 $kg_{COD}/m^3.d$) cannot be avoided. Furthermore, the relatively strong post-gassing running in parallel cannot either be avoided or reduced by using flocculation agents. Therefore, bulking sludge has generally to be avoided.

Organic, cationic flocculation agents are added at a dose rate of $3\,g\,m^{-3}$ at DM = 10–20 $g\,L^{-1}$, equivalent to 9.2 g flocculation agent/kg of DM or 1.5 g active agent per kg DM.

An intermittent intermixing in the reactor (minimal energy input, low shearing force stress) has a positive effect on the natural flocculation of aerobically activated sludge.

Flocculation agents prevent incrustation on metal parts of the plant (piping, heat exchanger, pumps, etc.).

In Table 4.14, the different process technologies are compared.

External phase separation for feedback of biomass By external phase separation and feedback of the concentrate into the digestion tower, an increase of total solids concentration in the outflow to 7–18 $g\,L^{-1}$ can be achieved. Different process technologies can be used for this purpose.

Table 4.15 Load per unit area.

Pressure head in the filter [bar]	*Load per area of the filter membrane [$m^3/m^2 \cdot h$]*
4.5	0.05–0.08
6.9	0.025

Sedimentation Tank

The area load w_s, given as volumetric flow rate related to the sedimentation tank area, should be at $w_s = 0.1–0.3\,m^3/m^2.h$ depending on the tendency of the sludge to sediment ($B_A = 5\,kg_{oDM}/m^2.h$). Should limestone be used for the pH regulation, a well-sedimenting sludge results. As a consequence, a higher area load can be used as compared to sludge with high organic load with bad sedimentation properties and high post-gassing tendency. Anaerobically active biomass can only be retained at $w_s < 0.1\,m\,h^{-1}$.

A 2- to 3- fold higher DM concentration compared to the inflow DM concentration can only be achieved at a return flow proportion of abt. 50–100%.

The sedimentation tank should have a minimal depth of 2.5–3 m.

Lamellar Settler

Lamellar settlers consist of a blocks of plates which are integrated into the settlers at a certain angle, so that the solids deposit between the plates over a short distance and then slip down the plates. The plates are made of PE, PP, PVC, glass-fiber-reinforced plastics, or stainless steel.

Lamellar settlers are categorized according to particle size and throughput of liquids. Because of the specific liquid conduction, all lamellas have to be loaded homogenously.[73] The total sedimentation area required should be divided into at least 2 structurally identical sedimentation units, in order to avoid interruptions during operation when doing maintenance and cleaning.

For the phase separation after anaerobic reactors, lamellar settlers have proven to be unstable, because the lamellas grow over. If the sedimentation areas and the sludge sumps are inclined less than 65° to the horizon, the sludge does not slip down.

Membrane Filter

In membrane filters, ultra-thin plastic membranes are put one on top of the other, resulting in a pressure-proof filtration layer with very fine holes. The suspension is pumped in at high pressure, forcing the liquid to flow through the membrane and the solids to settle on it.

73) Cp. JOU 27

The organic acids, however, are not filtered out because of their low molecular weight, increasing the COD value in the outflow.

Because of the large surface area of the membrane unit, the sewage water cools down significantly and has to be heated up again if necessary.

The membrane has to be cleaned if the surface load decreases from 0.025 to $0.014\,m^3/m^2$. A cleaning by caustic soda and hypochlorite is generally required once per month.

Membrane filters for phase separation in anaerobic plants could not gain acceptance because of economic constraints.

Flotation

Flotation by depressurization air of 3 bar results in a good separation effect, but it is expensive because of the air compression required. With a flotation lasting 15 min, the same solid retainment can be obtained as with an 8-h sedimentation process, which was found out by using a rice-starch-sewage water flotation. The liquid cannot be cleared completely by flotation. Therefore a separate sedimentation tank is required.

As a gas used in the process, air is well established. A gas mixture of CO_2 and N_2 with solubility similar to biogas is not suitable for the flotation of finely dispersed sludge because the CO_2 forms large bubbles, which ascend with high velocity.

Centrifuge

With the help of a centrifuge, the sedimentation time can be reduced to $1/6^{th}$, i.e., from 2 h to 20 min.

The liquid outlet however is contaminated with anaerobic sludge, so that a sedimentation step is mostly required afterwards. The contamination is caused by fragmentation of the sludge due to the shear forces.

4.4.1.2 **Operation of the digestion tower**

From the start-up period until the complete adaptation of the activated sludge, approx 7 weeks are required.

The methane reactor is first inoculated with sewage sludge from another plant. The starting volume load should be approx $B_R = 0.6\,kg_{COD}/m^3.d$. After about 3 weeks, a second inoculation is done, so that the volume load increases to $B_R = 3.0\,kg_{COD}/m^3.d$. It is afterwards subsequently increased by 50% per week. The load can also be increased every time the concentration of volatile fatty acids drops to a low level. Propionic acid is the leading substance for the concentration measurements.

After 7–10 weeks, a degradation of organic matter given as COD value of 95% at a volume load of $B_R = 11\,kg_{COD}/m^3.d$ can be obtained. In this way, a good degradation performance as well as a high volume load required for the anaerobic fermentation process can be maintained.

The course of fermentation can be monitored according to characteristic values, some of which are displayed in Table 4.16 together with their limits and application notes.

Table 4.16 Characteristic values of sewage sludge decomposition.[74)]

Characteristic value	*Remarks on the method*	*Sewage sludge*	*Partly fermented*	*Conditionally fermented*	*Fermented*
Smell	Subjective, depending on time	Strong	Less	Little	None
Ignition lost (content of oDM)	Often used, easy to apply, inaccurate	>90%	60%	45–55%	<45%
Ignition lost/ ignition residue		High	Medium	Low	Very low
Acetic acid in mg L^{-1}	Large measuring range	1800–3000	1000–2500	100–1000	<100
Fatty acids in mg L^{-1}	Large measuring range	>1000	–	300–1000	<300
BOD_5/COD-value ratio		>0.25	0.25–0.15	0.15–0.1	<0.1
Biogas yield		High	Little	Very little	Very little
Fat content	Difficult to apply	High	Little	Very little	Very little
Reduction of the relative ignition lost		High	Medium	Low	Very low
Content of carbon in the DM		High	Medium	Low	Very low
C/N-ratio	Not recommended	High	Medium	Low	Very low
C x H / ignition residue	Difficult to apply	High	Medium	Low	Very low
Volume of the sedimented sludge	Not precise enough	High	Medium	Low	Very low
H_2S formation	Inaccurate	High	Medium	Low	Very low

Other characteristic values include the acid formation capacity, the lead acetate test, the proportion of lipoid content to substances which can be extracted by ether, and the complicated thermo-analysis.

4.4.2 Industrial purification of sewage

Many different anaerobic processes have been developed for industrial water treatment. The objectives for these developments were:

- high concentration of biomass in the reactor.
- optimal contact between sludge and sewage.

In Germany, 150 anaerobic plants for the treatment of industrial sewage are operating. The decisive factor for the construction of these plants was mainly that anaero-

74) Cp. BOK 40

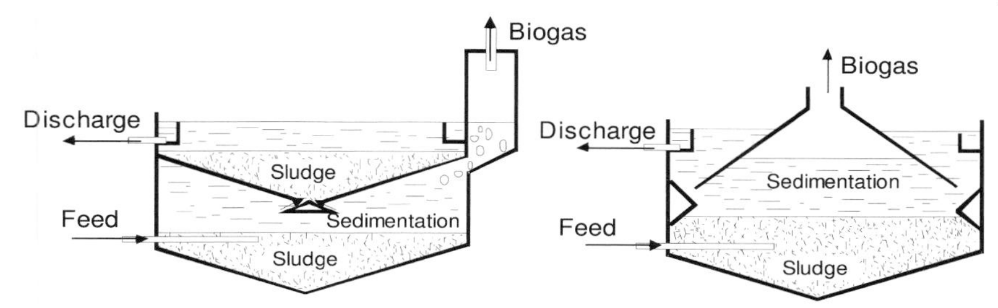

Figure 4.28 Common anaerobic sewage treatment plants:[75] Clarigester (left), UASB-reactor (right).

bic processes are able to purify very efficiently heavily loaded sewage sludge with $C_0 > 1500\,mg\,L^{-1}$. The production of biogas was only an additional advantage.

However, nowadays the production of biogas is coming more and more to the fore.

Simple anaerobic sewage purification plants still operate in the same way as they did more than 50 years ago (Figure 4.28).

4.4.2.1 **Process engineering and equipment construction**

There are two basically differing principles of equipment construction: one is the immobilization of microorganisms on a substrate of plastic or ceramic packing material, pumice stone, wood bars, or similar objects. The other is the formation of large granulates and flocculates sedimenting quickly.

In the second case, the sewage has to be carefully let in, preferably at still zones. Even in the upper part of the reactor the liquid has to be as still as possible to allow proper development of large granulates. In the region where the most and largest gas bubbles rise, the biomass has to be retained by a special installation so that it is not washed out by rinsing water coming from the region where the degradation takes place.

Alternatively, an additional purification step can be performed after sedimentation, aiming at recycling the biomass back to the reactor.

Sludge-bed bioreactor The most well-known sludge-bed bioreactors are: Upflow Anaerobic Sludge Blanket (UASB) Reactor, Expanded Granula Sludge-bed (EGSB) Reactor, Internal Circuit (IC) Reactor (Figure 4.29 and Table 4.17). The UASB-reactor was developed around 1980 from Letinga in the Netherlands.[76]

The **UASB-reactor** is made of a vertical cylindrical tank with a built-in three-phase precipitator for the effective separation of gas, liquids, and solids (sludge). The sewage is pumped in at the bottom of the reactor and flows to the top. Good mixing of the reactor's contents and contact with the microorganisms are ensured by the liquid stream and the development of biogas bubbles. Installations for the gas-liquid-solid separation are placed at the top of the reactor. The sludge settles out in the form of pellets onto the bottom.

The specific operation mode of the sludge-bed reactor results in good pelletization and good sedimentation of the sludge. Depending on size, rigidity,

75) Cp. BOK 58

76) Cp. BOK 28

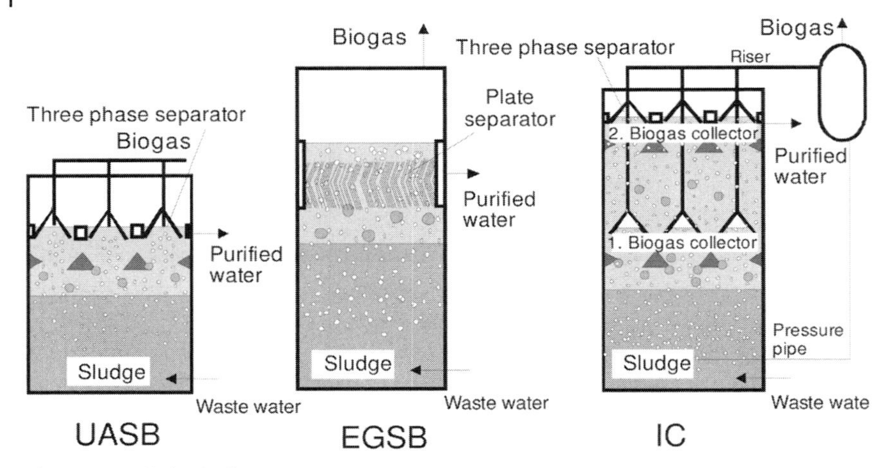

Figure 4.29 Sludge-bed reactors.

Table 4.17 Comparison of sludge bed reactors (throughput 125 m^3/h, organic matter as COD value: 5 kg/m).

	Units	*UASB*	*EGSB*	*IC*
Reactor volume	m^3	1250	n.a.	500
Reactor height	m	5		20
Volume load	$kg_{COD}/m^3/d$	12	n.a.	30
Residence time	h	10	n.a.	4
Circulations	–	1	n.a.	4.5
Liquid stream upwards at the bottom	$m\,h^{-1}$	1.0		27.5
Gas stream upwards at the bottom	$m\,h^{-1}$	0.9	<10	7.2
Liquid stream upwards at the top	$m\,h^{-1}$	1.0	<10	5.0
Gas stream upwards at the top	$m\,h^{-1}$	0.9		1.8

sedimentation velocity, specific CH_4-activity, and ash content of the sludge pellets, sedimentation velocities of 2–70 $m\,h^{-1}$ result with unobstructed sedimentation. The DM content of the sinking layer is relatively high at 75–150 $kg\,m^{-3}$. Consequently a large bioreactor capacity is required.

The UASB Reactor tolerates high solids concentrations without the risk of clogging. Special sewage, toxic substances in the sewage, or special process conditions can interfere with the development of pellets, consequently reducing the reactor performance and increasing the washing out. A storage tank with sludge pellets for the inoculation of a new mix should be ready in case the pellets are destroyed.

The **EGSB Reactor** is a modification of the UASB Reactor. It is run by recirculation of purified liquid with a higher upward velocity of the liquid. In this way, the contact time between granulate and sewage is intensified. A special realization of this concept is the BIOBED Reactor.[77] This is slender and tall and includes a

77) Cp. WEB 84

lamellar settler for the separation and feedback of the sludge. It offers the advantages:

- smaller reactor volume through volume loads up to 35 $kg_{COD}/(m^3.d)$
- lean tower construction methods with structural height from 10 to 20 m
- Lower investment costs because less material required for the reactor and its installation.

The **IC Reactor** is another modification of the UASB Reactor. It basically consists of two UASB Reactors arranged one on top of the other, with the same 3-phase separators at different reactor heights. The mixing itself is achieved by the ascent of the biogas from the separator to the headspace of the reactor through a gas pipe (riser). Consequently, a slight overpressure builds up in the headspace, forcing the biogas down through another gas pipe (downer), and from here again into the reactor. This allows reduced energy consumption. Forced circulation, for instance by pumping of the reactor's contents, is only required at start-up of the plant.

Assumptions for the operation of a sludge-bed reactor

One of the requirements for the operation of a sludge-bed reactor is that the inlet's COD value should not exceed 10 $g\,L^{-1}$. With higher loads, a dilution with reactor out-flow (recirculation) is required.

For nearly solids-free sewage, which is not prone to sludge or scum development, a pre-acidification of only 20–40% of the biomass is recommended. For all other sewage waters, with high solids content, a pre-acidification with subsequent removal of the acidifying sludge is recommended in order to increase the residence time in the reactor. This sludge should be stabilized separately by an aerobic process in a conventional intermixed reactor.

Generally, the calcium concentration has not to exceed 600–800 $mg\,Ca^{2+}\,L^{-1}$ to suppress carbonate formation. Furthermore, it has not to fall below 200–300 $mg\,L^{-1}$ Ca^{2+} ions, because otherwise the mechanical strength of the pellets diminishes.

Additionally, high NH_4-nitrogen concentrations of >1000 $mg\,L^{-1}$ and NaCl concentrations of >13–15 $g\,L^{-1}$ hinder the granulate development.

High fat concentrations (>50 $mg\,L^{-1}$) promote scum formation. The scum has to be removed from time to time and separately stabilized anaerobically.

Reactor design and ratings

The necessary reactor volume for the treatment of sewage or sludge is influenced by:

- the maximal daily load of dry matter expessed as COD
- the allowable area load of the three-phase separator
- the minimal sewage temperature
- the sewage concentration and composition (types of sewage ingredients and their biodegradability, presence of proteins, etc.)
- the permitted volume load at a specific sludge retention

- the targeted degradation performance and the targeted degree of decomposition.

The COD value is mostly reduced in the lower third of the bioreactor where the biomass sediments out.

The reactor volume V_R [m^3] can be calculated from the supplied volumetric flow V_G* [m^3/s], the load of the inflow c_0 [kg_{COD}/m^3], and the volume load B_R [$kg_{COD}/(m^3.d)$].

$$V_R = \frac{\dot{V}_G \cdot c_0}{B_R}$$

The sludge load B_{RS} [$kg_{COD}/(kg_{DM}.d)$] results from the volume load B_R and the biomass concentration X [kg_{DM}/m^3] in the reactor.

$$B_{RS} = \frac{B_R}{X}$$

The sludge age Θ [d] is calculated from V_R, X, the volumetric flow of excess sludge $\dot{V}_s$, [m^3/d] and the biomass concentration in the excess sludge c_S [kg_{COD}/m^3].

$$\Theta = \frac{V_R \cdot X}{\dot{V}_S \cdot c_S}$$

Table 4.18 summarizes all available findings for sludge-bed reactors with pelletized and flocculated sludge.

For only slightly contaminated sewage (COD value <1000 mg L^{-1}, $\vartheta > 25\,°C$), the required reactor volume is derived from the required residence time.

The surface load (wet surface of the separators) can also be decisive for the calculation apart from the volume load. The residence time is varied by the reactor height.

For UASB-Reactors the maximum permitted area load w_S amounts to:

- $w_S = 3\,m\,h^{-1}$ for water containing sewage a dissolved state with sludge
- $w_S = 1–1.25\,m\,h^{-1}$ for only partly dissolved sewage with granulated sludge
- $w_S = 0.5\,m\,h^{-1}$ for voluminous, flaky, activated sludge.

For the fermentation of sewage waters for which it is certain that solids are always previously separated and therefore all degradable substances are dissolved, the bioreactor height can reach up to 10 m and more (Table 4.19). For all other sewage waters a lower bioreactor height has to be chosen. For slightly contaminated sewage, the reactor height should be 3–5 m. A height of more than 7 m for sewage water without dissolved components leads to sludge losses.

Table 4.18 Applicable volume load for UASB reactors depending on the waste water composition.

Waste water concentration ($mg_{COD}\ L^{-1}$)	*Undissolved COD (%)*	*Applicable volume load B_R at 30 °C in $kg_{COD}/m^3 \cdot d$*		
		Granulated sludge	*Flaky sludge*	
			High DM decomposition	*Low DM decomposition*
Up to 2000	10–30	2–4	8–12	2–4
	30–60	2–4	8–14	2–4
	60–100	*	*	*
2000–6000	10–30	3–5	12–18	3–5
	30–60	4–6	12–24	2–6
	60–100	4–8	*	2–6
6000–9000	10–30	4–6	15–20	4–6
	30–60	5–7	15–24	3–7
	60–100	6–8	*	3–8
9000–18 000	10–30	5–8	15–24	4–6
	30–60	Uncertain at DM >6–8 g/l	Uncertain at DM >6–8 g/l	3–7
	60–100	*	*	3–7

* Application of the UASB process cannot be recommended under these conditions.

Table 4.19 Required residence time in h at different reactor heights at maximum tolerable average feed per area of $w_S = 3\ m/h$ and $1\ m/h$, under the assumption that only the hydraulic load is correct.

H_{BR} (m)	*t_{min}*	
	$w_S = 3\ m\ h^{-1}$	*$w_S = 1\ m\ h^{-1}$*
3	1	3
4	1.3	4.9
5	1.7	5.1
6	2	6
9	3	9
12	4	12

The distribution system, characterized by the number of inlets per reactor cross section or its inverse value, is adequate for the contact between sewage water and sludge (Table 4.20).

For slightly contaminated sewage waters, the area per inlet should be $0.5–2\ m^2$; for heavily contaminated ones $2–5\ m^2$.

Table 4.20 Reference values for the number of feeding points of UASB reactors.

Features of the sludge	B_R [$Kg_{COD}/m^3 \cdot d$]	***Area per feeding point*** [m^2]
Thick flaky sludge (>40 kg_{DM}/m^3)	<1	0.5–1
	1–2	1–2
	>2	2–3
Medium thick flaky sludge (20–40 kg_{DM}/m^3)	<1–2	1–2
	>3	2–5
Pellets	Up to 2	0.5–1
	2–4	0.5–2
	>4 kg	>2

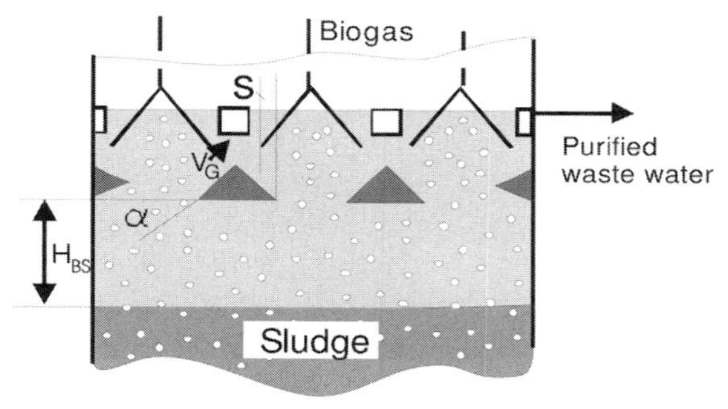

Figure 4.30 Three-phase separator in a UASB-reactor.

The three-phase separator in UASB, EGSB, and IC reactors is a rooflike installation (Figure 4.30).

Plate inclination:

$\alpha > 50$ to $60°$ for flocky sludge

$\alpha > 45°$ for pelletized sludge

Velocity of inflow:

$V_G = 3–5\,m\,h^{-1}$ for high-load sewage water

$V_G = 1\,m\,h^{-1}$ for low-load sewage water

Overlapping:

$s = 100$ to $200\,mm$

The area of the openings between the three-phase separators has to be between 15 and 20% of the bioreactor surface.

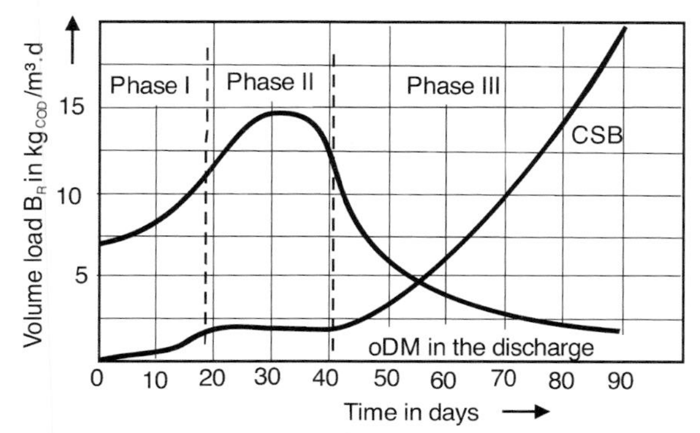

Figure 4.31 Starting phases of a UASB-reactor.

The height between the three-phase separator and the sludge-bed should be $H_{BS} = 1.5–2\,m$ at an overall reactor height of $H_{BR} = 5–7\,m$.

If sludge formation is expected, e.g., with sewage containing fats or proteins, spraying devices have to be placed above the water level.

Operation of a sludge-bed reactor

The start-up operation is divided into 3 phases, shown in Figure 4.31.

Phase I: Volume load $B_R < 2\,kg_{COD}/m^3 \cdot d$

The sludge-bed rises as a consequence of the starting gas formation. Very fine material is washed out with the sewage. Because of the growth of thread-like microorganisms, the sludge sediments less well. Scum develops at the end of phase I.

Phase II: Volume load $B_R = 2–5\,kg_{COD}/m^3 \cdot d$

As a consequence of the increasing volume load and gas formation, more material is washed out. Only the heavy parts of the sludge sink downwards. After ca. 40 days, the first sludge grains form (diameter up to 5 mm). Despite the heavy sludge losses caused by the washing out, the sludge load increases to $2\,kg_{COD}/m^3 \cdot d$ at constant volume load, because the specific sludge activity rises.

Phase III: Volume load $B_R = 2–50\,kg_{COD}/m^3 \cdot d$

If granulates are able to form better and as such the sedimentation ability of the sludge is improved, less material can be washed out. The volume load increases to its maximum value of $B_R = 50\,kg_{COD}/m^3 \cdot d$.

The granules can develop rapidly (6 months) especially when the sewage composition promotes the formation of natural biopolymers (extra-cellular polymers, polysaccharides, and proteins) or if residues of synthetic polymers are in the sewage.

The start-up phases can be shortened if the sludge-bed reactor is inoculated with sludge. The activation is carried out with 10–20 kg_{oDM}/m^3 activated sludge from another plant with low methane activity and low sludge index ISV at a sludge load of B_{RS} = 0.05–0.1 $kg_{COD}/kg_{DM} \cdot d$ in phase I and a sludge load of 0.6 $kg_{COD}/kg_{DM} \cdot d$ during phase II for the formation of pellets. The volume load should not be increased before all volatile fatty acids are degraded. Sludge difficult to sediment should be washed out and heavier sludge particles should be retained.

If pelletized excess sludge from another reactor operating for a long time is used, the sludge index ISV of the residual sludge should amount to approx 50 $mL g^{-1}$ after washing out the colloidal components during the first days of operation. The pelletized excess sludge should contain 75 kg_{DM}/m^3. Its ratio oDM/DM should be as high as possible.

The specific methane formation rate should be kept low at the beginning (below 0.6 kg CH_4 per $kg_{COD}/m^3 \cdot d$), so that no washout due to intense gas development takes place.

Pelletized sludge (DM ≈ 100 $g L^{-1}$) is available on the market at a price of approx 5000 US$/$Mg_{DM}$. During the starting phase, it is common to load the reactor up to approx H_{BP} = 1 m with pelletized sludge.

In the outflow of sludge-bed reactors, the dry matter content is $DM_{R,e}$ = 0.05–1.1 $g L^{-1}$ or $oDM_{R,e}$ = 0.02–0.6 $g L^{-1}$. This biomass is tolerated and even wanted, because flocky sludge has a negative influence on the formation of pellets.

Reactors with immobilized microorganisms All slowly growing microrganisms[78] of the anaerobic degradation process tend to immobilization. Microorganisms of the genus *Methanosaeta* grow particularly well on hydrophobic surfaces, because they are not covered with a barrier of adhesively bonded water molecules that prevent attachment to surfaces.

For this reason, reactors with immobilized microorganisms are equipped with packing material on which the microorganisms can grow as a thin layer, the so-called biofilm.

Biofilm

The thinner the biofilm on the packing material, the more effectively the plant works, because only the topmost, approx 1 mm thick biofilm layer is actively involved in the degradation of sewage water components. After a residence time of t = 154 days in an anaerobic reactor, a dense biofilm has grown on the carrier material. Tests with different plastic materials of the specific surface $O_{spec.}$ = 98–138 m^2/m^3 for the time specified above led to film thicknesses of 2–3 mm when synthetic, easily methanizable sewage consisting of 30% sugar and 65% ethanol was used.

The biofilm grows continuously (Figure 4.32). After about $1^1/_2$ years of operation, perhaps only ca. 65–70% of the total volume flows through or is intermixed, and 30–35% of the reactor's volume belongs to the stagnation space (skip zone). The biofilm has to be removed no later than this.

78) Cp. BOK 69

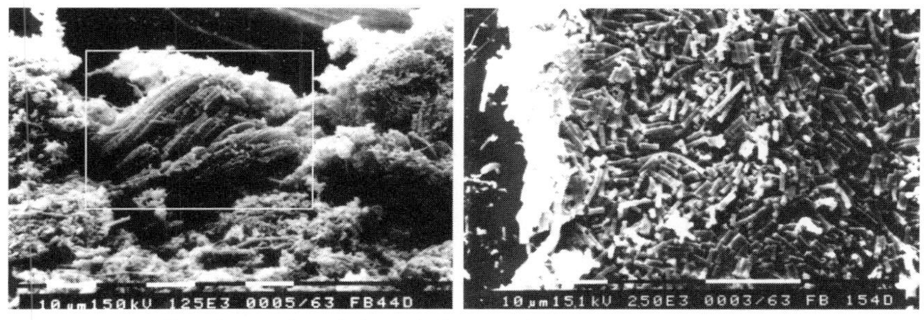

Figure 4.32 Biofilm after 44 days (left) and after 154 days (right) of anaerobic fermentation. After 44 days some filamentous growing *Methanosaeta* can be observed. After 154 days the typical bamboo-like segmented cell formation of *Methanosaeta* can clearly be seen. (SEM-photo: the length of a white bar equals 10 µm).[79)]

Should primarily carbohydrates (oligo- and polysaccharides) be degraded, microorganisms can rapidly grow and develop a slime shell guaranteeing good adherence. Other non-slime-developing microorganisms can then stick to this slime shell and so form a heterogeneous biofilm.

The form, surface, horizontal areas, and cavity volumes of the packing material determine the constitution of the biomass in the reactor. Randomly inserted material grows faster than axially flowed-through, packed packing material, but also forms easier skip zones with short-circuit currents and is more prone to clogging, leading to a lower elimination of organic matter as COD. By addition of flocculation agents (polyacrylamides) and calcium, the biofilm develops faster.

The packing material can be

- Random-fill packings or cubes.

To this category belong limestones, quartz, lava slag, activated coke, corals, shells, clay, expanded clay, plastics (polyurethane), ceramics, pumice stone, small stones, ordered and packed or filled plastic elements with various shapes, sizes, and surface structure (specific surface 70–400 m^2/m^3, porosity >95%), foils, glass marbles, glass ceramic, wire and plastic braidings, natural sponges, etc.

PVC is less suitable for the adhesion of microorganisms.

The following materials can also be used: carrier materials from brewer's barley, etc., (e.g., in the form of compressed plates), native fibers, e.g., hemp, flax, sisal, cotton, jute or cocoa, bamboo stick, wood plates, burr walnut, used tires. A natural fixed bed of cocoa fibers has proven to be superior to other carrier materials from the point of view of bacteria adhesion, degradation performance, homogeneity of the outflow, concentration, and costs.

Well established is also packing material of Siran™, with density $\rho_{FS} = 1\,g\,cm^{-3}$, open pore volume ε_{FS} = 55–70%, adjustable pores diameter d_{FS} = 10–400 µm, and usable colonization surface up to $O_{FSspec} = 90\,000\,m^2/m^3$ fill.

79) Cp. BOK 69

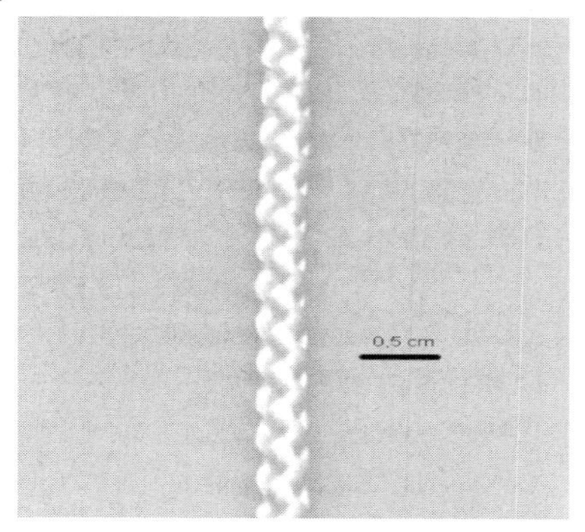

Figure 4.33 Packing material in the form of cord before installation in a bioreactor.

In a fluidized bed, spherical inert material with a large surface (O_{spec} = 3000 m^2/m^3, $\varepsilon = 80\%$), e.g., made of sand, activated coke, foamed plastic flakes, or plastics is used.

- Inserts in the form of webs or blocks.
- Strings

Vertically placed strings (Figure 4.33) of hydrophobic plastic strings are not only recommended for proofness against interlocking, large specific surface area (A_s > 100 m^2/m^3), and a cavity volume of 93%, but also because *Methanosaeta* settling is promoted by the hydrophobic surfaces.[80)]

Fixed-bed reactor, filter reactor, fixed film reactor In fixed-bed reactors such as anaerobic filters, PCR (Polyurethane-assisted Carrier Reactor) packing material is accumulated. The packing material takes up to 2/3 of the bioreactor height according to 10–20% of the bioreactor volume and its layer should be minimum 2 m thick.

In the fixed-bed reactor, packing materials with a coarse surface are provided for the microorganisms, on which they can immobilize particularly well. Packing material in general is characterized by its very large specific surface and very small flow resistance, so that clogging is prevented.

Fixed-bed reactors are particularly well suited for waste water, which contains few or no larger solid lumps.

The volume load of the reactors amounts to B_R = 5–100 $kg_{COD}/(m^3 \cdot d)$. It should be below B_R = 12 $kg_{COD}/m^3 \cdot d$. The degradation of organic substances in the waste water to COD values <1 kg_{COD}/m^3 is achieved. Without recycling of biomass, the normal DM concentration reaches oDM = 10–20 $g\,L^{-1}$, and with recycling it rises to 50 $g\,L^{-1}$.

Fixed-bed reactors can be operated both in upflow, i.e. substrate and developed biogas move parallelly from the bottom upward, and in downflow, i.e. substrate

80) Cp BOK 69

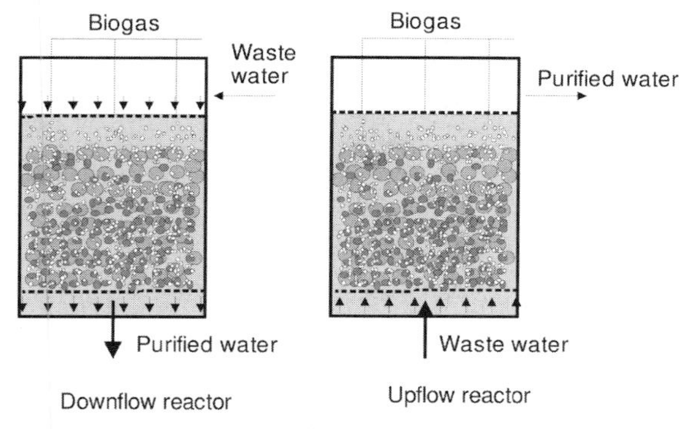

Figure 4.34 Technologies of fixed-bed reactor.

Table 4.21 Characteristic values for bioreactors with immobilized microorganisms.

	Upstream velocity (superficial velocity) [m s⁻¹]	*Expansion factor*	*Energy load [W/m³]*
Sludge floc reactor	1	0	5
Expanded-bed reactor	2–10	1.5–2	
Fluidized-bed reactor with Biolith[81]	10–30	5–10	5–15
Fluidized-bed reactor with sand			20–40

flows from the top downward and the biogas flows upward (Figure 4.34). For waste water with high organic load (DM_{COD} = 30–60 g L^{-1}) but low solids, upflow fixed-bed reactors (UAF = Upflow Anaerobic Filter) are used. It may happen that dead biomass tears away and contaminates the clear water. Fixed-bed reactors in downflow operation (DSFF reactor = Downflow Stationary Fixed Film Reactor) in contrast do not block so easily and therefore are used for solid-rich waste water.

Because of the large surface which can be settled by microorganisms, the residence time can be short. With a continuous or semi-continuous supply of waste water, the microorganisms in the fixed-bed reactor are held back for the most part and are not washed out.

Characteristic values for fixed-bed reactors are given in Table 4.21. During start-up, the velocity of the substrate should be below v_A = 0.4 m h^{-1} in the beginning and should rise slowly to ca. v_A = 1 m h^{-1}. In industrial scale plants the velocity of the upflow reachs v_A = 2 m h^{-1}. Higher velocities can cause losses of sludge. A circulation pump with adjustable drive is recommended, which is able to recycle 15–20 times the feed, in order to improve the flow through the entire reactor, to prevent blockage, and to keep the best biological conditions over the reactor height. The higher the concentration of the waste water, the higher is the circulation rate that should be chosen.

81) Biolith: sediments consisting of dead animals

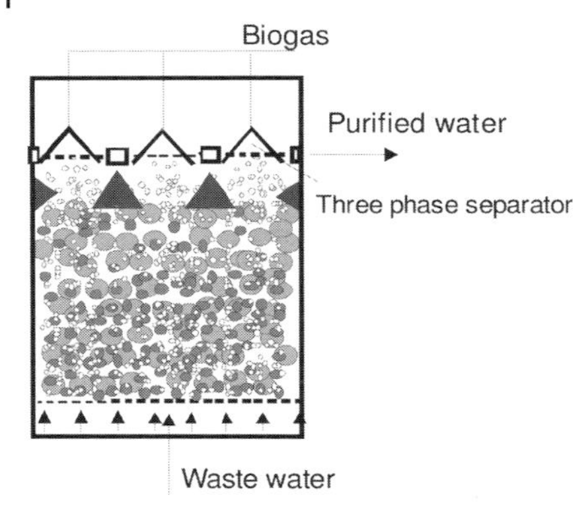

Figure 4.35 Fluidized-bed reactor.

By circulation of substrates with a concentration higher than oDM_{COD} = 8000 mg L^{-1}, the following can be achieved:

- The demand on alkalis and trace elements is reduced.
- The concentration of organic acids in the zone where the substrate meets the reactor wall is decreased.
- A slightly better cleaning is achieved.

If the biofilm is washed out or the supplied sludge is too strongly whirled up and thus biomass passes through the reactor and is to be found in the clear water, the waste water must later be cleaned and possibly retreated.

Expanded-bed reactor, fluidized-bed reactor

In the fluidized-bed reactor (Figure 4.35) and anaerobic attached film expanded-bed reactor, spherical packing materials are preferably used, covered over with microorganisms.

Well-approved materials especially suitable for growing microorganisms include Al_2O_3, PVC, clay, activated charcoal, bentonite, and sand with a specific surface of 1000–3000 m^2/m^3.

The packing materials are used on a perforated plate. Through the plate, substrate is pumped with such a velocity that the packing materials hover in the fluid. The high velocity is reached by the fact that substrate is circulated by a pump and the reactor is tall and narrow. Reactors are classified according to the upstream flow velocity.

In order to ensure sufficient matter transfer, the following power supply rates are necessary for continuous operation (according to experience). The installed capacity should be about twice as large, however.

Sludge floc reactors and expended-bed reactors are more suitable for weak to highly concentrated sludge (5–80 g L^{-1}). Without biomass feedback in the reactor, solid concentrations can be reached up to oDM_{COD} = 20–70 $kg_{COD}/(m^3 \cdot d)$.

Often fluidized-bed reactors are built in two stages in order to ensure a constant supply of acidified water (oDM_{COD} = 1500–3600 mg L^{-1}) into the second reactor.

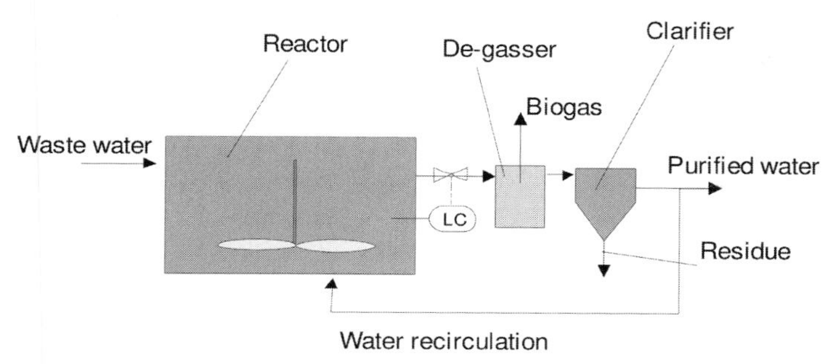

Figure 4.36 Contact process.

The contact between biomass and waste water is particularly intensive in the fluidized-bed reactor. There is no risk of clogging. Because of the high liquid flow locally, particularly high concentrations of poisoning materials are impossible.

Fluidized-bed and expanded-bed reactors are started up either with slowly added packing material and constant substrate flow rate or with maximum packing material and slowly increasing substrate flow rate. The smaller the amount of packing material, the shorter is the start-up period.

4.4.2.2 **Plant installations**

Contact Process (CP) (Figure 4.36) In the contact process, the bioreactor, a degassing device, and a clarifier with biomass recycling are combined.

This process is characterized by a low volume load and a residence time of 0.5–5 days. The procedure can be used for waste water with high concentrations of suspended solids (up to 40%) without risk of clogging. Clarification of the overflow could be difficult, because the biomass washed out does not have good sedimentation properties.

The contact process in general works with low biomass concentrations, long residence times for the sludge, and large reactors. The process is susceptible to changing environmental conditions and inhibiting materials.

Uhde-Schwarting process[82] (Figure 4.37) With this two-stage process, the entire biomass ($oDM_{COD} > 5000\,mg\,L^{-1}$) is passed through the methanation. The bioreactors are characteristic. They are "slim" and installed one behind the other. Each is equipped with several perforated plates. The process works with mesophilic acidification and thermophilic methane formation. Into the methane reactor a biomass retention system can be inserted. The reactors are mixed by sporadically circulating the substrate.

Hybrid reactor (UASB/Filters) (Figure 4.38) The hybrid reactor, also called CASB reactor (Captured Anaerobic Sludge Blanket Reactor) is a UASB reactor with a

82) Vgl. BRO 14

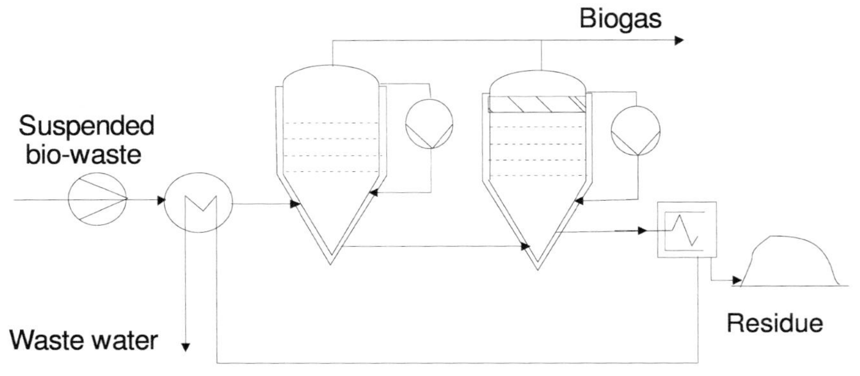

Figure 4.37 Uhde-Schwarting process.

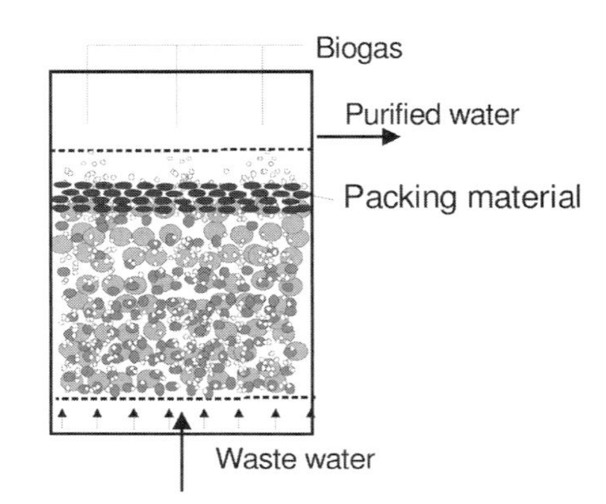

Figure 4.38 CASB-reactor.

packing material layer in the upper part, which works similarly to the three-phase separators described above. The packing material is chosen so that it does not clog, no channels are formed, and solids are not torn off.

Continuously-stirred tank reactor (CST reactor) (Figure 4.39) The CST reactor plant works in two stages, with a biomass feedback in the second stage.

The biogas plant is suitable for waste water with a high sludge concentration or extremely high concentration of soluble organics. It reacts sensitively to shock disturbances or poisonous substances in the waste water. Normally the CST reactor plant for the treatment of communal waste water is not economical.

There are complete plants with an upstream small conditioning tank for the addition of neutralization liquids, nutrients, and/or trace elements. These are necessary to adjust the optimum conditions for the degradation. The plant is equipped with a tank for the hydrolysis, and the bioreactor and can be built in a compact, completely encapsulated, and gas-tight design.

Loop reactor (Figure 4.40) Plants with a loop reactor instead of an agitated reactor are approved for the treatment of solid-rich, highly concentrated substrates. The

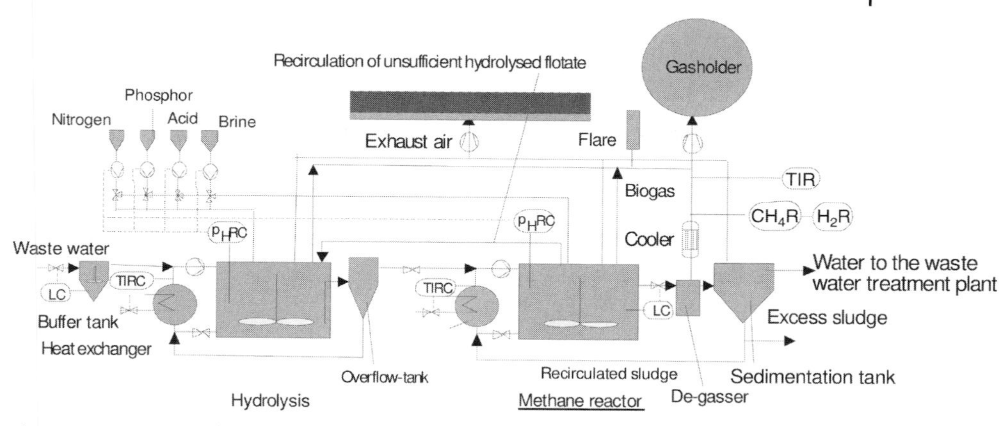

Figure 4.39 Two-stage bioreactor plant for industrial waste water.

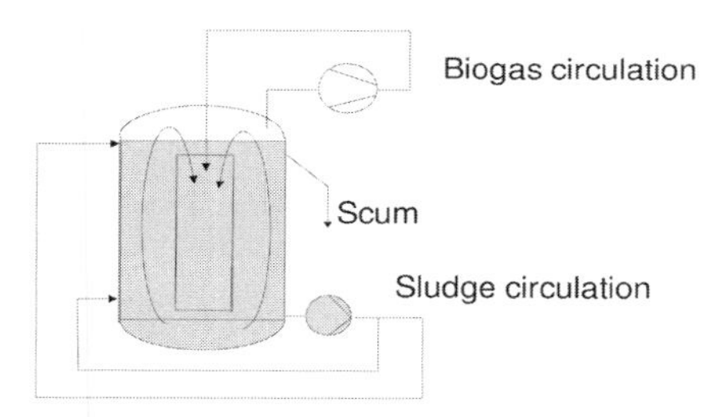

Figure 4.40 Loop reactor.

liquid is rotated by injected biogas or the air-lift effect, so that no scums develop. The formation of H_2S is suppressed by the biogas circulation, so that in many cases additional biogas desulfurization is not required.

4.5 Plants with separation of non-hydrolyzable biomass

Especially when planning two-stage plants, the separation of the biomass which cannot be hydrolyzed is to be taken into consideration in order to avoid operating problems in the methane reactor. Even the feedback into the hydrolysis stage may be useful in order to improve the leaching, especially if the biomass tends to form lumps easily.

In principle, there are two different technologies that are most often used to process bio waste from households and agriculture:

- Suspension process: the biomass is suspended in water and is then separated after the hydrolysis.
- Percolation process: the piled up biomass is sprayed with water, which percolates the stack.

4.5.1 Process of suspension

4.5.1.1 Process engineering and equipment construction

Figure 4.41 shows the flow sheet of a biogas plant for bio waste.

The biomass is continuously dehydrated by centrifuges applied during the hydrolysis stage. In this process the liquid phase is transferred into the methane reactor while the solid phase is recirculated for continuous hydrolysis. A series arrangement of several tanks allows the use of cultures of special bacteria and optimization of the degradation of organic substances. Usually about 40–50% of the organic substances dissolve easily within 5 days. Of these, about 40% are

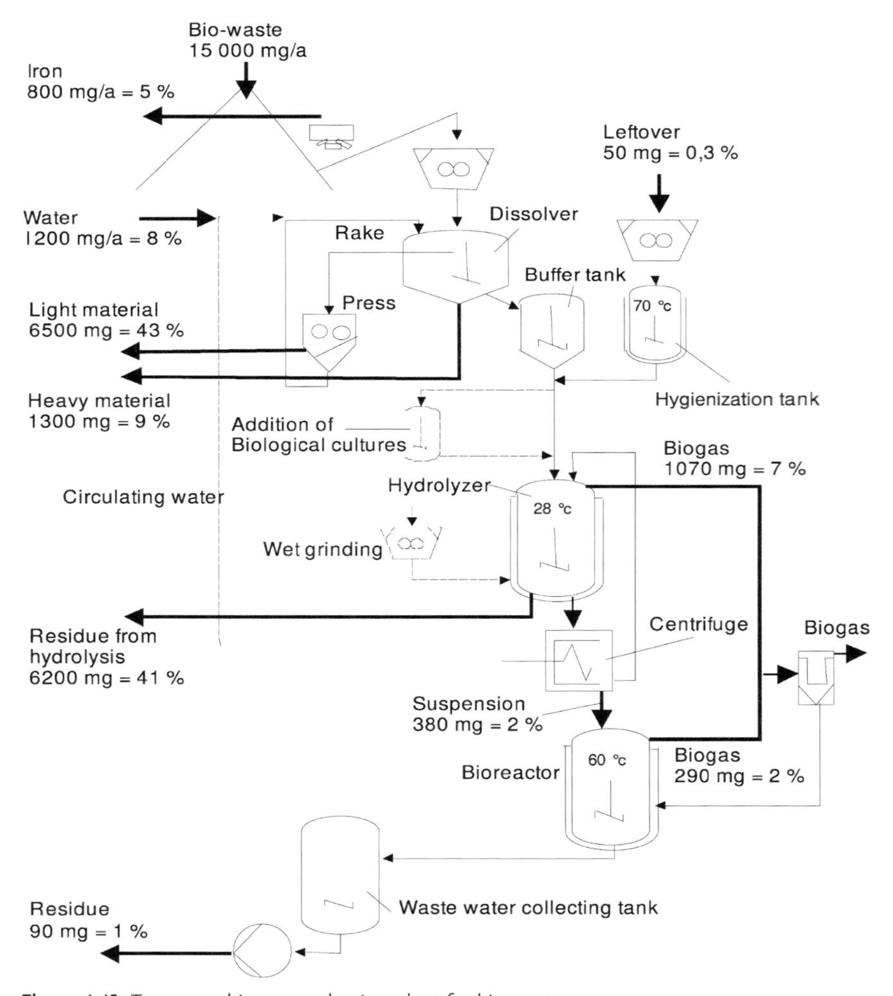

Figure 4.41 Two-stage biogas production plant for bio waste.

directly converted to hydrolysis gas containing 64% CH_4. The residue (any solids which are not decomposed) exits the second hydrolysis stage and is dehydrated and completely removed from the process. After 4 weeks of aerobic composting, it can be used as fertilizer.

Instead of using a separate methane reactor, the liquid phase can also be fed into the digestion tower of the sewage treatment plant.

Usually the methanation takes 2–3 weeks, depending on the temperature, the concentration of the substrate, the active biomass, and the targeted degree of degradation. In some cases, nutrient salt and trace elements may be added. The pH value should remain neutral.

Any liquid phase which is removed from the process is collected in a storage tank and recirculated for use as process water. Any excess water is purified within the plant itself and then circulated back into the sewage treatment plant. A biogas plant for 20000 $Mg\,a^{-1}$ bio waste produces 13000 $m^3 a^{-1}$ of waste water corresponding to 240 population units, while a plant for 30000 $Mg\,a^{-1}$ (360 population units) delivers about 20000 $m^3 a^{-1}$ waste water fed back to the sewage treatment plant.

It is possible to accelerate the process by wet grinding the biomass (as shown in Figure 5.56) with microorganisms (symbiosis of funghi, bacteria, and protozoa), which are present in the rumen of ruminants and able to dissolve specifically cellulose-containing substances within a short time.

The organic material is mashed with a specific buffer solution to a liquid mixture which is subjected to hydrolysis and the same cell digestion as that applied by the microorganisms in the rumen. Therefore it is important to adopt the same ambient conditions, e.g., temperature, pH, and mixing as those in the natural rumen. An internal recirculation through a refiner allows, in the rumen as well as in the fermentation process, further break-up of the material to facilitate the disintegration.

4.5.1.2 Efficiency

The costs for the fermentation of organic waste are not fixed but strongly vary depending on the different technical and structural requirements laid down by the operating authorities. Today in 2007, the overhead expenses for the treatment of 1 Mg bio waste amount to 70–2000 US$ Mg^{-1} for a plant producing 15000 $Mg\,a^{-1}$. With an increasing yield of the plant, the costs go down, e.g., for a plant with an output of 50000 $Mg\,a^{-1}$ the overhead expenses decrease to 50–90 US$ Mg^{-1}. Referring to the volume of biogas, the costs can be assumed with 37000 US$/($m^3$/h) of biogas.[83]

4.5.1.3 Plant installations

KCA process[84] (Figure 4.42) The KCA process is characterized by the automatic separation of foreign particles during the wet processing in a pulper and a drum screen. The whole plant is fully closed to avoid any pollutants getting into the environment. The hydrolysis happens in an aerobic way. The bioreactor is a vertical cylindrical fermenter with a centralized draught tube to recirculate the gas.

83) Cp. BOK 63

84) Cpl. WEB 63

Table 4.22 Products resulting from the fermentation of 1 Mg bio wastes (DM).

Products	*Amount*	*Features*
Biogas	90–200 m^3	50–70% methane, 27–43% carbon dioxide, traces of hydrogen sulfite, nitrogen, hydrogen, and carbon monoxide Calorific value 20–30 MJ/m^3
Residues in total	0.7–0.8 Mg	iron, heavy material, hydrolyzate
Hydrolyzate		Calorific value 30 000 kJ/kg Contains heavy metals in a problematical percentage

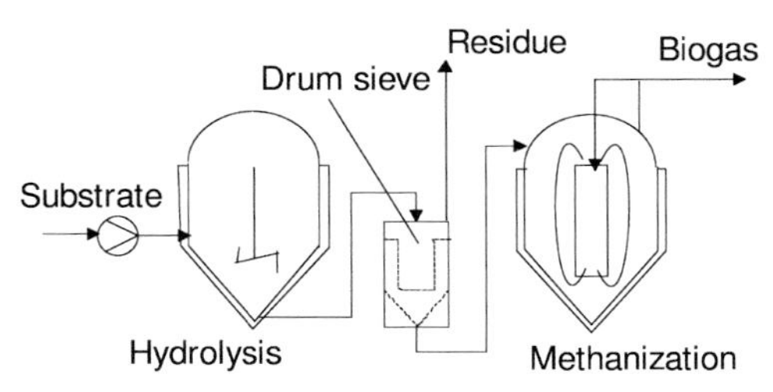

Figure 4.42 KCA process.

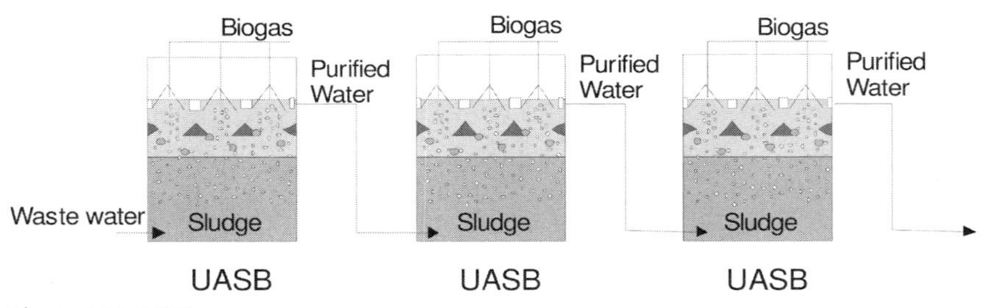

Figure 4.43 ABBR process.

Usually the residues of the fermentation are poor in foreign particles. Into the bioreactor are added chemicals[85] to prevent sludge flocculation and the formation of H_2S. In this way it is possible to achieve low concentrations of around $10 < H_2S < 100\,mg\,Nm^{-3}$.

Generally, the KCA process is preferably applied to co-fermented bio waste with waste water sludge and liquid manure.

Anaerobic baffled reactors (ABBR) (Figure 4.43) The ABBR process is simple, robust, and not affected either by a discontinous feeding of waste water, the introduction of toxic substances, or irregular high volumetric flow rates. It consists of

85) Cp. BRO 10

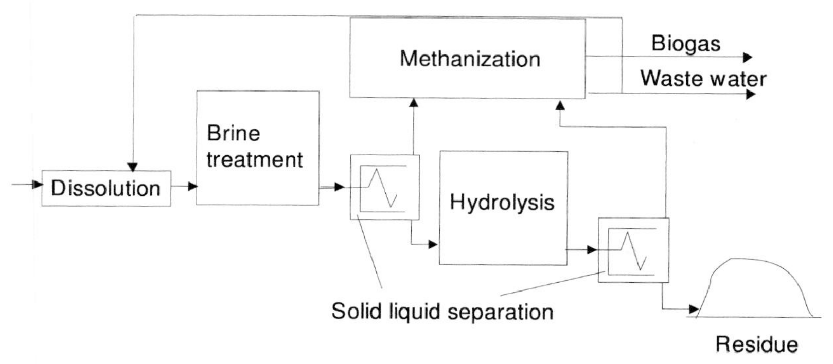

Figure 4.44 BTA process.

a cascade of UASB reactors, which makes it more costly compared to one single UASB reactor. The sedimented sludge is removed in each single cascade.

The BTA process (Figure 4.44) allows the solid substances to be well dissolved, resulting in a higher yield of methane and a better degradation of the residue.

In this process, the suspension of the substrate is hydrolyzed in a bioreactor equipped with a mixer. In a next step the liquid containing the hydrolyzed products is decanted from the solid phase until the targeted degree of hydrolysis is reached. The following fermentation is mesophilic and occurs like that in a fixed-bed bioreactor using plastic carrier material. The methanation of the purely dissolved organic material happens in less than 24h.

Several industrially implemented biogas plants run according to this principle.

4.5.2 Percolation process

The percolation process, also called the "leach-bed process", works with a very low water content (65%). The volume of the bioreactor can be small.

Industrial plants, e.g., for bio waste, are most commonly two-stage processes, while most plants in agriculture, e.g., for liquid manure or grass, work by a single-stage process.

Compared to the wet fermentation process, the volume load in the percolation process can be much higher, but the overall yield of biogas is between 10 and 20% lower, considering favorable conditions. If the biomass is piled up badly, the yield of biogas can even be much lower.

In future, however, the percolation process is expected to improve, and it should be possible with new plants to achieve the same gas yield as that with the wet fermentation technology.

4.5.2.1 Process engineering and equipment construction

Water-soluble metabolites are flushed out with water, which is continuously percolating the bulk goods. The phase separation happens either with cake filtration

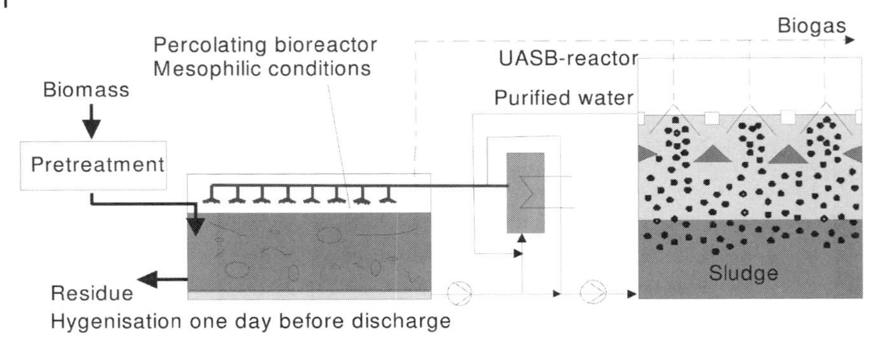

Figure 4.45 AN-Biothane process.

during the percolation or in a screw press, belt-type press, or centrifuge, often still combined with added precipitating agent. In the two-stage process, the wash water is then transferred into an anaerobic bioreactor. Such fermentation processes consist of a so-called “dry hydrolysis” and a “liquid methanation”. The suspension, rich in organic substrates, can be easily and well fermented, yielding biogas with up to 70% of methane. In the course of the hydrolysis the quality of the biogas deteriorates so that the final quality of the biogas mixture ends up the same as that from other plants.

If a single-phase process is applied, substrate and digested sewage sludge will be either piled up by turns or fresh material will be mixed with some fermented mass. A batch of liquid manure from cattle and horses, grass, and inoculation material provides a content of about 22% of dry matter, consisting of about 78% of organic dry matter.

Overall, the percolation serves to moisten the biomass to enable the methanation process to occur. Hence, the substrate is continuously reinoculated with bacteria to keep the fermentation going. It is possible to go without any percolation by providing a ratio of 1 : 1 of solids : inoculation material (chicken excrement).

4.5.2.2 **Plant installations**

AN/Biothane process (Figure 4.45) The AN/Biothane process was probably the first process to be developed for bio waste fermentation.

In this process the substrate is prepared, then carefully piled up in the percolation reactor and sprayed with a little water, which percolates through the biomass and removes all water-soluble substances. The wash water, containing all the dissolved substances, is roughly separated and fermented in an UASB-digester.

Prethane/Rudad-Biopaq process or ANM process (Figure 4.46) In the 1980s, this process was implemented twice on a large industrial scale (about $3000\,Mg\,a^{-1}$) for the treatment of bio waste (vegetable matter, slaughter waste, dung). In Germany it was called the ANM process, while in the Netherlandes it was named the Prethane™/Rudad™-Biopaq™ process.

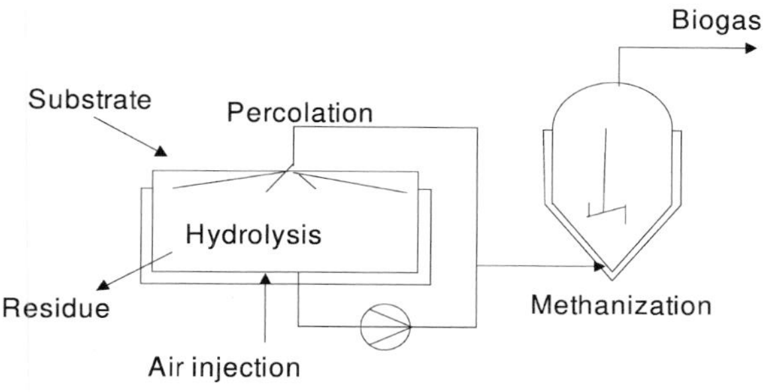

Figure 4.46 Prethane™/Rudad™-Biopaq™ process.

Two additional plants were implemented in Germany in the years 2000/2001 on the basis of the meanwhile gained experiences. Both plants were based on a two-stage process with an initial mechanical preparation stage. Their output amounts to 25 000 $Mg a^{-1}$ of total waste produced, since bio waste cannot be separately collected.

By the means of a rotating drum screen (sieve cutting of 140 mm), coarse particles with a high heating value and biologically degradable fine particles are separated, and metals are removed. About 96% of the degradable organic substances are removed by sieving with a 140 mm sieve and transferred to the percolator.

The fine particles in the percolation reactor are continuously circulated and strongly agitated for 2–4 days while being sprayed with about 2 $m^3 Mg^{-1}$ wash water from the top (related to the percolator supply) and cyclically blown through by air from the bottom ($V^* \approx 8.6 m^3/(m^3.h)$). In this way, soluble organic substances are removed from the biomass and organic compounds are hydrolyzed or acidified. Alternatively, the waste is suspended in water to give 10% of dry matter and is anaerobically hydrolyzed in a bioreactor at 35 °C up to a certain targeted degree of hydrolysis. The remaining solids are dehydrated by means of a screw press.

The total quantity of solution accumulating in the press and percolator is cleaned of sand and fibrous materials and injected into the fermentation reactor, e.g., into a UASB reactor, where the organic components are degraded for 2–6 days.

Part of the water flowing out of the reactor is re-used for percolation. From the other part, the nitrogen is removed in a NO_x-reduction plant. Afterwards, this water can be used as industrial water.

IMK process[86] In the IMK process, the acidification is done aerobically. The liquid with the dissolved substances is squeezed out and fermented anaerobically. For the separation of fine particles from the liquid stream, hydraulic cyclones are used.

The process is suitable for the utilization of all types of bio waste.

86) Cp. WEB 28

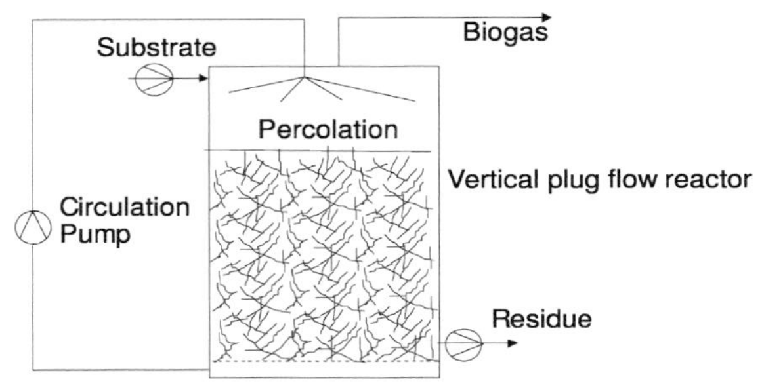

Figure 4.47 Dry anaerobic composting process.

Table 4.23 Process parameters.

Plant capacity	Mg a^{-1}	30 000
Residence time	days	
For methanation		10–20
For rotting		ca. 10
Spec. waste water accumulation	L Mg^{-1} Input	340
Spec. electrical energy consumption	kWh Mg^{-1} Input	64
Personnel requirement	Persons	6
Investment costs	Mio US$	14
Operation costs	US$ Mg^{-1} Input	90

Dry Anaerobic Composting (Figure 4.47 and Table 4.23) The DRANCO or Dry Anaerobic Composting of manure (Anacom) or ATF processes are one-stage processes in which substances dissolved during water percolation are methanized.

All three processes are very similar: The substrate preparation is performed by sieving and shredding of the waste. Afterwards, the particles smaller than 40 mm are heated up to the fermentation temperature of 55 °C and the substrate is moistened to DM = 25–45% in a special conditioning unit. The next step is a single-stage thermophilic fermentation in a vertical cylindrical bioreactor (residence time in the fermenter: 15–30 days). Fresh biomass is continuously loaded from the top of the reactor. It slowly flows to the bottom almost in a plug flow current. At the bottom, the biomass is continuously removed by a combination of floor scraper and discharging screw. While the biomass flows through the reactor, it is sprayed with circulation water.

In contrast to the Anacom process, in the ATF process the substrate is moistened to a DM concentration of 35–50%, so that no additional fresh water has to be added for the percolation step.

Prior to the rotting, the digested sludge is dehydrated by means of a screw press.

Figure 4.48 Four big size bioreactors for batch dry fermentation.[87]

Dry fermentation process in a stack In well-developed dry fermentation plants, gas-tight buildings serve as bioreactors, in which either piled up biomass[88] or biomass that has been filled into containers[89] is subjected to the fermentation process. The building is under slight overpressure and well heat-insulated. The doors are specifically equipped with special sealing lips to prevent oxygen from entering.

To achieve a throughput of 1000–1200 $Mg\,a^{-1}$, four drivable containers of about 100 m^3 volume each are necessary.

Fresh and fermented substrate in equal quantities are piled up together in layers.

Liquid is circulated over the piled-up biomass in a two-hour cycle through two distributors. The liquid percolates through the biomass, warms it up to 35 °C, and brings nutrients along, so that biogas forms. A double perforated floor in the reactor serves for the collection of drained off percolate.

The percolate is heated up in a separate tank with integrated heat exchanger or is conveyed through a tubular heat exchanger integrated in the floor. A floor heating system supplied with the exhaust heat of the CHP engines is enough to keep the substrate coming into the tank at a temperature of approx 65 °C after an initial cooling to 40 °C.

The formed biogas is withdrawn. The sulfur content of the biogas is very low at 10 $mg\,m^{-3}$. Desulfurization is thus not necessary in most cases.

The fermentation lasts for 2–4 weeks, depending on the substrate. The methane content in the biogas is relatively high at 80%.

A few days before the fermentation process is completed, the percolation is interrupted. The water is allowed to percolate off. Afterwards, the bioreactor is back-flushed with engine exhaust gas in order to avoid the development of an explosive

87) Cp. BRO 2
88) Cp. BRO 1
89) Cp. BRO 2

gas mixture. The biomass is dried out by injection of air through the perforated floor. Thus, the complicated afterfermentation under the roof is no longer required. The fermenter is then emptied and reloaded with fresh biomass.

Since gas generation and gas composition are not constant during a batch process, at least three bioreactors have to be run in parallel and out of phase.

Fermentation channel process (Figure 4.49) In this process, the biomass is placed in wire metal box wagons which are transferred into the channel by means of gas-tight sluices. The boxes are then pulled through the closed, completely gas-tight channel filled with liquid (bacteria and nutrients). The biomass remains in the channel until the methanation is finished. Two channels can be passed through sequentially to isolate the processes of hydrolysis and methanation.

Aerobic-Anaerobic-Aerobic process (3-A process) (Figure 4.50) The 3A process enables savings to be made in transportation costs, and includes three procedures which take place sequentially in the same bioreactor. First, the solids are subjected to an aerobic decomposition and dissolution of organic matters. In the second procedure, the biomass undergoes an anaerobic fermentation process followed by

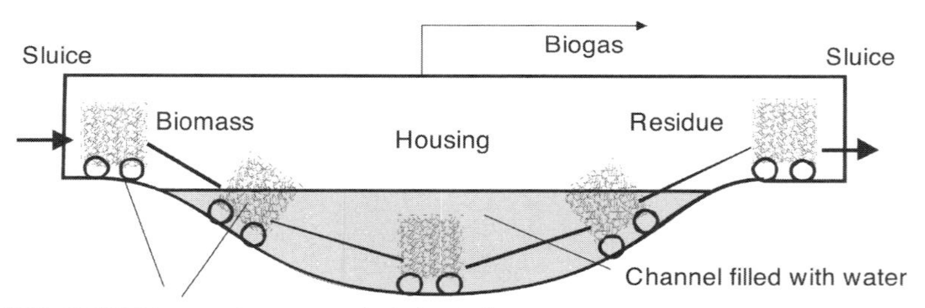

Figure 4.49 Channel process.

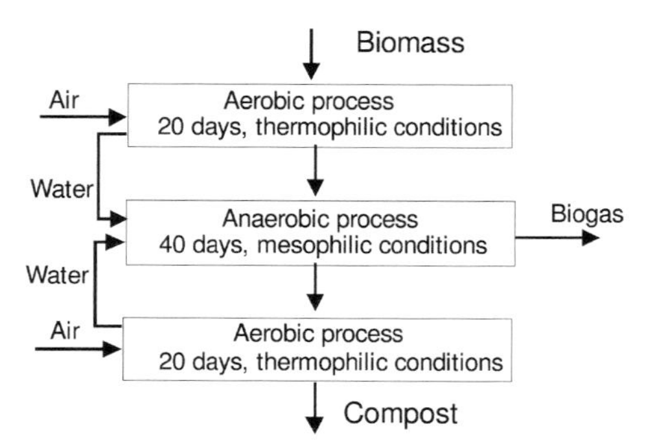

Figure 4.50 3 A process.

an aerobic rotting. One disadvantage to be mentioned here is the release of emissions during the composting, e.g., climatic gases and substances with an unpleasant smell.

Usually the first two procedures take 60–70 days. It is possible to improve and accelerate it by inoculating the biomass using methanogenics from the rumen.[90] In this way the aerobic pre-fermentation is reduced to 4 days and the anaerobic fermentation occurs in 15 days only. Between the two processes, the mass is inactivated by increasing the pH value to pH > 7.2.

4.6 Residue storage tank and distribution

Since the residue is not used immediately, it is transferred into the residue storage tank (Figure 4.51) at the end of the complete process of degradation. The size of the tank is usually determined by the necessary storage time given by the legal and technical requirements on the degree of fermentation the sludge needs to reach before it can be applied as fertilizer on agricultural fields.

Depending on the plant design and the storage conditions of the residue (residence time, degree of degradation, pH value, temperature, etc.) significant amounts of biogas (up to 25%) can still be generated during afterfermentation. This gas contain ammonia, odors, and H_2S besides the methane. In order to use these additional gas volumes it is recommended to design a gas-tight cover for the residue storage tank with a connection to the gasholder.

Figure 4.51 Storage tank and transportation of residue.[91]

90) Cp. PAT 1

91) Cp. WEB 36

5
Special plant installations

5.1
Combined fermentation of sewage sludge and bio waste

Experience[92] on the combined fermentation of sewage sludge and bio waste has been obtained in a plant in Germany (Figure 5.1).[93] The daily throughputs per inhabitant of processed waste water and bio waste are listed in Table 5.1.

The bio waste from different origins is collected separately and dumped together into a flat hopper from where a wheel loader transports it to a screw mill, from where it is then transported to the liquid preparation stage, consisting of two steps: dissolving the waste, and subjecting it to a treatment in a sieve drum. In this stage the bio waste is mashed by adding process water, and foreign particles (heavy- and light-weight substances) are released. All the mechanically treated and/or buffered waste is fed into the hydrolysis tank. From there the two methanation reactors, which operate in parallel, both in the mesophilic temperature range, are loaded with the substrate, always having to pass a hygienization stage in the inlet.

The generated biogas is transformed into electricity by means of two CHPs. The fermented sludge is dehydrated in decanters down to 28% DM. The residue has to be aerobically treated before it can be distributed to agriculture.

The exhaust air of the whole plant is purified and the process waste water is collected and added to the sewage water treatment plant.

The degradable load of organic dry matter as COD in the raw sewage sludge (60% of the total COD value) is the same as in the bio waste products from hydrolysis. The biomass products from hydrolysis would thus lead to a doubling of the biogas production during fermentation. The biogas yield increases in combination with co-fermentation: with 75% sewage sludge about the same volume of biogas is produced as with 4.5% fat sludge in the substrate.

The water from the sewage sludge fermentation is normally re-circulated to the waste water treatment plant. Because of the co-fermentation, the water from the fermentation step contains higher nutrients concentrations, so that a higher nitrogen elimination of 70–80% is achieved in the plant and up to 80% less precipitation agents are required, reducing the sewage sludge load by 5–10%.

92) Cp. WEB 10

93) Cp. WEB 55

Biogas from Waste and Renewable Resources. An Introduction.
Dieter Deublein and Angelika Steinhauser

ISBN: 978-3-527-31841-4

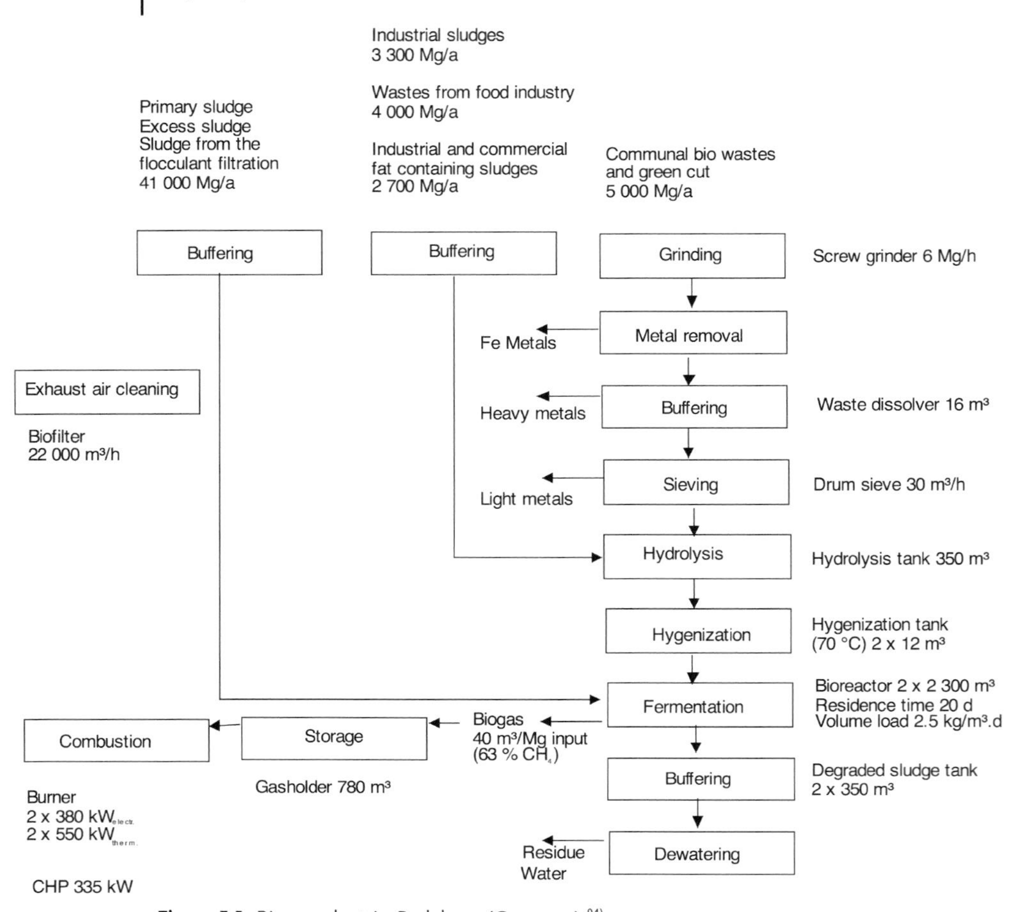

Figure 5.1 Biogas plant in Radeberg (Germany).[94]

Table 5.1 oDM_{COD} and N loads in waste water and in sewage sludge compared to possible loads in bio wastes and in the products from hydrolysis in a two-stage fermentation plant.

	Waste water $[g\ IN^{-1}\ d^{-1}]$	*Sewage sludge* $[g\ IN^{-1}\ d^{-1}]$	*Bio wastes* $[g\ IN^{-1}\ d^{-1}]$	*Products from hydrolysis* $[g\ IN^{-1}\ d^{-1}]$
oDM_{COD}	120	80	70	60
N	10	3–4	0.7	0.6

94) Cp. BRO 11

The considerable nitrogen loads in the efflux of the fermentation plants can be used selectively in the sewage treatment plant, e.g., during the night when the ammonium load of the waste water is low and as such does not have to be treated separately. By skilful control of the plant, it is possible to achieve a massive reduction of the plant's oxygen inflow.[95]

It is advisable to separate the hydrolysis of the bio waste and sewage sludge and perform only the methanation together. However, the shared methanation in the digestion tower should be accomplished only if the sludge is of very high quality, i.e., contains little heavy metal.

When sewage sludge and bio waste fermentation are combined, the whole plant can become more cost efficient, because many of the process steps like fermentation, gas exploitation, and dehydration can be carried out together and therefore only have to be up-sized.

The purified waste water of the plant can be used as cooling water for the combined heat and power station and as process water for starting the fermentation of the bio waste.

A higher degree of dehydration of the residue can be achieved when structured wastes are added, in comparison to dehydrating pure sewage water. As a consequence, the afterfermentation is much easier.

Sediments like broken brick and concrete, cullet, clay, stones, sand, etc., from bio waste preparation, which still contain some organic compounds, have to be deposited in a special waste disposal. By special treatments, it is possible to reduce the content of organic matter, so that it can be deposited in a less costly landfill.[96]

The combined treatment of bio waste and sewage sludge leads to two disadvantages, namely the formation of scum and grit layers. The hydrolysis provokes a viscosity reduction in the substrate in anaerobic processes. Consequently, more particles sediment.[97] The gases (CO_2, H_2S, CH_4) from hydrolysis simultaneously lead to scum formation by flotation. These developments have to be counteracted by mixing.

5.2 Bio waste plants

In Germany, 70% of the biogas plants for bio waste, operate based on wet fermentation with suspension of the wastes, while in 30% of the plants, stacked waste is leached by means of percolation.

Most plants work in batch mode (Table 5.2).

In all current processes, the bio waste is first accepted and then pretreated and subjected to a single-phase or multi-stage bioreactor plant. The biogas is utilized for power generation and the fermentation residue aerobically decomposed, while waste water is treated and purified (Figure 5.2).

60% of all biogas plants for bio waste are single-stage and 40% multi-stage plants (Table 5.3). All plants are equipped with one buffer tank for homogenization.

95) Cp. JOU 2

96) Landfill class I according to TASI = Technische Anleitung Siedlungsabfall

97) Cp. BOK 47

Table 5.2 Advantages of continuous and batch processes.

Continuous process	*Batch process*
High volume time yield	Cheap reactor
Normally well mixed bioreactor	All parts of the substrate are subjected to the same residence time – no risk of low hygiene
Easy to automate	Easy to operate
Continous production of consistent quality and quantity of biogas	
Low operating expenses	
Frequently a higher degree of decomposition	

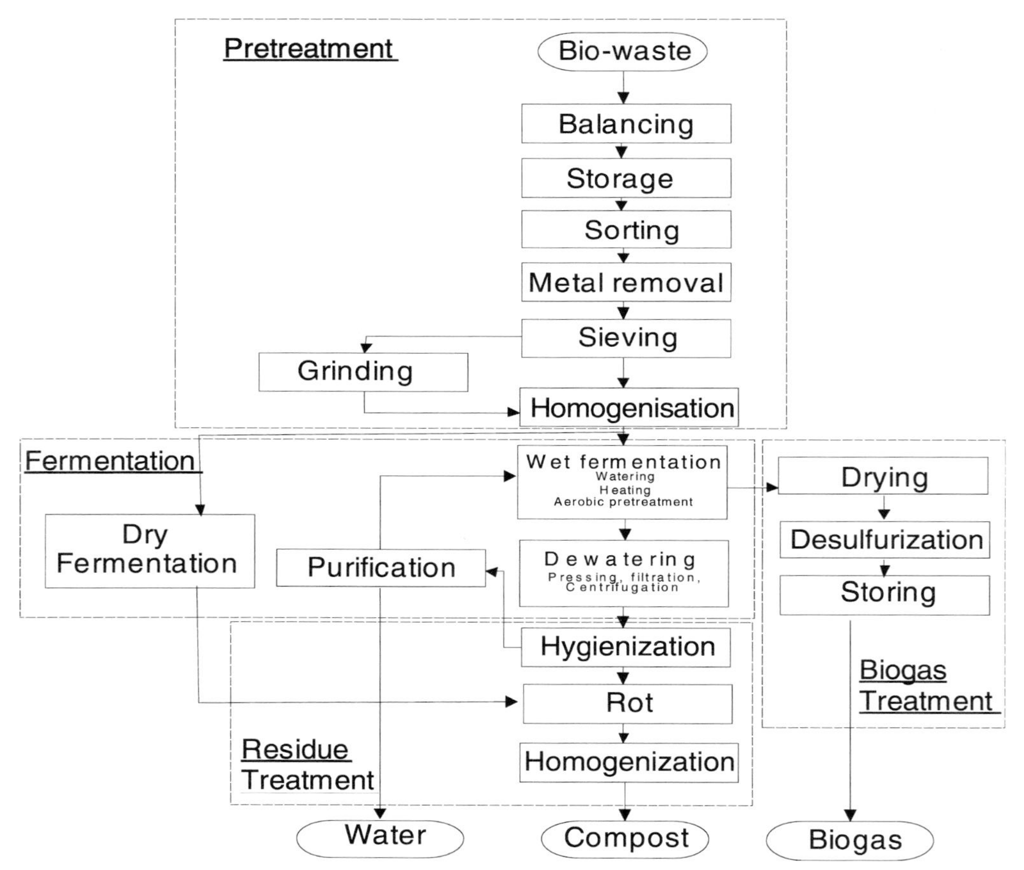

Figure 5.2 Basic flow sheet of a combined wet and dry fermentation process of bio waste.[98]

98) Cp. WEB 46

Table 5.3 Operating biogas plants for bio wastes.[99]

	Units	*Ravensburg*	*Bassum*	*Otelfingen/CH 1994*	*Hannover / Amiens*
System		Linde-BRV	Dranco	Kompogas	Valorga
Status		Pilot plant	Commercial plant	Commercial plant	Commercial plant – pilot plant for rotting
Parameters of anaerobic decomposition					
Capacity	Mg a^{-1}	n.a.	65000	n.a.	100000
Size of the communal waste	mm	<60 / <70 mm	<40 mm	<15 mm	<50 mm
Temperature		Mesophilic	Thermophilic	n.a.	n.a.
Residence time	days	26	(10)–21	15 (15–20)	n.a.
Degree of decomposition	% oDM	63–78	37	n.a.	n.a.
Bioreactor volume	M^3	n.a.	1280	n.a.	n.a.
Biogas yield	M^3/Mg TS	121–133	125	100	n.a.
Surplus of electric power	kWh_{el}/Mg	n.a.	200	135	n.a.
Waste water accumulation	L/Mg Input	n.a.	n.a.	300	n.a.
Parameter[100] of the residue					
AT_4	mg O_2/g DM	15–25	n.a.	10–25	7
TOC eluate	mg L^{-1}	200 to 300	n.a.	1000	330
Parameters of aerobic decomposition					
Amount of compost	m^3/Mg	n.a.	n.a.	0.54	n.a.
Period of aerobic decomposition	weeks	n.a.	8	3	7–8
Degree of aerobic decomposition	% oDM	12–43	ca. 20	n.a.	n.a.
AT_4	mg O_2/g DM	<5	n.a.	1–3	3
TOC eluate	mg L^{-1}	200–500	n.a.	110–170	240

99) Cp. BOK 46

100) The "breathing" oxygen consumption AT_4 is limited by law

Table 5.4 Comparison of anaerobic processes for waste water treatment.[101)]

Reactor	*Advantages*	*Disadvantages*	*$Kg_{COD}/m^3.d$*
Anaerobic lagoon	Decomposition of suspended solids over long periods Possibility to work as buffer Simple and cheap process	Requires big area required for at least 7–10 days residence time Potential heat loss Requires periodic sludge removal	<1–2
Continuously stirred reactor (CSTR) Anaerobic contact process (CP)	Long time of contact Good decomposition of the suspended material No plugging	High area and energy demand Biomass does not sediment well High amount of washout of active biomass; additional tank for sedimentation and recirculation required	1–5
Anaerobic Filter (AF)	Robust process; no negative effect of irregularities Good retainment of the microorganisms Low cost, no agitator required Plug flow	Precipitation of inorganics Filter plugging possible Short circuit stream possible High pressure drop High demand on construction Low stress load of the sludge Not suitable for high sludge concentrations	10–20
Upflow Anaerobic Sludge Blanket Reactor (UASB)	Low residence times of 48 h[102)] No plugging Natural mixing Good sedimentation of the sludge	High amount of washout of active biomass; recirculation necessary Problems when no granules appearSensitive with high concentrations of insoluble organic material	
Expanded Granular sludge Bed (EGSB)Internal circulation (IC) reactor	Effective mixing due to high velocity in the upward stream (3–10 $m h^{-1}$) Advantageous at low temperatures (5–20 °C), long-chain fatty acids and waste water containing organic poisons	High energy costs due to recirculation High amount of washout of active biomass Difficult formation of granules	

101) Cp. WEB 51

102) Cp. WEB 16

Table 5.4 *Continued*

Reactor	*Advantages*	*Disadvantages*	*$Kg_{COD}/m^3.d$*
Anaerobic Fluidized-bed Reactor	Good contact between biomass and waste water Easy and short start-up possible Insensitive to load variations	High investment and operating expenses Difficult maintenance High amount of washout of active biomass High stress load on the microorganisms Accident-sensitive with toxic feeds Adjustable velocity of sedimentation	
Anaerobic Hybrid Reactors (AHR)	Very effective due to the upward flow in the fixed bed High biomass concentration To be applied when certain miroorganisms have to be retained	Plugging and damage of the lower layers of microorganism possible Low throughput The activity of the methanogenic microorganisms in lower layers is higher than that in the upper levels Biogas must pass through the upper levels	10–20
Anaerobic Sequencing Batch Reactor (ASBR)	Low investment and operating expenses Low demand on area, technical equipment, and maintenance Low residence at high sludge concentrations Ideal for aftertreatment	Difficult sedimentation of the biomass Low volume load Batchwise operation	
Anaerobic Baffled Reactor (ABR)	Long contact time Advantageous when plug flow supports the anaerobic decomposition	Low volume load Not much experience available	

Some plants work with two bioreactors in parallel. Quite often the residue storage tank is designed as a gas-tight tank suitable for the afterfermentation.

With regard to the fermentation temperature, several options exist in industry: the systems can be operated within either a mesophilic (usually at 37 °C) or thermophilic (usually at 55 °C) temperature range. Regardless of the mode, the biogas yield and the degree of degradation of bio waste is about the same. On a scale of 1–5, the residue reaches a decomposition level of 2–3 and can be used as an excellent fertilizer after an aerobic composting.

Table 5.5 Biogas production from industrial plants[103] – collected values (the size of the reactor is given in m^3 volume per m^3 h^{-1} throughput of waste water).

Type of industry	*Waste water with degradable substances*	*Reactor*	*Residence time [d]*	*Degree of decomposition [%]*	*Volume load [$kg_{COD}/m^3.d$]*
Sugar beet processing	Washing, pressing, and other waste water	UASB 0.23 $m^3.h/m^3$	n.a.	n.a.	9.4–10.0
Sugar production[104]	Alcohol	Completely mixed UASB, 0.23 $m^3.h/m^3$	5	97	2.14
			0.55	87	8.18
			0.75	90	6.9
			0.10	90	40
					8.4–9.4
Milk processing	Whey from processing	n.a.	n.a.	n.a.	n.a.
Distilleries	Mash Yeast	Completely mixed contact, anaerobic filters	17	57	1.6
			8	84	2.75
			7.2	77	6.6
Slaughterhouses	Waste water with rumen content	n.a.	0.6	65	6
Fish processing		n.a.	n.a.	n.a.	n.a.
Fruit juice	From washing, trasportation, sludges	0.155 $m^3.h/m^3$	n.a.	n.a.	13.85
Pectin production	Starch (wheat, maize, potato)	Contact UASB	3.8	72	2.38
			0.8	95	12
Breweries, malthouses	From washing, cleaning, and cooling	Contact UASB 0.008–0.064 $m^3.h/m^3$	4.8	80	13.5
			0.45	85	6
Potato industry	From washing, transportation, pressing, cleaning	UASB 0.040–0.125 $m^3.h/m^3$	0.33	80	6
					8.1–16.9
Cheese factories		Anaerobic filters	5.6	85	8

103) Cp. BOK 16

104) Cp. BRO 12

Tinned food industry	From washing, bleaching of fruits or vegetables	Contact	9.52	85	1.37
Vineyard	Mesh	Completely mixed UASB	15	81	2.5
Citric acid production[105)]		n.a.	n.a.	n.a.	n.a.
Yeast		Fluidized bed	0.5	77	10
		$0.034–1.250\,m^3.h/m^3$	0.12	75	29
Enzyme		Anaerobic	2.3	78	6.4
Fat, oil, and margarine[106)] production	From washing, extraction, filtration	n.a.	n.a.	n.a.	n.a.
Tea and coffee production	Residues from extraction	n.a.	n.a.	n.a.	n.a.
Pharmaceutical and cosmetics industries	Culture broth Mycelium sludge Washing water	n.a.	n.a.	n.a.	n.a.
Paper and board industry[107)]	From production of pulp, paper, waste paper, board	Completely mixed UASB	10	97	1.75
		$0.034–0.300\,m^3.h/m^3$	0.2	70	7
		$0.057–0.200\,m^3.h/m^3$			7.9–12.5
					9.3–14.7
Wood processing	From pulp and hard board production	n.a.	n.a.	n.a.	n.a.
Leather production		n.a.	n.a.	n.a.	n.a.
Wool laundry		n.a.	n.a.	n.a.	n.a.
Insulin production		$0.193\,m^3.h/m^3$	n.a.	n.a.	13.7

105) Cp. BOK 70
106) Cp. JOU 19
107) Cp. BOK 38

In Table 5.4 different technologies for anaerobic waste water treatment are compared. Table 5.5 shows some applications.

5.3
Purification of industrial waste water

5.3.1
Process engineering and equipment construction

During the past 15–20 years, such reactors as the contact process reactor, the UASB reactor, anaerobic filters, and fluidized-bed reactors, were most preferred. Nowadays, this is changing, since more and more often technologies like the hybrid reactor, the ESB reactor, or the IC reactor are used.

The ASBR reactor or ABBR reactor, used for the purification of industrial waste water, is still under development.

5.3.2
Plants for industrial waste water fermentation

In the past the anaerobic purification of industrial waste water was not widespread and was only applied quite rarely, but, today, there is a trend toward a more frequent implementation of this technology.

Usually the UASB reactor is used for substrates with a low waste water load of about $oDM_{COD} = 1\,g\,L^{-1}$. Especially the waste water from the paper, brewing, or potato processing industries meets this criterion and is treated in such reactors provided it is free of floating solids, oils, or fats.

Part VI
Biogas to energy

Biogas is a promising renewable source of energy. It can be directly converted into electrical power, e.g., in a fuel cell. It can be burnt, releasing heat at high temperature. It can be burnt in a CHP for the simultaneous production of heat and power. Finally, it can be fed into the natural gas network for energy saving purposes or it can serve as fuel for vehicles, being distributed by gas stations.

Often the biogas has to be transported over long distances and has to be purified before it can be further utilized.

Biogas from Waste and Renewable Resources. An Introduction.
Dieter Deublein and Angelika Steinhauser

ISBN: 978-3-527-31841-4

1
Gas pipelines

Biogas pipelines are subject to the same regulations as those for natural gas,[1] e.g., branches attached to the main biogas pipes must always be smaller in diameter than the main pipes. The flow rate should not exceed $20\,m\,s^{-1}$ at pressures above 16 bar in branch pipes or $5\,m\,s^{-1}$ at pressures below 16 bar. Under certain conditions, long-distance pipelines have to be heated, or the gas which is to be transported has to be preheated. Otherwise the temperature would fall below the dew point, because the gas temperature drops about 0.4–0.6 °C per bar pressure drop in the pipeline and corrosive condensate would develop.

If gas pipes are embedded in the soil, they must be absolutely corrosion-resistant toward all gas components. They must withstand all influences from the outside, e.g., corrosion in contact with soil or liquids or, e.g., pressure from passing trucks. Furthermore, the pipes must tolerate internal pressure to the extent that welding seams do not break and leak.

Biogas pipes can be of steel, high-grade steel, brass, red brass, polyethylene (PE hp), or PVC. High-grade steel like 1.4571 should be preferred to normal stainless steel, because attachments of sulfur could result in pitting corrosion. Chloridic corrosion has also to be avoided. Copper is not suitable as pipe wall material, because it is not resistant against ammonia in the biogas. Commercial PVC pipes are not permitted, since they could break. Normal plastic pipes are not sufficienty corrosion-resistant. They are only allowed to be mounted outside closed areas below and above ground if they are protected against mechanical and thermal damage. Vertically mounted plastic pipes may only be attached to walls if these are massive constructions. Welded steel pipes are recommended. The welds have to be done by certified welders. The steel pipes often are lined inside with an epoxy resin layer to prevent corrosion. Outside they are protected with a bitumen layer (4–6.5 mm thick). After welding the removed layer has to be repaired perfectly.

When transporting humid biogas, the pipes must be protected against freezing and must be equipped with condensate traps at their lowest point. All gas pipes have to be mounted slightly inclined to enable drainage. Exhaust gas pipes must diccharge at least 0.5 m above the highest point of the building.

Pipes have to be painted in accordance with the local regulations. In Germany, for example, biogas pipes have to be painted yellow according to the DIN 2403

1) Cp. LAW 16

Biogas from Waste and Renewable Resources. An Introduction.
Dieter Deublein and Angelika Steinhauser

ISBN: 978-3-527-31841-4

standard. In all trenches a warning plastic band has to be placed above the biogas pipe.

Valves, safety devices, sight glasses, covers for openings, and other components exposed to the biogas must be designed according to the state-of-the-art and must have been approved by the authorities. They must be operable from a safe place under safe conditions only by authorized people. Condensate traps and safety devices such as over/negative pressure safety devices must always be accessible.

The quality-controlled manufacture and suitability of all valves and other equipment built into the pipeline must be certified by the deliverer. Pipe joints must be installed longitudinally force-locked.

2
Biogasholder

In principle, biogasholders cannot be too large. The larger the gasholder volume, the better the variation in the biogas production can be balanced, and the less biogas has to be burnt in the flare, i.e. the smaller is the energy loss. But biogasholders are very expensive and have their limits. Therefore, in agricultural biogas plants, the biogasholders are in general small. Often recommended are biogasholders which can take the daily biogas rate of the plant in order to prevent trouble in unexpected situations.

In sewage treatment plants, the size of the sewage gasholder varies normally between 0.75 and 1.5 times the daily produced biogas rate. The smaller value is applied to larger plants. But the actual size depends on the utilization of the sewage gas.

If the biogas is used only for the heating of the bioreactor and other tanks of the plant, a biogasholder is not a real essential. When the gas is used in a CHP and the produced current is fed into the power network, the biogasholder should be designed to take half of the average daily gas production rate. When the power consumption of the biogas plant itself is mainly covered by the produced current, the biogasholder should take up three quarters of the daily accumulated biogas. Only when the biogas power station is used to cover peak loads must the biogasholder be able to store the daily biogas output.

Biogasholders (pressure <0.1 bar) in general have to be installed,[2] maintained, and operated so that the safety of the operators and other personnel is assured (see Part IV).

2.1
Biogasholder types

The following classes of biogasholders can be distinguished (Table 2.1).

2.1.1
Low-pressure biogasholder

Most (i.e. 80%) of the installed biogasholders are of the low-pressure type. Many consist of a biogas bag of plastic foil. Others are constructed of steel.

2) Cp. LAW 16

Biogas from Waste and Renewable Resources. An Introduction.
Dieter Deublein and Angelika Steinhauser

ISBN: 978-3-527-31841-4

Table 2.1 Categories of gasholders.

Pressure		*Usual sizes [m³]*	*Design*
Low pressure	10–50 mbar	5–2	Water cup gasometer
	0.05–5 mbar	10–2000	Bioreactor cover or biogas bag of foil
Medium pressure	5–20 bar	1–100	Steel pressure tank
High pressure	200–300 bar	0.1–0.5	Steel flasks

The plastic foils used are manufactured from plastics resistant to UV, weather, fungus, microbes, and biogas, e.g., tearproof, high-strength, on both sides PVC-coated polyester tissues. They have high break stability and are extremely durable.

Sometimes the bags are just laid on a baseplate beside the bioreactor. Simple canopies prevent damage to the plastic foils and assure long-lasting weather protection. Preferably, the bags are protected on all sides in a special steel tank, a wooden housing, or equivalent. Often plastic covers on top of the bioreactors are used as gasholders.

Plastic bags have the disadvantage that they can easily be perforated and have often to be repaired.

Some biogasholders consist of special constructions which give the gasholders their permanent form.

Double-membrane biogasholders

Double-membrane biogasholdes for low-pressure storage consist of three high-strength, fabric-strengthened membranes. They are fastened to a reinforced concrete foundation with a hot-dip-galvanized steel ring. The bottom membrane seals the gasholder to the foundation. The interior membrane takes up the biogas, being more or less strained depending on the amount of biogas. The exterior membrane protects the interior membrane and is held under constant tension by air, which is blown into the space between interior and exterior membranes by an explosion-proof ventilator. Thus, a strong, rigid outer skin is formed, which resists the weather and stabilizes the construction. An over-pressure safety valve protects the biogasholder against positive pressure.

If the double membrane biogasholder is installed as a cover on top of a bioreactor, a timber construction or a wire net is necessary above the bioreactor, so that the interior membrane cannot fall into the reactor and cannot be damaged by the agitators or other equipment.

Biogas bags

The plastic biogas bag is sometimes installed in a frame with guide bars, so that the gas bag can be loaded by weights in order to enable smooth filling and

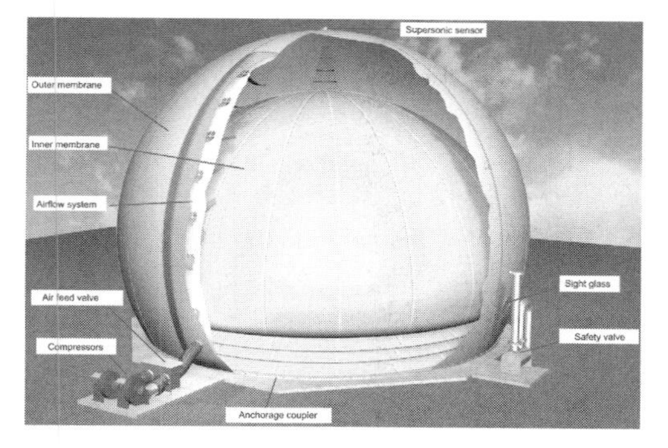

Figure 2.1 Gasholders.
Double membrane gasholder (pat.)[3] (top left); Bag bio gasholder (top right)
Enclosed theomoplastic foil gasholder (middle left); Theomoplastic foil gasholder on a bioreactor (middle right)
Cube gasholder (below left); Steel tank gasholder (below right)

3) Cp. WEB 22

emptying. Such gas bag installations, called gas cushions, can be provided with up to 5 cushions, one above the other, of 25 m^3 volume each.

Biogas bags benefit if the interior gas rate and thus the gas pressure does not vary too much.

In the hanging biogasholder, with a volume of 30–2000 m^3, the bag is hung in a special device to the top of a steel tank and is stretched by a carrying ring, so that it can be blown up and collapsed smoothly under all operational conditions.

Other low-pressure biogasholders

A simple biogasholder consists of two open cylindrical steel tanks, one fitting into the other. The tanks are connected by a gastight membrane, so that the volume of the tank can vary. The upper half can sink into the lower half when the tank is emptied.

The floating-dome biogasholder is formed by inverting an open plastic or steel drum above another tank filled with water, so that the edge dips into the water. When the biogasholder is empty, the edge dips deep into the water, otherwise only a little. If the biogasholder is integrated in the bioreactor, the substrate functions as sealing liquid.

2.1.2
Medium- and high-pressure biogasholders

Biogasholder of medium pressure are smaller in size but require increased operating expenditure for the compression and expansion of the biogas.

Pressurized biogasholders are made of steel and are subject to special safety requirements, because of the explosion risk. Small pressurized tanks are cylindrical and suitable also for high pressures.

Pressurized ball gas tanks of steel are to be found in sewage sludge fermentation plants.

2.2
Gas flares

A biogas flare (Figure 2.2) prevents the escape of unburnt biogas into the atmosphere and therefore is essential for environmental protection and is even prescribed by law in many countries. Of course, the biogas flare should be used as seldom as possible, since otherwise usable energy is lost. It is preferable, for example, to switch the biogas flow over to another consumer.

The gas pipe connecting the flare to the biogas plant must contain the following devices: an isolation device operated manually, an emergency shut-down device which interrupts the biogas feed automatically, a flame trap as a safety device, and at the flare an automatically operating ignition mechanism, flame control equipment, and lightning protection.

Figure 2.2 Gas flare for high temperature (1050°C) with hidden flame.

The flare must be installed at a height such that its opening is at least 4 m above ground and at least 5 m away from buildings, traffic ways, and storage for inflammable materials, and is outside defined zones for explosion protection.

Either after or before the biogas is stored in a biogasholder it has to be prepared in many cases.

3
Gas preparation

Biogas is not absolutely pure, but contains droplets, dust, mud, or trace gases. All this contamination has to be removed, depending on the further utilization of the biogas.

Solid particles in the biogas and sometimes oil-like components are filtered out of the biogas with the usual dust collectors.

Sludge and foam components are separated in cyclones. The separation can be improved by injecting water into the biogas before the cyclone. Process water can be used.

Following such equipment, gravel pots (used as prefilters and for dehydrating) and cartridge filters (fine filters) are often used for the removal of coarse contamination in the biogas from the bioreactor. The gravel pots separate pollutants at their internal surfaces and serve at the same time as flame traps.

For the removal of trace gases, scrubbing, adsorption, absorption, and drying are applied among other techniques. Since trace gases in the biogas impair the lifetime of catalysts severely, catalysts are less suitable for the decontamination of biogas.

Table 3.1 indicates some requirements on the purity of the pretreated biogas. The values originate from standards and regulations or are the result of detailed attempts and tests. For the case, that the biogas is just burnt, e.g., in a gas burner, no obligations exist for the purity of the biogas. But the exhaust air behind the burner might to be decontaminated.

The removal of trace gases is carried out stepwise:[4)]

1. Rough separation of hydrogen sulfide in the bioreactor or a separate scrubber
2. Removal of traces of hydrogen sulfide
3. Separation of carbon dioxide and other biogas components
4. Dehumidification (if the carbon dioxide removal is a dry gas process, drying must be carried out before step 3).

The rough desulfurization and dehumidification steps are conducted in nearly all biogas plants. The methane enrichment by the removal of carbon dioxide and

4) Cp. WEB 27

Biogas from Waste and Renewable Resources. An Introduction.
Dieter Deublein and Angelika Steinhauser

ISBN: 978-3-527-31841-4

Table 3.1 Biogas qualities required for different applications in Europe.[5]

Gross calorific value and gas components	*Gas motor* [6),7)]	*Fuel cellMCFC*[8),9)]	*"Green gas" for vehicles ISO/ DIS 15403*	*Addition to natural gas according to DVGW G 260*
Gross calorific value	n.a.	n.a.	No minimum value	8.4–13.1 kWh/m^3
CH_4	Minimum ca. 430 mg/Nm^3 (60% by volume)	n.a.	>96%	No minimum value
Hydrogen sulfide H_2S	<200 mg/Nm^3 (0.013% by volume)	<0.1 mg/Nm^3	≤5 mg/Nm^3	<5 mg/Nm^3
Total sulfur without odorizing agents	<2200 mg/Nm^3*	0.1 mg/Nm^3	<120 mg/Nm^3	≤30 mg/Nm^3
COS	n.a.	<0.14 mg/Nm^3	n.a.	n.a.
Mercaptan sulfur	n.a.	n.a.	<15 mg/Nm^3	≤6 mg/Nm^3
CO_2	ca. 60 mg/Nm^3	n.a.	<3%	No upper limits
O_2	n.a.	<1%	<3%	≤3% dry net/≤5% humid net
Hydrocarbons	n.a.	n.a.	<1%	<Dew point (at the relevant pressure/ temperature)
Water	<80% rel. humidity	Dew point <15 °C	<0.03 g/Nm^3	<Dew point (at the relevant pressure/ temperature)
Oil vapors (<C10)	3000 mg/Nm^3*	n.a.	<70–200 mg/Nm^3	n.a.
Oil vapors (>C10)	250 mg/Nm^3*			
Glycol/methanol	n.a.	n.a.	Technically free	n.a.
Dust	<10 mg/Nm^3*		Technically free <1 μm	Technically free
Particle size	3–10 μm	1 μm size		
NH_3	<30 mg/Nm^3*	<400 $mg\,m^{-3}$	n.a.	n.a.
Polysiloxanes[10)]	0.2 mg/Nm^3	n.a.	n.a.	n.a.
Chlorine	100 mg/Nm^3*	0.1 mg/Nm^3	n.a.	n.a.
Fluorine	50 mg/Nm^3*	0.01 mg/Nm^3	n.a.	n.a.
Heavy metals	n.a.	0.1 mg/Nm^3	n.a.	n.a.
CO	n.a.	n.a.	n.a.	n.a.
Hg	n.a.	30–35 mg/Nm^3	n.a.	n.a.

* This is related to the CH_4 content.

5) Cp. BOK 50
6) Cp. WEB 105
7) Cp. WEB 105
8) Cp. WEB 30
9) Cp. BOK 44
10) Cp. WEB 25

other biogas components is only necessary if the biogas is to be fed into natural gas and/or used as fuel for vehicles.

3.1
Removal of hydrogen sulfide

Hydrogen sulfide in the fermentation gas impairs the lifetime of pipework and all installations for the utilization of biogas. It is toxic and strongly corrosive to many kinds of steel.

When the hydrogen sulfide-containing biogas burns it is converted into sulfur oxides, which on the one hand corrode metallic components and on the other hand acidify the engine oil, e.g., of the engine in the CHP.

In order to prevent damage of the CHP and other equipment, e.g., heat exchangers and catalysts, hydrogen sulfide must be removed from the biogas or at least reduced.[11),12)]

For trouble-free operation of the CHP, limit values of 100–500 mg/Nm^3 H_2S may not be exceeded (equal to 0.05% by volume), depending on the recommendations of the manufacturer of the CHP. Short peak loads above these limits can occasionally be accepted. In general, low values affect the lifetime of all plant positively.

For the removal of hydrogen sulfide from biogas, different chemical, physical, and biological procedures are possible. The target is a residual content of 20 mg/Nm^3 hydrogen sulfide in the decontaminated biogas. This value is difficult to achieve with biological and physicochemical procedures.[13)] Therefore, combinations are frequently used, e.g., a biological process for rough gas cleaning and adsorption for final cleaning.

Generally, the desulfurization procedure should be always selected in relation to the H_2S concentration, the mass flow of sulfur, and also the possibilites for disposal of the residues from the gas cleaning. Table 3.2 offers a decision-making aid.

3.1.1
Biological desulfurization

Biological desulfurization is mostly applied for the reduction of the H_2S content. The process is effective at concentrations up to 3000 mg Nm^{-3}.[14),15)] In the biological desulfurization process, the hydrogen sulfide is absorbed in water and then decreased biologically. Microorganisms of the species *Thiobacillus* and *Sulfolobus*, which are omnipresent and therefore do not have to be inoculated especially, degrade the hydrogen sulfide.

11) Cp. BOK 49
12) Cp. BOK 71
13) Cp. PAT 4
14) Cp. DIS 9
15) Cp. BOK 63

Table 3.2 Decision-making aid for the selection of procedures for the removal of hydrogen sulfide (++ very suitable, + suitable, – less suitable, – not suitable; comment in parentheses means under certain conditions).

Technology	*Investment costs*	*Operational costs*	*Air intake to the biogas required*	*Rough/fine decontamination*	*Remarks*
Internal biological desulfurization	–	–	Yes	Very rough	Low dynamic in the change of load, corrosions hazard in the bioreactor
Percolating filter plant	+	–	Yes	Rough	Blocking hazard at low air intake
Bioscrubber plant	++	–	No	Rough	
Sulfide precipitation	+	++	No	Rough	
Ferric chelate	(+) ++	–	Yes	Rough	
($Fe(OH)_3$ – bog iron ore)	(+) ++	+	(Yes) No	Fine	Fire hazard
Fe_2O_3	(+) ++	+	(Yes) No	Fine	Fire hazard High costs of disposal High costs of disposal
Activated carbon – KI, K_2CO_3, $KMnO_4$	–	++	No	Fine	Removal as hazardous waste
Zinc oxide	–	++	No	Fine	High process temperature
Surfactant	+	+	No	Fine	Only for small gas rates
Absorption at glycol and ethanolamine	++	++	No	Fine	
Algae	–	+	Yes	Fine	
Direct oxidation	+	++	Yes	Fine	Lifetime of the catalyst is short

The procedure results in biogas which is roughly desulfurized but still suitable to be burnt in gas engines or in agricultural biogas plants. In plants where bio wastes are fermented, it may be possible to remove only 50% of the hydrogen sulfide.

The decomposition of H_2S to form sulfate and/or sulfur occurs according to the equations:

$$2\,H_2S + O_2 \rightarrow 2\,S + 2\,H_2O$$

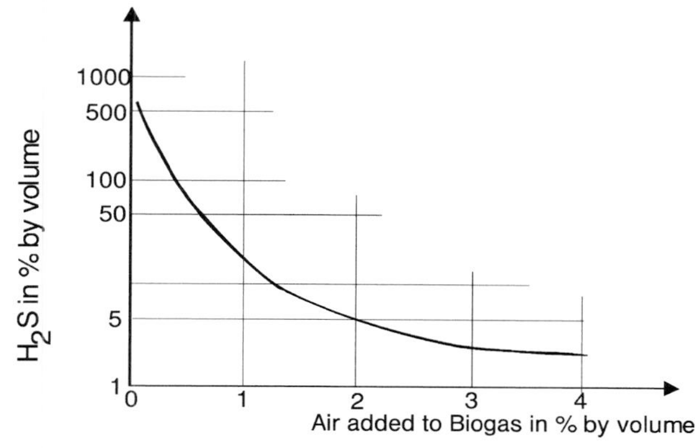

Figure 3.1 Correlation between hydrogen sulfide content in biogas and air flow into the bioreactor.

$$2\,S + 2\,H_2O + 3\,O_2 \rightarrow 2\,H_2SO_4$$

The direct reaction of H_2S to sulfate is also possible:

$$H_2S + 2\,O_2 \rightarrow H_2SO_4$$

Usually 75% of the H_2S in biogas is biologically decomposed to pure sulfur and the rest to sulfate. If sulfur-producing organisms dominate the decomposition, then a pH value of ca. 7 is generated in the proximity of these microorganisms. But if microorganisms which produce sulfuric acid prevail, a pH value of 1–2 can be reached.

For these reactions, the microorganisms need carbon and inorganic salts (N, P, K) as nutrients as well as trace elements (Fe, CO, Ni) in addition to H_2S. These nutrients are to be found in substrates, mostly in sufficient quantity.

The H_2S-attacking microorganisms are aerobic and therefore require oxygen. Air added at a rate of 4–6% of the biogas is sufficient (Figure 3.1). Because of the risk of explosion, the air dosage must be limited, giving a maximum air concentration of 12% by volume in the biogas. The air dosage is to be fixed in such a way that even with a malfunctioning of the air rate adjustment no higher rates of air can be supplied. In the air inlet to the gas space a non-return valve is necessary.

The microorganisms also need sufficient surface (which is moistened with substrate) for immobilization. About $1\,m^2$ surface for the desulfurization of $20\,m^3\,d^{-1}$ biogas at 20 °C is necessary.

Immobilization in the bioreactor

In an easy and economical manner, the biological H_2S-degradation can be accomplished by achieving immobilization inside the bioreactor on wall and cover surfaces, which are above the substrate level. Particularly in smaller agricultural plants this kind of immobilization is preferred. If this surface is not sufficient, plates or cloths have to be hung in the upper space of the bioreactor.

With this kind of immobilization, the injected air certainly causes disturbances of the fermentation process, because the methanogens are obligatory anaerobic, so there can be lower biogas yield and formation of new H_2S, because the sulfur remains in the bioreactor.

Therefore, larger plants with a capacity above 200 kW_{el} are equipped with trickling filters or scrubbers.

Trickling filter

In trickling filter installations, the biogas is fed into an external container which is filled with packing materials and through which nutrients diluted in water are circulated. The hydrogen sulfide-attacking microorganisms live in the circulated nutrients and are immobilized on the surface of the packing material. Under certain conditions, biogas decontamination is facilitated by inoculating sewage sludge or specially bred cultures.

The contaminated biogas is first mixed with 8–12% air and then fed into the trickling filter from the bottom, so that it flows upwards through the packing material.

The circulation of the nutrient solution can occur either continuously or at intervals. With continuous flushing, the optimum rate of nutrient solution is 0.5–0.75 $m^3/m^2.h$. In order to prevent high concentrations of sulfur or sulfate in the system, particularly with long running facilities, the sulfur has to be rinsed and the nutrient solution has to be continuously replaced. The removed sulfur-containing nutrient can be used as fertilizers.

In trickling filters, up to 99% of the sulfur can be removed, and a biogas purity of less than 75 mg/Nm^3 sulfur is achievable.

The installation of trickling filters is obviously more expensive than immobilization in the bioreactor, but less so if additional surfaces have to be inserted into the bioreactor. The advantage of the trickling filters is the steadier operation, because contact of oxygen with the methanogens is completely avoided. A real alternative to trickling filters is scrubbers.

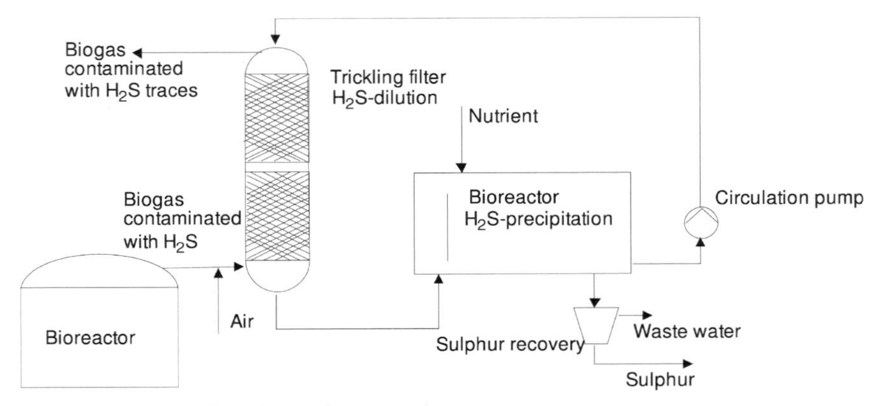

Figure 3.2 Trickling filter plant with separate bioreactor.

Bioscrubber

Bioscrubbers consist of two columns, through which diluted caustic soda at a concentration of ca. 20% and containing the required nutrients is circulated. Through one column flows the biogas, leaving its H_2S content with the caustic soda by forming sodium sulfide or sodium polysulfide.

$$2\,NaOH + H_2S \rightarrow Na_2S + 2\,H_2O$$

or by forming NaHS

$$H_2S + NaOH \rightarrow NaHS + H_2O$$

Through the other column flows air, so that the immobilized microorganisms remove the H_2S from the caustic soda and regenerate the brine.

$$2\,NaHS + O_2 \rightarrow 2\,NaOH + 2\,S$$

$$2\,NaHS + 4\,O_2 \rightarrow NaHSO_4$$

The elementary sulfur settles and is withdrawn at the bottom of the column.

Because of the sulfate formation, the caustic soda must be partly removed and neutralized, in order to prevent the acidification of the scrubber.

Bioscrubber plants are particularly suitable to decontaminate biogas which carries very high hydrogen load of up to 30 000 mg m^{-3} at a medium to high rate. When the biogas does not contain too much H_2S, low concentrations like 75–150 mg/Nm^3 hydrogen sulfide are achievable in the purified biogas. Disadvantageous are the high costs of the two scrubbers, and the fact that only very soluble contaminants can be removed from the biogas as well as hydrogen sulfide.

Favorable conditions for the removal of hydrogen sulfide are temperatures at 28–32 °C and humidity. Therefore, bioscrubbers as well as trickling filters should be heated in winter and cooled in summer.

Of course, the problem of corrosion arises in bioscrubbers and also in trickling filters due to the low pH value. This problem can be handled as follows:

- by selecting species of microorganisms which form sulfur as a product of their metabolism. The sulfur settles down in the bioreactor and is to be found in the residue, whereby its fertilizing value increases and the otherwise possible problem of disposal of the residue is avoided.
- by constructing the bioscrubber or the trickling filter from glass-fiber reinforced plastic and by using packing materials of polypropylene.

3.1.2 Sulfide precipitation

For the fixation of sulfur, Fe^{2+} ions in the form of iron(II) chloride ($FeCl_2$) or Fe^{3+} ions in the forms of iron(III) chloride or iron(II) sulfate are admitted, what leads to the precipitation of stable iron(II) sulfide and sulfur, which remain in the residue.

$$Fe^{2+} + S^{2-} \rightarrow FeS$$

$$2\,FeCl_3 + 3\,H_2S \rightarrow 2\,FeS + S + 6\,HCI$$

For sulfide precipitation, only an additional mixing tank and a dosing pump are necessary. However, fresh iron salt must be constantly admitted, which costs ca. 100.00 US$ per Mg.

The process is applied predominantly in sewage water treatment plants. If 3–5 g $FeCl_2/m^3$ is added to waste water, the content of hydrogen sulfide can be lowered at best to less than 150 mg/Nm^3.

For the decontamination of biogas, however, considerably higher $FeCl_2$ dosages are necessary. In a co-fermentation plant with a yearly throughput of 3300 Mg substrate (two thirds is liquid manure from pigs, one third is co-substrates like flotates from grease removal tanks, mayonnaise, or gelatine), the hydrogen sulfide content in the contaminated biogas of 3400 mg/Nm^3 can be reduced to less than 30 mg/Nm^3 by a 4.2-times stoichiometric overdosing of iron(II) sulfate.[16] But it also shows that H_2S concentrations can be achieved far below the required values indicated in Table 5.1.

The procedure is therefore often applied despite the high operating cost for the iron salt.

3.1.3
Absorption in a ferric chelate solution

In ferric chelate solutions iron(III) ions (Fe^{3+}) are reduced to iron(II) ions (Fe^{2+}),[17] a process in which hydrogen sulfide is oxidized to elementary sulfur.

$$2\,Fe^{3+} + H_2S \rightarrow 2\,Fe^{2+} + S + 2\,H^+$$

The equipment consists of a vessel containing the solution of the chelating agents or ligands with iron(III) ions at a concentration of 0.01–0.05% by weight, into which the biogas and some air are injected.

If the biogas contains some oxygen for the regeneration of the Fe^{3+} ions, one container is enough for the solution. If oxygen cannot be accepted in the biogas, a second container is required through which the solution is circulated. In the first container the biogas is desulfurrized. In the second container the solution of Fe^{3+} ions is regenerated by injected air.

Chelate is needed so that the iron(II) ions do not react spontaneously to iron sulfide and/or iron hydroxide.

The elementary sulfur concentrates at the bottom of the container and has to be removed from there from time to time.

16) Cp. BOK 37

17) Cp. DIS 11

The chelate, which is regenerated by converting iron(II) ions to iron(III) ions by oxygen and water, can be further used for precipitations.

$$4\,Fe^{2+} + O_2 + 4\,H^+ \rightarrow 4\,Fe^{3+} + 2\,H_2O \text{ (Oxidation)}$$

The process is known as the LO CAT process.[18] Most of the hydrogen sulfide (99.9%) can be removed with it when the sulfur content of the biogas is high (in the range of 500–30 000 mg/Nm^3. Mercaptans can sometimes be removed, but not COS and/or CS_2.

If too much air is injected, thiosulfates are formed and the pH value sinks. Dependent on the CO_2 content of the biogas, it is possible that carbonates will result. In both cases, neutralization of the solution could be necessary before circulating it.

3.1.4
Adsorption at iron-containing masses

In this procedure, hydrogen sulfide is adsorbed onto iron(III) hydroxide ($Fe(OH)_3$), also known as bog iron ore, and/or adsorbed onto iron(III) oxide (Fe_2O_3). Both procedures run similarly and are dry desulfurization processes. For the preparation of sewage gas and biogas desulfurization by adsorption onto iron(III) hydroxide, ($Fe(OH)_3$) is preferred.

H_2S is converted at the iron(III) hydroxide surface or iron(III) oxide surface to iron(III) sulfide and water

$$2\,Fe(OH)_3 + 3\,H_2S \rightarrow Fe_2S_3 + 6\,H_2O$$

or

$$Fe_2O_3 + 3\,H_2S \rightarrow Fe_2S_3 + 3\,H_2O$$

Ferric oxide or hydroxide masses are stacked layer by layer in a tower desulfurizer either as impregnated steel wool or as impregnated wooden chips or wooden pellets. The biogas is fed at the bottom at low pressure and a temperature of 15–50 °C. It should be humid, so that it does not carry water from the mass.

For the regeneration of the iron(III) hydroxide air is admitted to the biogas, so that the iron(III) sulfide reacts with oxygen and water and elementary sulfur is formed.

$$2\,Fe_2S_3 + 3\,O_2 + 6\,H_2O \rightarrow 4\,Fe(OH)_3 + 6\,S$$

For continuous running with high gas flow rates, e.g., 200 m^3 h^{-1}, at least two tower desulfurizers should be operated working alternately.

18) Cp. WEB 118

Table 3.3 Characteristic figures for a desulfurization tower.[19]

Biogas rate	*Ca. 1000 m^3/day*
H_2S content before the tower	1–2 g m^{-3}
H_2S content after the tower	0–20 mg m^{-3}
Ferric oxide consumption	2.2 $m^3 a^{-1}$
Process air consumption	1.0 $Nm^3 h^{-1}$
Diameter of the tower	1–3 m
Height of the tower	2–3 m
Material	1.4571
Pressure drop	25 mbar
Operational costs	13 US$/1000 Nm^3 biogas

With each regeneration, sulfur is deposited at the surface of the iron(III) hydroxide, so that the surface area is reduced. This is available for the decomposition of the hydrogen sulfide, and the iron(III) hydroxide is slowly made ineffective. The iron(III) hydroxide can be loaded up to 25% by weight with sulfur. Wooden chips can be loaded maximally with 200 g H_2S and wooden pellets with 500 g H_2S per 1 kg. The regeneration of the iron(III) hydroxide can be carried out at most ten times, and the regeneration of iron(III) oxide maximally three times. Afterwards the tower filling has to be renewed.

The adsorption process is used at H_2S contents of 150–7500 mg Nm^{-3} in the biogas. An H_2S concentration of less than 1.5 mg Nm^{-3} in the decontaminated biogas can be achieved if the crude biogas does not contain too much impurities. Otherwise an H_2S concentration of less than 150 mg Nm^{-3} can be guaranteed (see Table 3.3).

For the regeneration it should be noted that the oxidation occurs exothermically and that the adsorbent can catch fire when it is well saturated with sulfur.

If iron hydroxide sludges are used, hydrochloric acid can be formed, which must be neutralized in the discharge.[20]

3.1.5 Adsorption on activated charcoal

Molecular hydrogen sulfide adsorbs at the surface of activated charcoal if the biogas is oxygen-free and is of medium or high H_2S concentration. However, the efficiency of decontamination is generally not sufficient.

Therefore, the activated charcoal is impregnated with catalysts, whereby the reaction rate of oxidizing the H_2S to elementary sulfur is increased. Several impregnating agents are available.

Impregnation with potassium iodide (KI) at a concentration of 1–5% by weight based on the activated charcoal is reasonable only in the presence of oxygen and

19) Cp. WEB 53

20) Cp. JOU 12

water. H_2S dissolves in the water layer on the activated charcoal and reacts there with oxygen at low temperatures (50–70 °C) and an operating pressure of 7–8 bar.

$$2\,H_2S + O_2 \rightarrow 2S + 2H_2O$$

The catalyst potassium iodide (KI) also prevents the formation of sulfuric acid, since the oxidation potential for this reaction is too low.

Biogas can be desulfurized by this procedure to a value of less than $5\,mg\,Nm^{-3}$.

When impregnating with potassium carbonate (K_2CO_3) at a concentration of 10–20% by weight based on the activated charcoal, potassium sulfate is formed in the presence of water vapor and oxygen at a temperature above 50 °C. The potassium carbonate adsorbs at the activated charcoal.

$$H_2S + 2\,O_2 + K_2CO_3 \rightarrow CO_2 + H_2O + K_2SO_4$$

Another approved impregnating agent is potassium permanganate ($KMnO_4$).

The sulfur adsorbs at the activated charcoal. Loadings of up to 150% by weight sulfur based on the unloaded activated charcoal are possible. The loaded activated charcoal must be replaced and disposed of. Regeneration with hot gas and/or with superheated steam at temperatures above 450 °C is possible. However, even then a certain residual load remains in the charcoal.

3.1.6 Chemical binding to zinc

In small agricultural plants, biogas with low hydrogen sulfide concentration can be produced passing it through a zinc oxide (ZnO) cartridge according to the equation:

$$ZnO + H_2S \rightarrow ZnS + H_2O$$

The sulfur remains chemically bound inside the cartridge, which has to be replaced from time to time. Even carbonyl sulfide (COS) and mercaptans can be removed with zinc oxide, when both are previously hydrolyzed to H_2S.

$$COS + H_2O \leftrightarrow CO_2 + H_2S$$

Using zinc oxide cartridges, sulfur in biogas can be reduced to values of less than $1\,mg\,Nm^{-3}$, exceptionally to $0.005\,mg\,Nm^{-3}$.

3.1.7 Surfactants

Hydrogen sulfide can be removed absorptively by surfactants, when the biogas is passed through the surfactant foam. However, this procedure is applicable only with small gas rates, since the foam must always be kept permeable.

3.1.8
Passing the biogas through an algae reactor or addition of sodium alginate

The idea of connecting an algae reactor to the discharge pipe of the bioreactor was disclosed to the patent office,[21] but was not realized for economic reasons.

On the other hand, the addition of sodium alginate[22] to the substrate appears very promising. Sodium alginate is won from sea algae and is usually added today to animal feeds. It is recommended to add to the substrate in the bioreactor a quantity at a dilution of 1 : 10 000; e.g., to inoculate a bioreactor of 650 m^3 volume with 65 kg sodium alginate. Thereby the biological procedures are activated in such a way that the biogas yield increases, NH_3 production decreases, and the H_2S concentration falls below the value of 20 mg Nm^{-3}.

3.1.9
Direct oxidation

When the H_2S concentration is above 15 000 mg m^{-3} and the biogas has a high sulfur load (of several Mg per day), catalytic direct oxidation can be economical if executed in a modified Claus's procedure. In this procedure the sulfur is released in liquid form. The reaction runs similarly to adsorption onto special metallic catalysts, but requires a preheated biogas/air mixture at a temperature above 100 °C.[23]

3.1.10
Compressed gas scrubbing

The good solubility of H_2S in water makes it possible to clean the biogas by means of scrubbing. In order to minimize the scrubber volume, the biogas should be compressed before processing it. The procedure is described in detail in the next chapter, because carbon dioxide can be separated from the gas in the same way.

When applying this procedure, small quantities of H_2S can also be removed. It is however very energy intensive and needs about 10% of the electric current generated from the biogas.

3.1.11
Molecular sieves

Molecular sieves take up and very selectively separate molecules with certain characteristics, e.g., size or shape. Best approved are molecular sieves designed especially for the separation of methane, which is retained in the pores of the sieves. All other impurities flow through unhindered such as CO_2, H_2O, or H_2S. Pure methane is removed from the molecular sieves. Only ca. 10% of the methane is lost. The molecular sieves can be used again after regeneration. The procedure is described in detail in the next chapter.

21) Cp. PAT 3
22) Cp. PAT 4
23) Cp. JOU 17

3.2 Removal of the carbon dioxide

Methane-enriched biogas is biogas with a methane concentration of more than 95%. To reach this concentration, CO_2 has to be removed, i.e., the gas volume has to be reduced by approximately 40%.

The procedure for the carbon dioxide removal (Table 3.4) has to be chosen according to the following criteria:

- minimum required concentration
- low consumption of absorbing or adsorbing material; i.e. high load, easy regeneration, chemical and thermal stability
- low flow resistance (low viscosity, large pores)
- no environmental impact
- easy availability and low price.

Depending on the capacity of the plant, costs arise as shown in Figure 3.3.

3.2.1 Absorption[24),25)]

Methane and carbon dioxide are differently bound to liquids. In water as a scrubbing agent the acidic components in the biogas such as CO_2 are more easily dissolved than hydrophobic, nonpolar components such as hydrocarbons. The physical absorption can be explained by different van der Waals forces of the gases, and the chemical absorption by different covalent binding forces.

The solubilities of different biogas components in water are shown in Table 3.5.

An absorber for scrubbing (Figure 3.4) with pressurized warm water consists of a high column filled with packing material which is percolated with the fresh water. The compressed biogas at 10–12 bar is supplied at the bottom of the column. Flowing upwards, it passes through the packing material and thereby transfers its CO_2 to the warm water (5–25 °C). The biogas leaves at the top of the column with a CH_4 concentration of more than 95%. The loaded water is discharged at the bottom. Although an even higher CH_4 concentration is reached if fresh water is continuously supplied to the column, the water is mostly regenerated. Therefore the loaded water is released through a buffer tank into the desorbing column, which is under atmospheric pressure. Here the CO_2 evaporates from the water. The application of vacuum or higher temperatures would facilitate the desorption process. The CO_2 from the desorbing column can be collected and further used, e.g., in a liquefied form. Besides CO_2, the warm water takes up traces of H_2S and other impurities in the biogas. Only oxygen and nitrogen cannot be removed from the biogas by the water scrubbing process.

The lower the temperature of the scrubbing water and/or the higher the pressure in the column, the more CO_2 can be taken up by the water. A reduction in temperature from 25 °C to 5 °C gives double the plant's capacity.

24) Cp. BOK 43 **25)** Cp BOK 57

Table 3.4 Technologies for the decontamination of biogas[26] (++ very suitable, + suitable, – less suitable, – not suitable).

Technology	Costs[27]		Removed contamination	Temp. [°C]	Pressure [bar]	Remarks
	Investment	Operational				
Absorption						
In water	+	+	Dust, CO_2, H_2S	5–25	10–12	Often applied
Physically	+	+	CO_2	<40	10–20	For the production of L Gas
N-Methyl-pyrrolidone	+	+	CO_2, H_2S	<40	10–20	Purisol scrubbing
Methanol	+	+	CO_2, H_2S, HCN, COS	<40	10–20	Rektisol scrubbing
Polyethylene glycol dimethylether	+	+	CO_2, H_2S, COS, CS_2	<40	10–20	Selexol process
Tetrahydro thiophenedioxide	+	+	CO_2, H_2S, COS	<40	10–20	Sulfinol scrubbing
Methylisopropylether	+	+	CO_2	<40	10–20	Sepasolv process
Chemically	+	++	CO_2, H_2S	<40	20–30	Regeneration not possible
K_2CO_3 (10% in water)	+	++	CO_2	<40	20–30	Cold potassium scrubbing
K_2CO_3 (15–30% in water)	+	++	CO_2, H_2S	<40	20–30	Hot potassium scrubbing
NaOH (8% in water)	+	++	CO_2, H_2S	<40	20–30	Cold lye scrubbing
NH_3 (5% in water)	+	++	CO_2	<40	20–30	Ammonia scrubbing
Alcazid M in water	+	++	CO_2, H_2S	<40	20–30	Alcazid scrubbing
Methanolamine	+	++	CO_2, H_2S, COS, Mercaptans	<40	20–30	Amisol scrubbing
Monoethanolamine (10–20% in water)	+	++	CO_2	<40	20–30	MEA scrubbing
Methyldiethanolamine (10–25% in Wasser)	+	++	CO_2, H_2S	<40	20–30	MDEA process
Adsorption with pressure or vacuum changes						
Zeolite	++	–	CH_4, N_2	<40	10–12 or 1	Low selectivity between CH_4 and N_2
Carbon	++	–	CO_2, H_2S, COS, H_2O, O_2, NH_3 and Hg	<40	10–12 or 1	Often applied
Membranes	++	++	All	<40	30	For the production of L Gas
Cryogenic processes	++	++	CH_4	<–80	200	

26) Cp. BOK 63

27) Cp. BOK 37

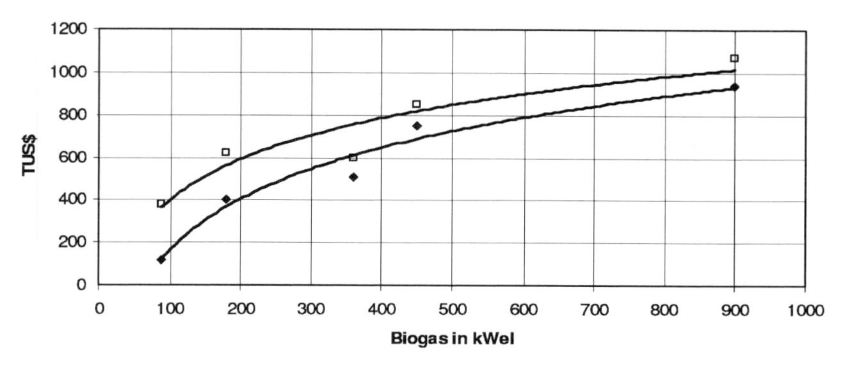

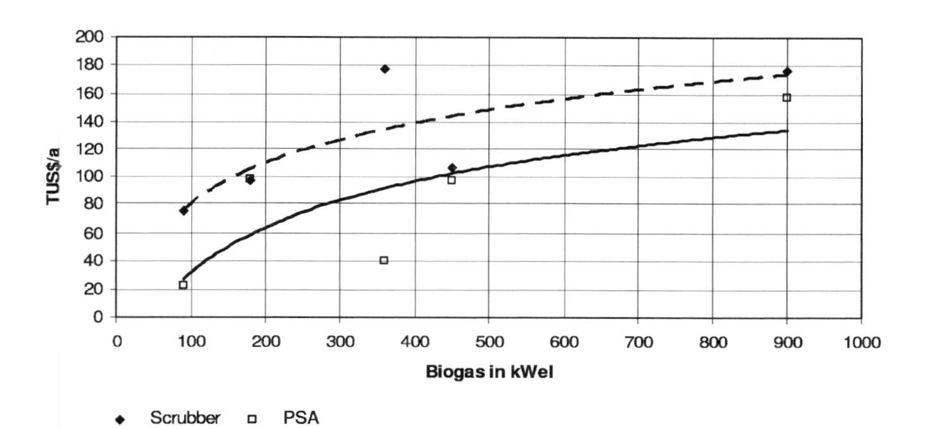

Figure 3.3 Investment costs (top) and operational costs (below) in TUS$ (TUS$ = 1000 US$).

Table 3.5 Solubility of biogas components in water.

Biogas component	*Solubility in water at 1 bar partial pressure of diluted gas [mmol/kg.bar)]*	
	0 °C	*25 °C*
Ammonia	53 000	28 000
Hydrogen sulfide	205	102
Carbon dioxide	75	34
Methane	2.45	1.32

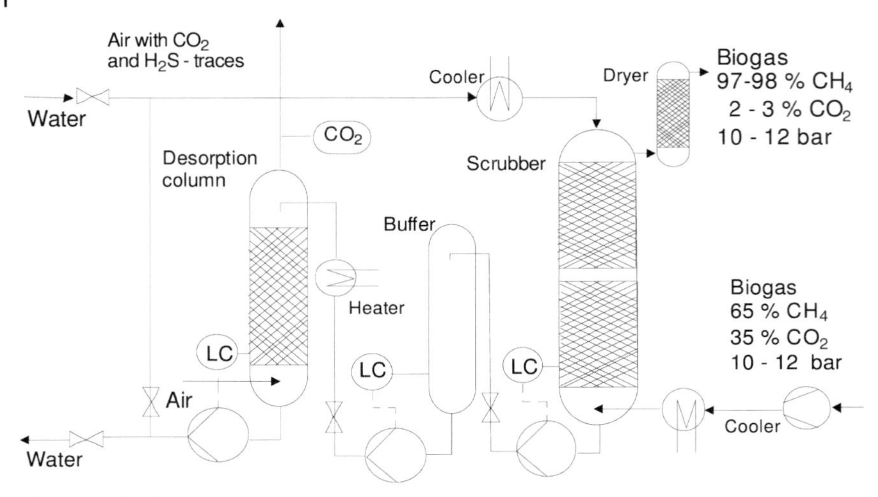

Figure 3.4 Facility for scrubbing with water under high pressure.

The advantage of physical absorption in solutions instead of pressurized water is the fact that more CO_2 and H_2S can be taken up. The columns can be smaller in size. Less circulated liquid is required, and the power consumption for the circulation of the liquid is decreased.

Well-accepted absorbents include mixtures of dimethyl ether and polyethylene glycol (trade name, e.g., Selexol), in particular because they are non-toxic and non-corrosive. In such plants an absorption pressure of 20–30 bar is applied.

It cannot be denied that the regeneration generates some problems if the biogas contains H_2S and COS, because these are more easily absorbed than CO_2 and more difficultly desorbed. For desorption of both, even vacuum is not sufficient. The absorbent has to be boiled under pressure, because polyethylene glycols have a high boiling point (200–350 °C).

Another physical absorbent is cold methanol (trade name, e.g., Rectisol). With it biogas can be decontaminated even when it contains only traces of CO_2 and sulfur (H_2S and COS) of ca. 0.1–1 mg Nm^{-3}. The process works very economically with an everywhere available detergent. Sulfur compounds and CO_2 can be separated selectively, so that both can be further used without additional steps of decontamination.[28)]

3.2.2
Absorbents based on glycol and ethanolamines

Even higher loads and selectivities are achieved with chemical absorbents (Figure 3.5). Chemical absorption into alkaline solvents occurs under medium or low partial pressure. Absorbents are mainly amines such as monoethanolamine

28) Cp. WEB 119

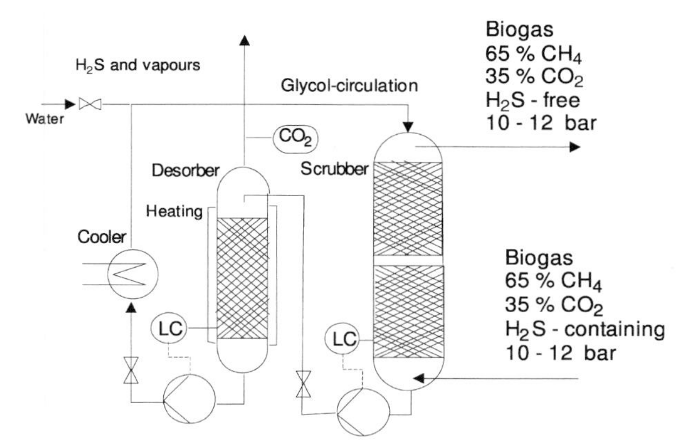

Figure 3.5 Chemical scrubber plant with glycol.

(MEA), diglycolamine, diethanolamine (DEA), triethanolamine and methyldiethanolamine (MDEA) or a hot potassium carbonate solution. At present, the best approved and most established procedure is the separation of CO_2 with MEA. If additionally hydrogen sulfide has to be removed, then MDEA is preferable.

The absorptive power of amines is impaired by impurities in the biogas, e.g., solid particles, SO_2, NO_x, and oxygen. The precleaning of the biogas from such impurities is therefore a prerequisite for chemical absorption.

Normally, desorption of such chemical absorbents is not possible, so that they have to be disposed of.

For cleaning exhaust gas streams which contain organic materials, absorbing liquids that work satisfactorily in industrial-scale installations include those based on glycols,[29] e.g., polyethylene glycol dimethyl ether (polyglykol DME) and polyethylene glycol dibutyl ether (polyglykol DBE). The polyethylene glycol dialkyl ethers with higher molecular mass have low vapor pressure, low viscosity, and good solubility for many organic materials and acidic gas components, e.g., H_2S, CS_2, COS, CH_3SH, and NH_3.

For the continuous removal of CO_2 from biogas with a flow rate of 0.5 to 300 000 $m^3 h^{-1}$ from an absorption plant, an amount of 0.5–250 $Mg h^{-1}$ circulating absorbent would be necessary. For desorption, the glycols are heated up to 130 °C at 10 mbar, which entails a considerable loss of energy.

3.2.3 Adsorption with pressure swing technology (PSA)

With a PSA (Pressure Swing Adsorption) plant, very pure CH_4 can be obtained containing only up to 0.1 $mg\,Nm^{-3}$ impurities. TSA (Temperature Swing Adsorption) is more seldom performed, but is also suitable for the removal of CO_2, H_2S, COS, H_2O, CO_2, NH_3, and Hg.

29) Cp. WEB 38

Adsorbents can be activated charcoal, zeolite molecular sieves, and carbon molecular sieves (CMS). Carbon molecular sieves consist of hard coal, which is finely comminuted, pre-oxidized with air, mixed with pitch, and extruded to form shaped products in order to get a maximum internal surface for the adsorption of carbon dioxide. Zeolite molecular sieves are naturally or synthetically hydrated aluminum silicates containing mono- or polyvalent metallic ions. They are characterized by the fact that they can emit water of crystallization without changing their structure and replace it by other materials such as CO_2. The free spaces left by the emitted water of crystallization are all exactly identical, so that only molecules of a specific shape can be adsorbed, and thus the molecular sieves work very selectively.

However, zeolite molecular sieves unfortunately do not have high selectivity for nitrogen and methane, so that despite many advantages of zeolites for the cleaning of biogas, carbon molecular sieves are preferred.

Since zeolite molecular sieves can be hydrophilic and carbon molecular sieves hydrophobic, the biogas must be dried before the methane enrichment.

The lifetime of the molecular sieves is theoretically unlimited if the biogas does not contain contaminants, e.g., oil from the compressor.

Usually adsorption plants consist of four adsorption columns, which are operated as follows:

First the biogas is compressed in an oil-free compressor up to 10–12 bar, whereby it warms up. Therefore it has to be cooled down to below 40 °C. It is than fed into adsorber 1. It passes through the adsorber from the bottom to the top, whereby it is cleaned. When adsorber 1 is loaded, i.e. the surface of the molecular sieves is covered with CO_2, the crude biogas flow is switched over to adsorber 2. The residual biogas in adsorber 1 is released to adsorber 3, which is under vacuum. The remaining gas, mainly CO_2, in adsorber 1 is exhausted to the environment. The adsorber eventually is flushed with air. Than the adsorber 1 is evacuated to a vacuum of ca. 100 mbar in order to keep the residual load as low as possible. Now adsorber 1 is available for adsorption. The crude biogas flow is now switched from adsorber 2 to adsorber 3. The residual biogas from adsorber 2 is released to adsorber 1. Because the loading of the columns does not always last the same time, the fourth adsorber column is necessary for bridging purposes. The duration of this cycle depends on the size of the columns and the size of the whole adsorption plant.

With such plants with four adsorption columns, a methane concentration of 95% can be reached. Higher concentrations are possible with intermediate flushings, so that plants with six adsorbers would then be necessary.

When methane with a purity of 99% is desired and the impurities consist mainly of carbon dioxide, the enriched methane is the main stream and the impurities are retained in the molecular sieves.

When the biogas contains much nitrogen and oxygen, the methane is adsorbed and is recovered during desorption. For special biogas compositions, two-stage processes are required.

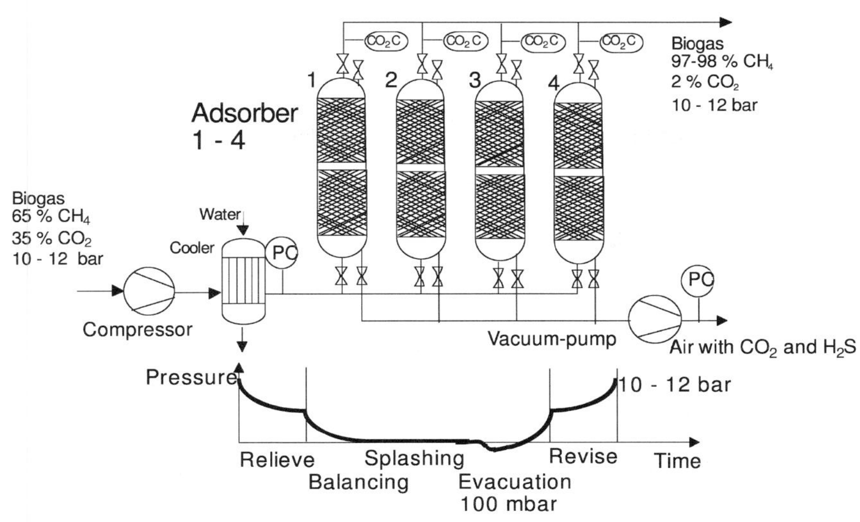

Figure 3.6 Adsorption plant with pressure swing technology.

3.2.4
Adsorption with pressure swing technology (VPSA) under vacuum

Vacuum pressure swing adsorption plants (Figure 3.6) work in principle like the pressure swing adsorption plants and use the same adsorbents.

The special economy of the VPSA technology is derived from the lower operating pressure. The biogas is only slightly compressed (<1 bar) by a ventilator and does not need to be pretreated or precleaned. Thus, the difficult oil and water separation is avoided. The pressure swing is achieved with a vacuum pump. The substantial advantage of these plants is the extremely low energy consumption.

Industrial plants operate with up to four adsorbers. Adsorption and desorption occur at approximately same temperature.

3.2.5
Diaphragm technology

In principle, gases such as CH_4, CO_2, and impurities in the biogas can be separated because of the different permeability of diaphragms. There are pore diaphragms, with which a differential pressure is responsible for the gas passage, and diffusional diaphragms, through which the gases must diffuse.

With high flow rates, the differential pressure can be up to 30 bar for both types of diaphragms. With small flow rates vacuum and pressure differences of less than 1 bar are preferred.

For the cleaning of biogas, diffusional diaphragms have become generally accepted. The biogas components pass through the diaphragm according to their

Table 3.6 Manufacturers of diaphraghms.

Membrane material	*Producer*
Polyethersulfone	Bayer, BASF, Monsanto
Cellulose acetate	Grace
Polyetherimide	General Electric
Hydrin C	Zeon
Pebax	Atochem
Polyacrylate	Röhm
Polydimethylsiloxane	Wacker, GKSS
Polyhydantoin	Bayer

molecular structure; so, e.g., hydrogen sulfide 60 times faster and carbon dioxide 20 times faster than methane. It has to be noted that always a considerable part of methane passes through the diaphragm and gets lost with the impurities.

In order to achieve a high throughput at low energy consumption, the pressure differences over the diaphragm must be kept small and thus the thickness of the diaphragm. Very thin membranes of 0.2–10 µm consist of various layers – always a thicker layer with bigger pores below a thinner layer with smaller pores. The diaphragms can be mounted either in plate form or in cushion form or as hollow fibers.

Diaphragm technology is still very new. The diaphragm materials, listed in Table 3.6, have been proved to be suitable to separate CH_4 and CO_2 according to manufacturers' data.

The material of the diaphragm decides the selectivity. It is possible to separate carbon dioxide, sulfur dioxide, and hydrogen sulfide selectively in single-stage or two-stage installations. The separated biogas components are absorbed in a solution, this process being called the wet diaphragm procedure, or they are used in the dry form, when the process is called the dry diaphragm procedure.

The dry diaphragm procedure is very costly and seldom used.

The wet diaphragm procedure seems to be more economic for cleaning biogas. Caustic soda solution is used as the solvent for H_2S and amine solutions for CO_2. The wet procedure works at low pressures. Problems with the procedure result from the still insufficient selectivity of the diaphragms and from aggressive substances in the biogas, which destroy the diaphragm material so that it has to be replaced very frequently.

Methane contents of up to 80% can be reached in a single-stage process. Higher biogas purities are achievable, but only with much more expensive multi-stage plants. The decontaminated biogas leaves the plant under high pressure, so that it can be fed directly into the natural gas network.

Diaphragm technology for the cleaning of biogas seems to be reasonable only at flow rates above $500\,m^3h^{-1}$. Pilot plants have been built in Sweden and Switzerland.

3.2.6
Mineralization and biomineralization

With these procedures, CO_2 is separated by chemical reactions, e.g., with CaO (quicklime) to form calcium carbonate ($CaCO_3$), which can be used for constructing houses.

However it has to be remembered that quicklime is manufactured by "burning" lime, a process which liberates one molecule of CO_2 per molecule of CaO produced, and this contributes to the environmental impact.

3.2.7
Cryogenic biogas purification

Cryogenic or low-temperature biogas purification[30] is a procedure which has not yet been sufficiently tested. It is expected to deliver methane of a quality suitable for vehicles.

After compression to approximately 200 bar and liquefaction of the biogas, the impurities like H_2S are adsorbed onto molecular sieves (Figure 3.7). The liquefied gas mixture is than separated by means of low-temperature distillation, often called "very low-temperature rectification" at approximately 30 bar, because the cooling causes an appropriate reduction of the pressure. The technology of the separation is based on the different boiling points of the biogas components: at a pressure of 50 bar CH_4 is liquefied at –80 °C and CO_2 at +15 °C.

CO_2 and about 80% of the CH_4 are drawn off in liquid form, the remaining 20% of the CH_4 being in gaseous form. The liquefied methane is stored at –161 °C,

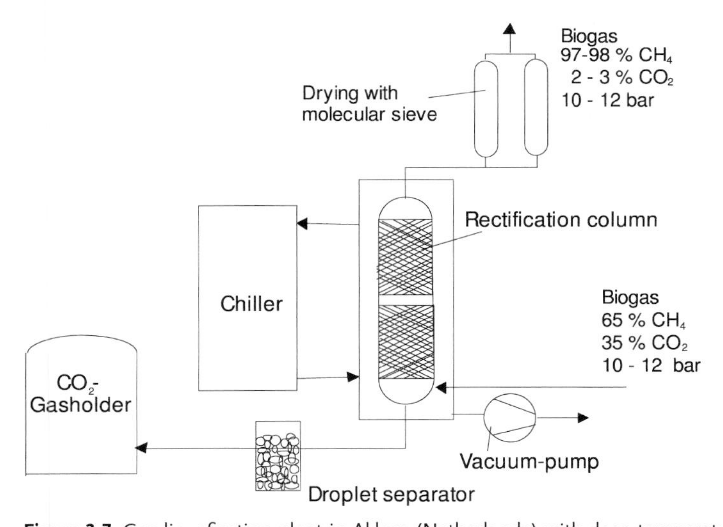

Figure 3.7 Gas liquefication plant in Aklam (Netherlands) with deep-temperature rectification.

30) Cp. BOK 41

when atmospheric pressure is applied, and the CO_2 in pressure vessels at 20 bar and ambient temperature.

The advantage of the liquefication of biogas is the high purity of the cleaned biogas. However, because of the high power consumption the procedure is very costly.

3.3 Removal of oxygen

Too high oxygen content in the biogas can only occur exceptionally. This oxygen can be removed using desulfurization procedures.

Adsorption processes, e.g., with activated charcoal, molecular sieves, or diaphragm technology are also applicable.

3.4 Removal of water

Biogas must have a relative humidity of less than 60% to prevent the formation of condensate in the transporting pipelines. This condensate, particularly in combination with other impurities, could corrode the pipe walls. Often the biogas has to be dried before further cleaning, e.g., by means of downstream absorption plants.

In order to be allowed to feed biogas into the natural gas network, a fine drying process is categorically required according to many European regulations.

Biogas can be dried by compression and/or cooling of the gas, by adsorption at activated charcoal or silica-gel, or by absorption, mostly in glycol solutions.

After compression to pressures up to 12 bar, which is necessary for many biogas decontamination procedures, the biogas leaves predried, when the condensate is removed from the compressor.

By lowering the dewpoint to 5 °C, biogas with a relative humidity of $\varphi < 60\%$ at normal temperature is obtained and corrosion can be prevented. In 60% of all agricultural biogas plants with power generators, the biogas is dehumidified by cooling it in a biogas pipe at least 50 m long enbedded in the soil. This pipe is inclined at ca. 1° to the horizontal and equipped with a condensate trap at its lowest point, this being protected against freezing and installed so as to be easily accessible.

Landfill gases are sometimes cooled down to 2 °C or even to −18 °C by a refrigerating machine in order to lower the dewpoint to 0.5–1 °C. After separation of the condensate, the landfill gas is heated to ambient temperature.

For drying of biogas in an adsorption process, SiO_2 is mostly used, although activated charcoal or molecular sieves are used also. For continuous running, a minimum of two adsorbers are necessary. While one is being loaded at a pressure of 6–10 bar, the other is desorbed with hot air at 120–150 °C. Vacuum or recirculation of a part of the still hot dry biogas also results in sufficient desorption.

For the absorption process, glycol or triethyleneglycol are appropriate as absorbents. They scrub out not only water vapor but also long-chain hydrocarbon gases out of the biogas. Two absorbers are normally used in industrial scale installations.

For fine drying, only adsorption onto activated charcoal is a suitable procedure. Activated charcoal separates not only water but also other trace impurities from the biogas.

Adsorption is preferably applied to small to medium flow rates of biogas ($<100\,000\,m^3h^{-1}$). Absorption is better suited to large flow rates. Both procedures are very costly and are only adopted when it is essential for the utilization of the biogas.

3.5 Removal of ammonia

When liquid manure and in particular wastes from fish processing or the food industry are used as substrates, ammonia can occur in considerable quantity, depending on the stability of the process of fermentation. Actually, ammonia is formed at high pH values from ammonium, which is formed in the liquid manure. Therefore, the ammonia formation can be avoided by a suitable operation of the plant.

The removal of the ammonia should be combined with other biogas cleaning procedures. When ammonia is passed through a slightly acidic solution, it remains in this liquid in the form of ammonium.

3.6 Removal of siloxanes

Siloxanes, which create abrasion in engines due to the formation of silica, can be removed from the biogas by means of adsorption onto activated charcoal, activated alumina, or silica gel. Since also other biogas components like water vapor, etc., are adsorbed, these should be removed at an earlier stage by more economical procedures in order to use the maximum potential of the adsorbents for the siloxane separation.

A simple cooling can also be sufficient for the removal of siloxanes,[31] but the purity of the biogas so obtained is less than that achieved by the adsorption process.

31) Cp. WEB 25

4
Liquefaction or compression of the biogas

Depending on the utilization of the biogas, the methane must be either compressed or liquefied. Both processes occur in analogy to the compression or liquefaction of natural gas.

In vehicles, either CNG (compressed natural gas) or LNG (liquefied natural gas) is used as fuel, the latter being liquefied under very low temperatures (–162 °C). CNG and LNG can be replaced by purified methane from biogas.

4.1
Liquefaction

Liquefied gas is advantageous for fueling vehicles because of its drastically reduced volume and its resulting higher power density. Comparatively small tanks are necessary and the cruising range is large.

Depending on the biogas flow rate, different liquefaction procedures are applied. For very large flow rates, combinations of the different possible procedures are chosen, e.g., turbine, Joule Thomson, and magnetocaloric procedures. All procedures work using compression for liquefaction and subsequent irreversible expansion in throttling valves or partly reversible expansion in expansion machines. For liquefaction, the methane has to be cooled down to –162 °C. The cooling medium is liquefied nitrogen (–196 °C). The methane gas leaves the cooler as a clear liquid at a temperature of –162 °C.

The thermodynamical process is shown schematically in Figure 4.1. The methane is compressed (from point 1 to point 2 in the figure) almost isothermally, e.g., in a multi-stage compressor, to a supercritical pressure, which for methane is about –82.59 °C. Then the methane is cooled down (from point 2 to point 3 in the figure) in counter-current heat exchange with cold gaseous methane. At point 3, methane starts to expand isenthalpically to ambient pressure (from point 3 to point 4 in the figure). Thereby a fraction of saturated liquid methane z and a fraction of dry saturated gaseous methane (1–z) are formed, from which the cold is transferred in the counter-current heat exchanger. The temperature differences necessary for the heat transfer in the heat exchanger can be seen in the figure. The temperature differences are shown as ΔT_E and ΔT_A. The energy balance for the heat exchanger is to be calculated.

Biogas from Waste and Renewable Resources. An Introduction.
Dieter Deublein and Angelika Steinhauser

ISBN: 978-3-527-31841-4

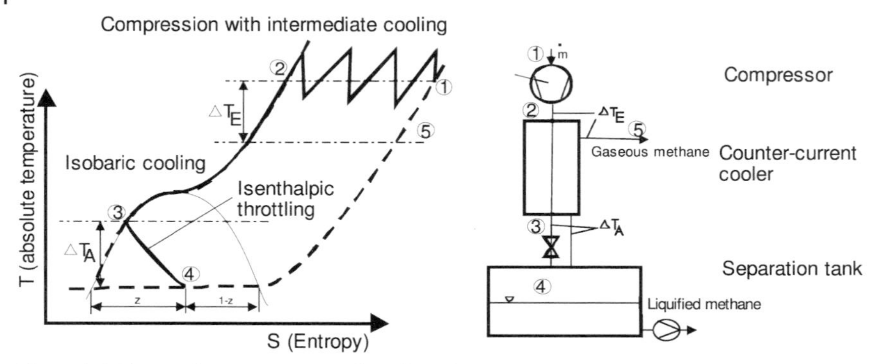

Figure 4.1 Thermodynamic process of gas liquefaction.

$$(1-z)\cdot \dot{m}\cdot h_5 + z\cdot \dot{m}\cdot h_{4'} = \dot{m}\cdot h_2$$

The fraction of liquefied methane can be calculated from

$$z = \frac{h_5 - h_2}{h_5 - h_{4'}}$$

The above describes continuous steady-state operation of the plant. During start-up, no cold methane is yet available. Under these conditions, the compressed biogas gradually cools down because of the integral throttling effect until the final mode of operation is reached.

A liquefaction plant with a capacity of 4.5 Mg d^{-1} costs about 15 Mio US$. The costs depend strongly on the steel grade used, and can be roughly estimated according to following equation:

$$K_1/K_2 = (0.6 \text{ to } 0.7)\cdot(\dot{M}_{G1}/\dot{M}_{G2})$$

K stands for the investment costs and $\dot{M}_G$ for the capacity of the plant. The overall costs are attributable to planning (ca. 10%), components (ca. 60%), and construction and erection (ca. 30%).

4.2 Compression

To compress methane from biogas the same compressors can be used as those for compressing natural gas, i.e. reciprocating piston compressors, rotary piston compressors such as sliding vane compressors, liquid ring compressors, rotary screw compressors, and turbo compressors such as radial and axial compressors (Table 4.1). Diaphragm compressors are in principle technically too complex and therefore uneconomic for methane gas compression.

Table 4.1 Biogas compressors.

Compressor	*Throughput*	*Pressure head*	*Remarks*
Reciprocating piston compressor	Small to medium	High (>10 bar)	High efficiency, economic, abrasion-sensitive
Rotating piston compressor	Small to medium	High	High reliability, low abrasion with contaminated gases
Sliding vane compressor	Medium	Medium to high (three stages)	
Liquid ring pump	Medium	Small to medium (two stages)	
Rotary screw compressor	Medium	Medium to high (two stages)	
Centrifugal compressor	Medium to high	Very high (300 bar)	Low abrasion
Axial flow compressor	Small to medium	High	Low abrasion

The necessary specific compression work w_t is calculated for isothermal compression, which can be assumed for a housing-cooled compressor, theoretically according to

$$w_t = R_{CH4} \cdot T \cdot \ln(p_2/p_1).$$

R_{CH4} stands for the specific gas constant for CH_4 and T stands for the absolute temperature.

For calculation of the power consumption P_A of the compressor, the flow rate and the compression efficiency must be considered:

$$P_A = \frac{\dot{m} \cdot W_t}{\eta_K}$$

where $\dot{m}$ is the flow rate of biogas and η the efficency including all hydraulic and mechanical losses. The efficiency can be estimated for small compressors to be $\eta_K = 50\%$.

The logarithmic relationship between necessary compressor power and compression ratio shows clearly that the pressure at the inlet determines the power. Thus, compression from 1 to 10 bar needs the same power as a compression from 10 to 100 bar. This influences the investment costs and even more the operational costs, i.e. the costs for the gas or power consumption of the compressor itself. Biogas compressors are designed to include several stages, in which the first stage produces a pressure of >10 bar.

For small plants (in the range of $10\,kW_{el}$), costs are up to $5000\,US\$\,kW_{el}^{-1}$. With increasing compressor size (in the range of $250\,kW_{el}$), the costs decrease to $500–750\,US\$\,kW_{el}^{-1}$.

Most standard compressors are oil-lubricated except the rotary piston compressor. Some oil (1–5 $mg\,m^{-3}$) is therefore to be found in the compressed biogas. Specially designed compressors do not need oil lubrication, but suffer from gas leakages and extraordinary wear. Traces of compressor material could sometimes contaminate the biogas.

5
Utilization of biogas for the generation of electric power and heat

Biogas can be used either for the production of heat only or for the generation of electric power.

When current is obtained, normally heat is produced in parallel. Such power generators are called combined heat and power generation plants (CHP) and are normally furnished with a four-stroke engine or a Diesel engine. A Stirling engine or gas turbine, a micro gas turbine, high- and low-temperature fuel cells, or a combination of a high-temperature fuel cell with a gas turbine are alternatives.

Biogas can also be used by burning it and producing steam by which an engine is driven, e.g., in the Organic Rankine Cycle (ORC), the Cheng Cycle, the steam turbine, the steam piston engine, or the steam screw engine. Another very interesting technology for the utilization of biogas is the steam and gas power station.

Figure 5.1 shows the range of capacities for the power generators which are available on the market as pilot plants or on an industrial scale. The efficiency indicates the ratio of electrical power to the total energy content in the biogas. Efficiency figures are given for different manufacturers. Small-capacity engines can result in lower efficiencies than high-capacity engines.

The generated current and heat can supply the bioreactor itself, associated buildings, and neighboring industrial companies or houses. The power can be fed into the public electricity network, and the heat into the network for long-distance heat supply. Vehicles can sometimes be driven by the power or the heat.

5.1
Supply of current to the public electricity network

The mode of operation of a gas engine varies in principle according to whether it

- covers peak load
- covers basic load
- supplies its own needs and only feeds the surplus into the network.

The mode of operation is determined by local conditions, especially the price of electric power.

Biogas from Waste and Renewable Resources. An Introduction.
Dieter Deublein and Angelika Steinhauser

ISBN: 978-3-527-31841-4

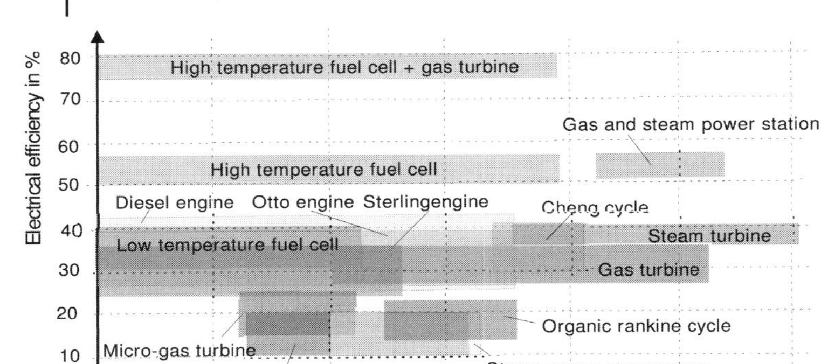

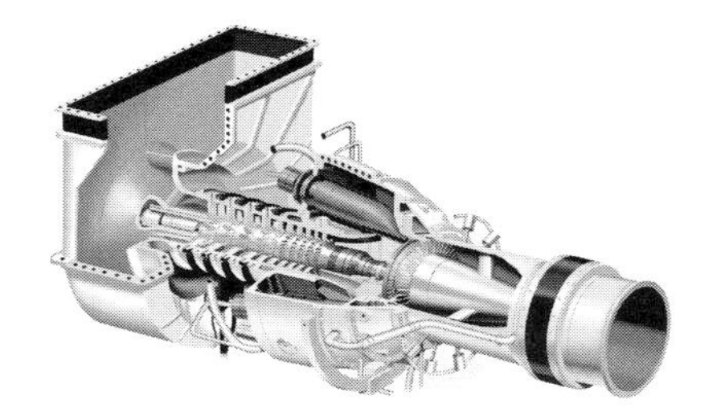

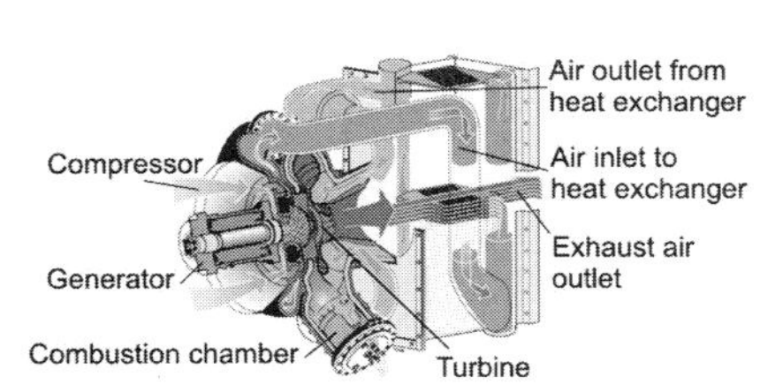

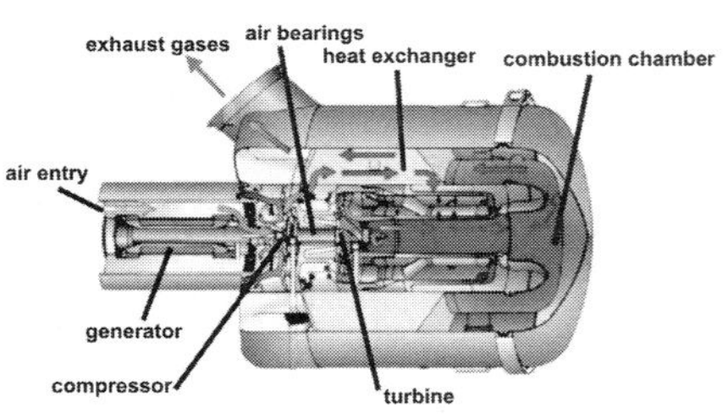

Figure 5.1 Power generators – survey (1st row).
CHP (2nd row left); Gas turbine (2nd row right) Siemens SGT-100
Molten carbonate fuel cell (biogas plant in Leonberg/Germany)
(3rd row left); Diesel engine (3rd row right)
Micro gas turbines: Turbec 100 (4th row left); Capstone
(4th row right)

Different plant designs are needed for covering a constant basic load and for covering peak loads for certain periods of the day only. Peak load covering requires complex and expensive gasholders for longer periods and larger and more expensive power stations.

The worldwide ongoing system of promoting renewable energy, as from biogas, does not especially consider whether the power is generated for basic or for peak load and at what time of day the current is fed into the network. Therefore biogas plants are normally designed to cover the basic load, although the produced power depends on the activity of the microorganism and therefore varies.

Biogas plants are usually constructed at places, where the power network is not available and special effords are required to connect the CHP to the public power network.

Usually special adjustments with synchronizing control, switching devices, network failure control and compensation of short-circuit power, power failure and wattless current.

5.1.1 Generators

For power generation, asynchronous or synchronous generators can be used. The application of an anynchronous generator is only reasonable if the generated electrical power is less than 100 KW because of the necessity for wattless current. However, even then they are not chosen for isolated operation.[32] Much more often, synchronous generators are used.

5.1.2 Current-measuring instruments

For feeding the current into the public electricity network, a transformer with attached voltage measurement is necessary. Outgoing and incoming current are measured in a so-called 4-quadrant enumerator. For this, a mean voltage measurement is essential.

5.1.3 Control of the synchronization

The electrical power unit may be connected to the public electricity network only if the generator voltage is adapted to the net; i.e. mains voltage must be provided by all three phases and must reach a value in all three power lines above the tripping value of the voltage decrease protection. This is controlled by the low-voltage protection of the current net failure registration. Nevertheless, the connection of the CHP is always slightly retarded.

32) i.e. without feeding current into the public power network

The connection of an asynchronous generator happens automatically at a rotational frequency of 95–105% compared to a synchronous generator. The connection of a synchronous generator is done when the following three conditions are fulfilled:

Variation of the voltage:	+/−10% of the nominal voltage
Variation of the voltage frequency:	+/−0.5 cycles per second
Difference in the phase angle:	+/−10°.

The frequency, the voltage and the phasing of the net and the generator are measured by appropriate sensors and controlled digitally by a synchronization device. The generator is controlled and adjusted within the limits of tolerance by signals in the range of 0–10 V. Only then does the PLC (programmable controller) release a signal for connection, with a slight time delay.

5.1.4 Switching devices

For connection of the generator to the electricity network, off-load switches should be used, in particular for capacities up to 100 KW circuit-breakers with upstream safeguard load disconnecting switch and for capacities above 100 KW circuit-breakers with electric drive.

5.1.5 Network failure registration

The generator is disconnected when the values under-run or over-run the limiting values stipulated below:

Low-voltage protection:	80% of the rated voltage
High-voltage protection:	110% of the rated voltage
Decreasing frequency protection:	49 cycles per second
Increasing frequency protection:	51 cycles per second.

For monitoring the frequency and the voltage, a three-phase relay is used. With synchronous generators, in place of the frequency monitoring a vector jump relay is used, which registers much faster disturbances like short breaks in the network due to its very short response time.

During short breaks (duration of 150–500 ms) the synchronous generator runs freely. With return of the mains voltage a wrong phase position could occur, which could lead to the destruction of the synchronous generator. The vector jump relay compares the current phase positions between the generator and the network (Figure 5.2). If these deviate by more than ca. 10°, then within 100 ms a signal is given to cut off of the generator connection. Thus, out-of-phase connection of the network to the generator is prevented.

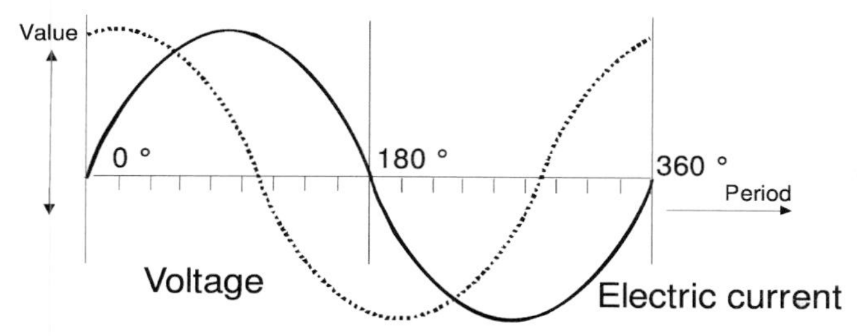

Figure 5.2 Phase difference between electric current and voltage.

5.1.6
Short-circuit protection

Generators usually produce a short-circuit current of 1.5–4 times the nominal current. Fuses or switches are not available on the market, which react to short-circuit current. Only special current guards can measure the short-circuit current. These have an adjustable time of retardation, so that they do not react when load is connected.

5.1.7
Wattless current compensation

Synchronous generators produce wattless current due to a phase shift. The phase shift $\cos\varphi$ of a generator, which is to be connected to the electricity network, should be within the limits of 0.9 capacitively to 0.8 inductively in the case of withdrawal and supply of power.

Generators are equipped with an automatical $\cos\varphi$ controller even for varying power generation. Additional compensation is not necessary.

With asynchronous generators, compensation capacitors are necessary. These may not be switched on before connecting the generator and have to be switched off at the same time as the generator is disconnected. The capacity of the capacitors should be ca. 75% of the rated output (in kVA) of the generator.

5.2
Heat

The economics of a biogas plant are highly dependent on the utilization of the heat produced from the biogas (Figure 5.3). It must be borne in mind that the heat is produced over the whole year and not only in the winter, when it can be easily used. The heat could be used, e.g., for

- heating swimming pools and/or industrial plants
- heating stables as for the breeding of young animals under infrared emitters

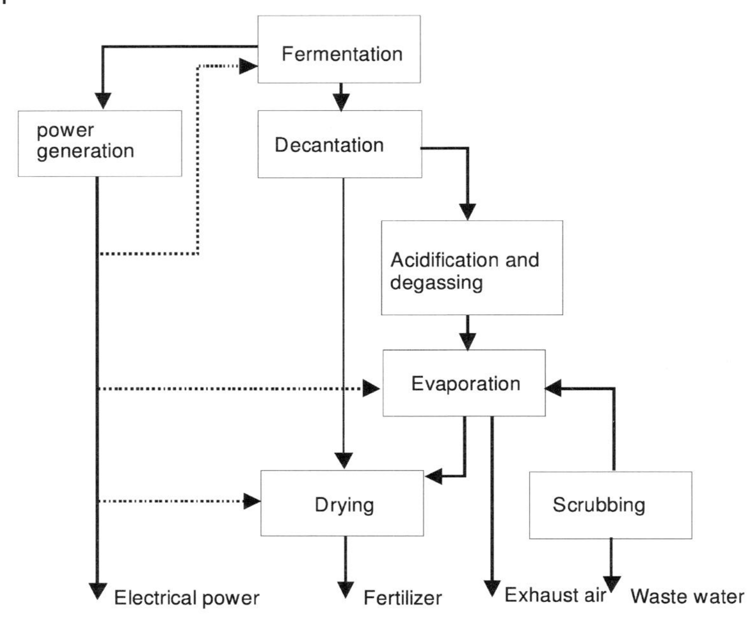

Figure 5.3 Utilization of heat generated in the biogas plant itself (dotted lines).

- the treatment of products, e.g., the conversion of liquid manure to fertilizer,[33] or for drying processes, e.g., of hay.
- heating greenhouses
- cleaning and disinfection of the milking equipment
- Transformation of warmth in cold e.g. for milk cooling

A remarkable step forward could be the use of the heat for operating an absorption refrigerator.[34] The cold could then be used in agricultural plants for cooling purposes, e.g., for stables or for milk.

The heat generated, e.g., in a CHP, partly escapes with the exhaust gas and has to be recovered in the exhaust heat exchanger for further use. But it can only be regained to a certain extent, because the exhaust gas is at a minimum temperature of 120–180 °C. The heat is partly transferred to water in the cooling water heat exchanger. The heat of the lubricating oil is mostly dissipated to the engine cooling water. But heat losses of the entire biogas plant and radiation losses of the CHP cannot be avoided.

Engines with a turbo charger are usually equipped with an intercooler or with a mixture radiator if they are gas engines. Depending on the design, the heat generated therein is transferred to the cooling water or to a separate water cycle.

The water in the cycle, which transports the heat from the biogas burner to consumers, is normally heated to 90–130 °C and flows back to the burner at a temperature of 70–110 °C.

33) Cp. BRO 7

34) Cp. WEB 88

5.3
Combined heat and power generator (CHP)

CHPs are very common in biogas plants. In parallel to the generation of current, a more or less high percentage of heat is developed in CHPs, depending on the power generator technology.

Approximately 50% of the CHPs installed in biogas plants in Europe run with four-stroke engines and about 50% with ignition oil Diesel engines. More modern technologies like fuel cells or micro gas turbines are very seldom to be found.

The total efficiency, i.e. the sum of the electrical and thermal efficiencies, is within the range 85–90% with modern CHPs. Only 10–15% of the energy of the biogas is wasted. But the electrical efficiency (maximum 40%) is still very low: from 1 m^3 biogas only 2.4 KWh electric current can be produced.

The equipment of a complete CHP includes

- a generator set consisting of drive unit and generator
- a waste gas system
- a ventilator for the supply of the combustion air on the one hand and on the other hand for the removal of the radiant heat of the engines, generators, and pipework
- a sound-damping hood
- an automatic lubricant supply.

5.3.1
Engines

As mentioned above, engines which differ widely in their technology are available for CHPs. As a decision-making aid, the engine types are listed in Table 5.1 with some characteristic performance figures. These have improved during recent years following development work inspired by the worldwide boom in biogas plants. Some manufacturers have already exceeded the figures given here.

5.3.1.1 Generation of electricity in a four-stroke gas engine and a Diesel engine (Figure 5.4)

Today's four-stroke biogas engines were originally developed for natural gas and are therefore well adapted to the special features of biogas. Their electrical efficiency normally does not exceed 34–40%, as the nitrogen oxide output NO_x has to be kept below the prescribed values. There are, for example, four-stroke engines with electrical efficiencies above 40% working with a recuperator (Figure 5.5). The capacity of the engines ranges between 100 KW and 1 MW and the lifetime is given as ca. 60 000 h.

The engine operates at 1500 rpm and consists of

- Engine block with crankshaft, crankshaft bearings and seals, housing, piston rod, piston with piston rings, cylinder, oil sump, flywheel housing
- Cylinder head with cylinder head gasket, cam shaft, valves, tappet, rocker arm.

Table 5.1 Characteric values of power generators[35),36)] (* relating to 15% O_2 in the exhaust gas, ** compressed biogas, so that ignition oil is not needed).

Feature	*Four-stroke engine*	*Gas-Diesel engine***	*Ignition oil Diesel engine*	*Stirling engine*	*Fuel cell*	*Gas turbine*	*Micro gas turbine*
Range of capacity (kW_{el})	<100	>150	30–1000	<150	1–10000		30–110
Spec. investment-costs	Medium	Medium	Medium	High	Very high	Medium	Low
(US$/$kW_{el}$)	1200	1300	1300	1600	n.a.	1200	600–900
Spec. maintenance costs (US$/$kWh_{el}$)	High	Low	High	Very high	Very high	Very low	Very low
Electrical efficiency	30–40%	35–40%	32–40%	30–40%	40–70%	25–35%	15–33%
Decrease of efficiency at	High	Low	Low	High	Very low	Very high	Low
partial load; given value for 50% load	84%	84%	84%	84%	96%	ca. 75%	88%
Cooling water temperature	110 °C	110 °C	110 °C	60 °C	n.a.	210 °C	300–500 °C
Revolutions per minute	1500 U/min	1500 U/min	1500 U/min	1500 U/min	0	100 000 U/min	100 000 U/min
Pressure ratio	10 : 1	20 : 1	20 : 1	5 : 1	n.a.	5 : 1	5 : 1
Controllability of the power/heat ration	Not possible	Not possible	Not possible	Not possible	good	Very good	Very good
Weight	Medium	Medium	Medium	Medium	High	Low	Low
Lifetime	Medium	Medium	Long	Long	Very short	Long	Long
Noises	Medium	Loud	Loud	Medium	Silent	Silent	Silent
Emissions NO_x	High	High	Carbon black	Very low	Very low	Low	Low
			600–700 mg/Nm^3*		3 mg/Nm^3 flue gas*	25 mg/Nm^3 flue gas*	20 mg/Nm^3 flue gas*
			600–700 mg/Nm^3*		3 mg/Nm^3 flue gas*	25 mg/Nm^3 flue gas*	20 mg/Nm^3 flue gas*
Alternative fuel in case of shortage of biogas	Liquid gas (gasoline)	Liquid gas	fuel, petroleum, (vegetable oil)	Any	Natural gas	Natural gas	Natural gas, biogas, Kerosene, fuel oil
Minimum heating value	Medium	Medium	High (10–30% ignition oil)	Any	Any	Low	Low

35) Cp. WEB 58

36) Cp. WEB 57

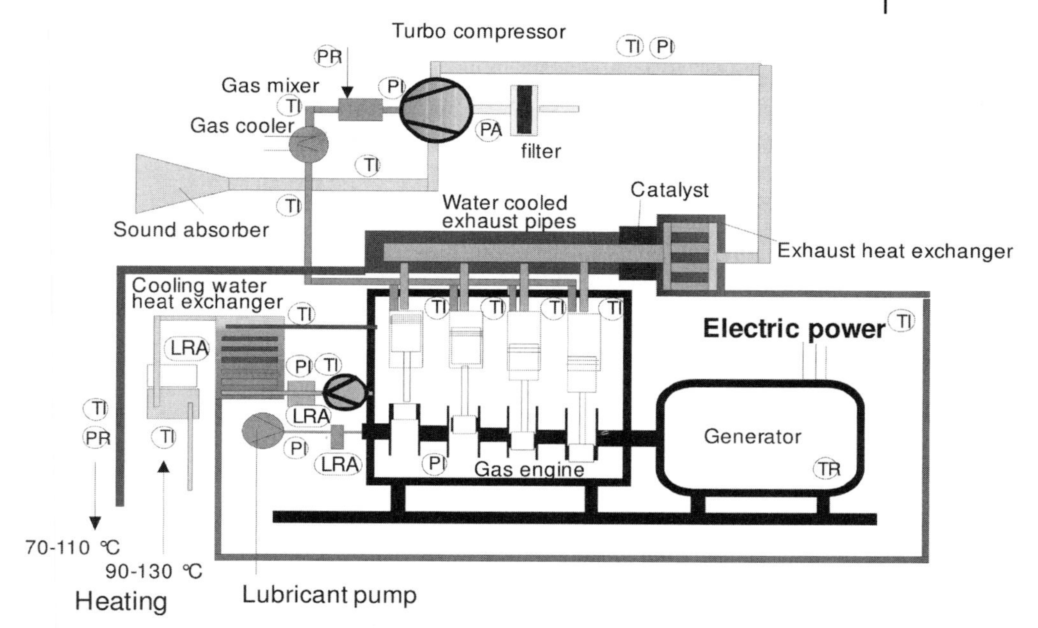

Figure 5.4 CHP instruments.
Control instruments (for cooling water temperature at motor inlet and outlet, cooling water temperature at the exhaust pipes, cooling water pressure, heating water temperature inlet and outlet, pressure of the lubricant, pressure in the crank case, pressure in the exhaust pipe, charging air pressure, gas mixture temperature before and after cooler, exhaust temperature after each cylinder, after turbo compressor, and after heat exchanger, voltage of the lambda probe on both sides, pinging signal).
Contactor for (level of lubricant, level of cooling water, contamination of the intake filter).
Digital indicator for (temperature of the electric coil, low voltage of the starting battery, missing heating water, overpressure of heating water, level of lubricant, level of cooling water, over-current for all electrical motors, minimum gas pressure, tightness of all gas pipes).

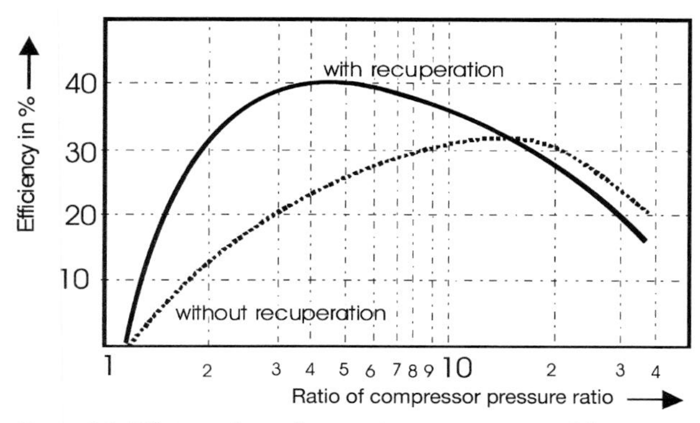

Figure 5.5 Efficiency depending on the pressure ratio of the compressor.

Medium- and large-sized engines have several single cylinder heads, small engines usually one cylinder head for all cylinders. Into the cylinder head replaceable valve seat rings are shrunk as valve guides. Lubrication of the bearings is by a pressure feeding system operating by means of a gear wheel-driven oil pump.

Engines over ca. 60 KW have wet, replaceable cylinder liners made by centrifugal casting. With the smaller engines the cylinders are fitted into the engine block (dry liners). During a complete overhaul these are drilled out and provided with new cross grinding (honed), so that oversize pistons can be used.

The air/fuel mixing is accomplished in the gas engine by a suction nozzle depending on the drawn-in air volume. Because of the outside regulation of the mixture, gas engines have longer response times to nominal and actual values of the rpm or the capacity. This is to be considered, particularly when the engine is operated in isolation from the electrical network.

Four-stroke biogas engines are used preferably with precompression of the gaseous fuel (turbocharger). By precompressing and subsequently cooling the mixture, the efficiency can be increased by the factor 1.5.

In general the efficiency decreases with an increase of the CO_2-concentration in the biogas. Because of the CO_2 content of the biogas and the consequently increased anti-knock properties, the compression ratio can be raised technically from 11 to 12.5 (as with a propane gas engine), whereby the efficiency rises ca. 1–2%.

Four-stroke biogas engines working in CHPs are equipped with digitally operated spark ignition, actuated by capacitor discharge, which do not have wear parts and which deliver a high-energy, time-exact ignition, whereby low waste gas emissions and long service lives of the spark plugs result.

The microprocessor-controlled ignition makes it easy to adjust the engine to different kinds of gas or varying gas quality. The timing of the ignition can be changed, e.g., depending on the analog signal of a methane analyzer or a knocking monitor. Usually a power adaption is made at the same time.

Four-stroke biogas engines often run in the lean-burn range (ignition window $1.3 < \lambda < 1.6$),[37] where the efficiency drops. The efficiency of lean-burn engines with turbocharger is ca. 33–39%. The NO_x emissions can be reduced, however, by a factor of 4 in comparison to ignition oil Diesel engines, and the limiting values[38] can be met without further measures. Regulation is then made by a butterfly valve in the suction pipe.

The CO content in the exhaust gas must be kept below 650 mg Nm^{-3} according to European regulations. This can be achieved by cooling the exhaust gas below ca. 400 °C in water-cooled collectors, because at lower temperatures the oxidation from hydrocarbons to CO (a post-reaction in the tail pipe) is slowed down. But the lower temperature brings the efficiency down to 27–35% before the turbocharger. A low CO value in the exhaust gas can also be achieved with an oxidation catalyst. The catalyst makes an activated charcoal filter essential in the suction pipe to the CHP to retain catalyst poisons (siloxanes, sulfur compounds). The efficiency rises ca. 3%, and emissions of SO_2 (formed by reaction of sulfur and oxygen to SO_2 in the engine) are prevented in the exhaust gas.

37) λ = air/fuel ratio

38) Cp. LAW 31

Since the engines can tend to knock with varying gas qualities, a methane content of at least 45% should be ensured.

All parts of the engine which come into contact with sulfur compounds and can corrode (copper, chrome), must be corrosion-protected, using best available materials which are easily replaceable. The corrodibility can be decreased by using specially designed bearings (Sputter bearings) instead of ball bearings.[39)]

In order prevent the drastic reduction of oil change intervals and considerable wear of the cylinder heads due to the sulfuric acid in the biogas, special lubricating oils are used which are ash-poor and ensure a long-term high alkalinity. Additionally, CHPs designed for biogas are equipped with large lubricating oil tanks (200 L) to allow a higher capacity for impurities and and a longer running time of the oil. Regular oil analyses are indispensable, depending on the sulfur content of the biogas. Regular means at intervals of 160–2000 h, on the average every 465 h.

In small agricultural plants, ignition oil Diesel engines with ca. 4 dm^3 piston displacement are frequently installed. These engines are more economical and have a higher efficiency than four-stroke engines in the lower capacity range. But their NO_x-emissions are high. Their lifetime is given as 35 000 h of operation.

In general, gas Diesel engines work by direct injection, because pre-chamber engines develop hot places, resulting in uncontrolled spark failures with biogas. Because of the internal formation of gas mixtures Diesel engines can be faster controlled.

The ignition oil Diesel engine is operated ideally at a air/fuel ratio of $\lambda < 1.9$. The efficiency is then up to 15% better than in a four-stroke engine.

Biogas is usually inhomogeneous. High temperatures of combustion can lead to increased NO_x emissions, low temperatures to incomplete combustion and unburnt carbon in the exhaust gas. The engine can knock because of a too early self-ignition of the mixture with high methane contents. These problems are exacerbated by NH_3 in the biogas. The variation of the biogas quality can be compensated by varying the feed of ignition oil. For an adequate biogas quality, feeding of 10–18% of ignition oil is recommended. When the methane content in the biogas is low, more ignition oil must be added.

If the Diesel engine remains a so-called self-ignition engine, the fuel can ignite a biogas air mixture which is sucked in, equivalent to a strong ignition spark. The limited values for NO_x can then be kept at lean burn with small ignition oil feed. Otherwise the NO_x-value can only be kept low with an catalyst.

Operation with mineral ignition oil requires special storage and in Germany monitoring by the customs. The ignition oil consumption, the operational data of the engine, and the quantity of electricity fed into the electricity network must be registered continuously. If the operator of the plant sets a value on environmental protection, vegetable oil and/or biodiesel (rapeseed methyl ester) can be fed instead of mineral ignition oil. If then the regulation of λ is replaced by a gas supply technology, an even leaner air/fuel ratio compared to the Diesel process can be used. Advantages of renewable ignition oils are the lower carbon monoxide emissions, the sulfur-free exhaust, and the biological degradability. Only if biodiesel (RME) or vegetable oil is used as ignition oil, in Germany a special promotion is granted.

39) Cp. WEB 117

5.3.1.2 **Generation of electricity in a Stirling engine** (Figure 5.6)
An alternative to the commonly used four-stroke engine and/or the Diesel engine is the Stirling engine. The efficiency of the Stirling process is closest to that of the ideal cycle.

In the Stirling engine, heat and cold must be supplied alternately to the working medium once per cycle. Because of the slow-acting heat transfer through the walls of the combustion chamber, external heat exchangers are installed in which a special driving gas is heated and cooled. The driving gas moves between two chambers, one with high and one with low temperature.

In stage I, the driving gas is to be found uncompressed in the cold chamber. When the driving piston moves upwards (see Figure 5.7), the driving gas is compressed (stage II) and then forced to the warm chamber. On its way it is heated in an external heat exchanger. The heated driving gas presses the displacement piston downwards by its expansion pressure (stage III). The displacement piston is taken downwards with the driving piston (stage IV). Then the displacement piston moves upwards again, forced by the rhombus gear with which both pistons

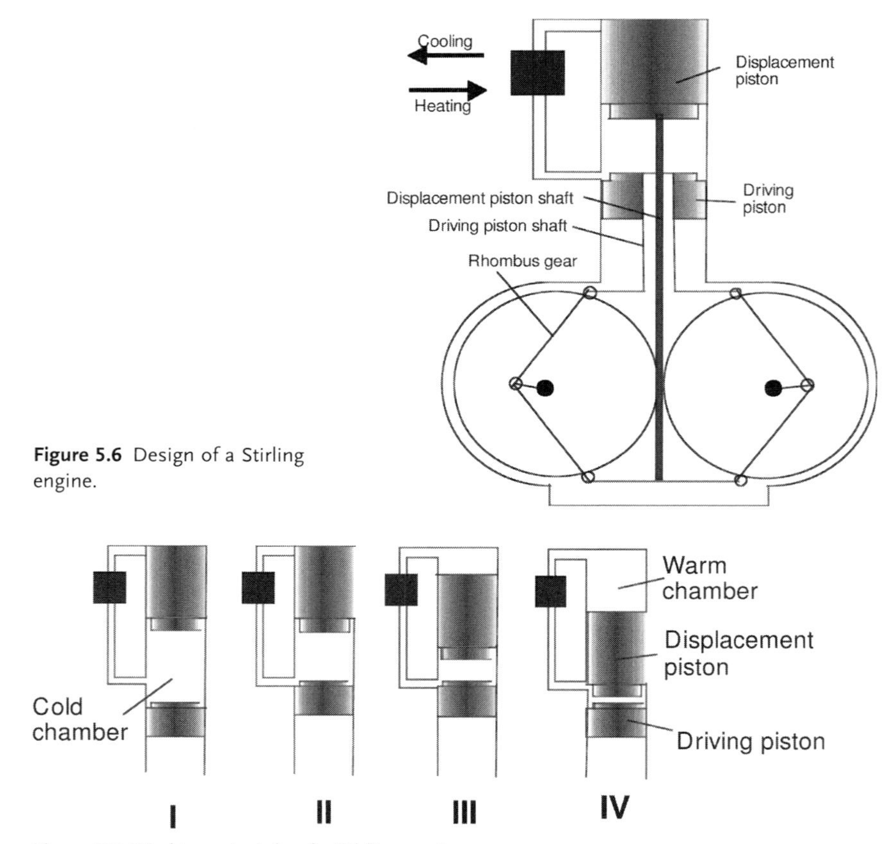

Figure 5.6 Design of a Stirling engine.

Figure 5.7 Working principle of a Stirling engine.

are connected. The displacement piston presses the driving gas downwards back to the cold chamber. On its way it is cooled in a cooling heat exchanger. Then the cycle starts again.

The rotation of the rhombus gear can be used to run a power generator.

The Stirling engine was recommended for power generation for many years, but was seldom realized on an industrial scale because of technical problems in details. Industrial scale installations are not known in which power is generated from biogas in Stirling engines.

5.3.1.3 Generation of electricity in a fuel cell

Compared to combustion engines, the fuel cell converts the chemical energy of hydrogen and oxygen directly to current and heat. Water is formed as the reaction product (Figure 5.8).

In principle, a fuel cell works with a liquid or solid electrolyte held between two porous electrodes – anode and cathode (Figure 5.9). The electrolyte lets pass only ions and no free electrons from the anode to the cathode side. The electrolyte is thus "electrically non-conductive". It separates the reaction partners and thereby prevents direct chemical reaction. With some fuel cells, the electrolyte is also permeable to oxygen molecules. In this case the reaction occurs on the anode side. The electrodes are connected by an electrical wire.

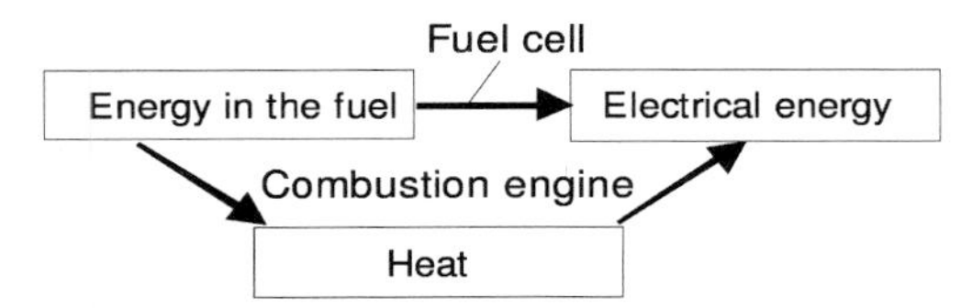

Figure 5.8 Energy conversion in a combustion engine in comparison to a fuel cell.

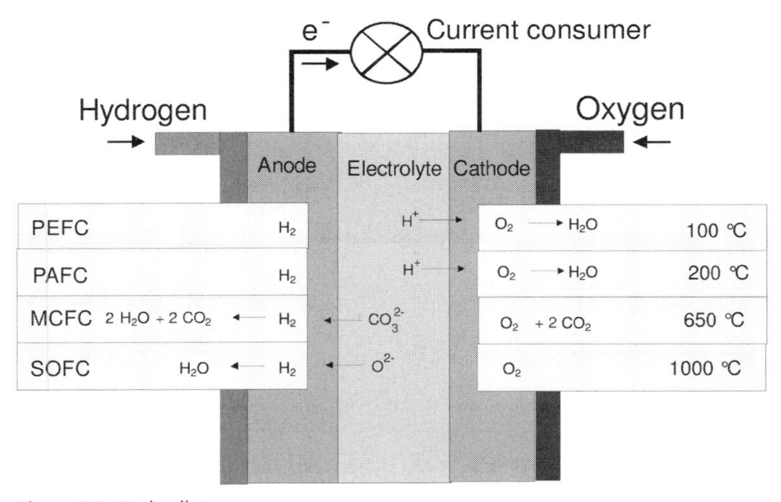

Figure 5.9 Fuel cell types.

Table 5.2 Industrial scale and pilot plant fuel cells for biogas.

Type	*Start-up year*	*El. power [kW]*	*Biogas from*	*Gas pretreatment*
PAFC	2000[40]	200	Sewage plant	Gas decontamination, reforming
MCFC	2000[41]	0,3	Pilot plant	
SOFC	2001	1	Agricultural plant	s-stage decontamination
MCFC	2003[42]	0.3	Sewage plant	Chemical pretreatment
MCFC	2003	n.s.	Agricultural plant	
MCFC	2003[43]	1	Sewage plant	
PEFC	2003[44]	0,25	Agricultural plant	Gas decontamination, reforming, fine decontamination
MCFC	2003[45]	0,3	Agricultural plant	Biological desulfurization
MCFC	2004[46]	0,3	Landfill gas	Chemical pretreatment
MCFC	2005	250	Sewage gas	Separation of sulfur, halogenes, siloxanes, water vapour
MCFC	2006	250	Biogas from waste fermentation	Separation of sulfur, halogenes, siloxanes, water vapour

Both reaction partners are continuously fed to the two electrodes. The molecules of the reactants are converted into ions by the catalytic effect of the electrodes. The ions pass through the electrolyte, while the electrons flow through the electric circuit from the anode to the cathode. Taking into account all losses, the voltage per single cell is 0.6–0.9 V. The desired voltage can be reached by single cells arranged in series, a so-called stack. In a stack, the voltages of the single cells are added.

Depending on the type of fuel cell, the biogas has to be purified, especially by removing CO and H_2S, before feeding the fuel cell.

Only a small number of fuel cell plants, mostly pilot plants, are in operation for the generation of electricity from biogas (Table 5.2).

Figure 5.10 shows the flow patterns of complete fuel cell plants. Decontamination of the crude biogas is absolutely necessary with all fuel cells, which work in a temperature range up to 200 °C. Subsequently, the methane from the biogas must be reformed to H_2. The hydrogen is fed into the stack. That part of the hydrogen which does not pass through the electrolyte is after-burned and used for the generation of heat.

In the MCFC cell, carbon dioxide serves as the cooling medium for the biogas on the anode side. Therefore, less excess air is necessary on the cathode side. The blower can be designed with a lower capacity. CO_2 slows the kinetics on the anode side to an insignificant extent, but accelerates it on the cathode side, so that a higher efficiency (ca. 2%) is observed when using biogas instead of pure hydrogen.[47]

Another type of fuel cell is the SOFC plant, shown in Figure 5.11, for providing houses with current and heat.

40) Vgl. WEB 64
41) Vgl. WEB 76
42) Vgl. WEB 81
43) Vgl. WEB 49
44) Vgl. WEB 7
45) Vgl. WEB 76
46) Vgl. WEB 76
47) Cp. WEB 67

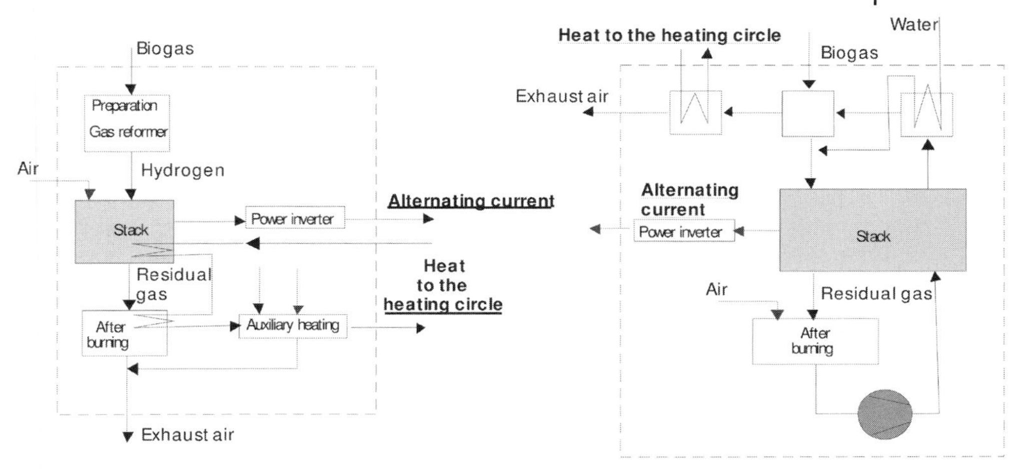

Figure 5.10 Flow chart of fuel cell plants: PAFC (left); MCFC (right).

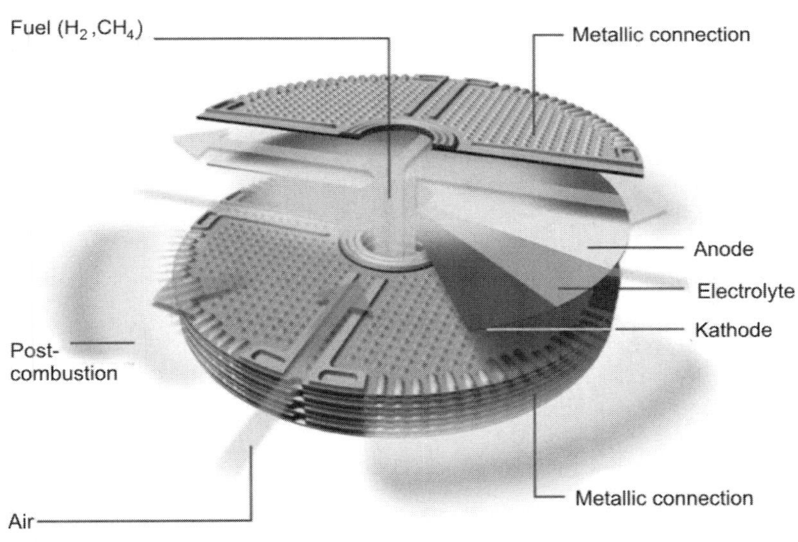

Figure 5.11 SOFC.

Pilot plant tests have shown that in SOFCs operated with biogas a lower efficiency (max. 5%) is obtained in comparison to an SOFC fed with natural gas.

Table 5.3 shows the state-of-the-art of the fuel cell technology. Although fuel cells show significant advantages in comparison to other "engines" for generating current from biogas, there are still many doubts about their breakthrough on the market because of the high costs. Their main advantage is their very high electrical efficiency. It is expected that the assumed considerable decrease in costs when fuel cells are manufactured on a production basis will push their commercialization.

Table 5.3 Fuel cell types.

Abbreviation	***AFC***	***PEFC***	***PAFC***	***MCFC***	***SOFC***
	Alkaline Fuel Cell	***Polymer Electrolyte Membrane Fuel Cell***	***Phosphoric Acid Fuel Cell***	***Molten Carbonate Fuel Cell***	***Solid Oxidized Fuel Cell***
Electrolyte	30% KOH	Membrane from polymers (Nafion, Dow)	Concentrated H_3PO_4 absorbed in plastic fleece (Si-PTFE)	Molten Li_2CO_3/K_2CO_3 bound in $LiAlO_2$ matrix	$Zr(Y)O_2$ with yttrium-doped zirconium oxide
Temperature (°C)	60–90	60–120	130–220	650	800–1000
Fuel gas	Pure H_2, hydrazine	H_2, methane, methanol,	Methane, H_2, natural gas, coal gas, biogas	Methane, coal gas, biogas,	Methane, H_2, coal gas, biogas
Gas	Oxygen	Air (3–4 bar)	Air	Air and CO_2	Air
Yield (W/cm^2)		0.6	0.2	0.1	0.4
$\eta_{el.}$ (%)	62	60 (H_2), 40 (CH_4)	40	50	55
Status of development	Well engineered	Marketable in niches	Marketable for some years Tested with biogas	Marketable for some years Tested with biogas	Marketable for some years Tested with biogas
Advantages/disadvantages		Dynamic Fast start-up, insensitive to load changes	Less dynamic Insensitive to CO	Hardly dynamic Nickel and nickel oxide catalysts usable instead of platinum Heat of reaction directly usable for gas reforming.	Not dynamic Long lifetime No catalysts required Heat of reaction directly usable for gas reforming Coupling possible with gas turbine

Current problems	CO_2^- incompatible due to reactions with caustic potash	The low temperature retards the reaction – a platinum–catalyst is required. Platinum is CO-sensitive especially at low temperatures	Overflow of acid and wetting of the electrodes Drying	Highly corrosive melt Deterioration of the cells during colling after shut down Short lifetime	Tightness of the planar cells
Attempted developments		Saving of catalyst material Minimizing of the losses of the ionic conductors Higher temperatures (180 °C)	Scale-up of the plants and amplification of the market	Longer lifetime of minimum 40 000 h Lowering costs	Lowering of the manufacturing cost Lowering the working temperature Other materials
Applications	Today's very seldom applied	Vehicles, small CHPs, rechargeable batteries	Small to medium-sized CHPs	Medium to large-sized CHPs, generators fired with coal	CHPs, combi power stations

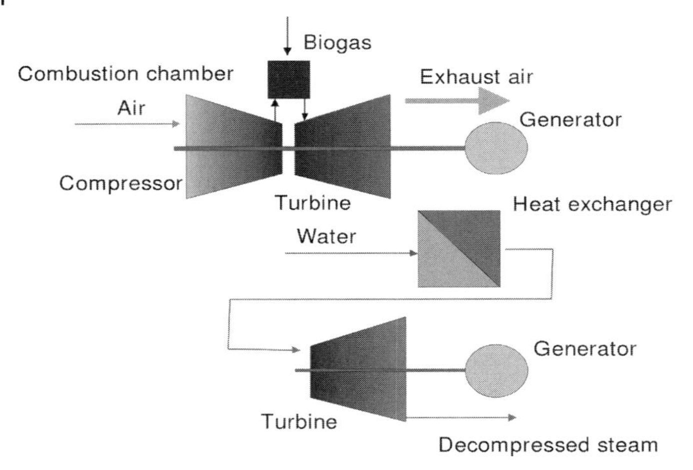

Figure 5.12 Gas turbine process with heat recovery in a steam turbine downstream.

5.3.1.4 Generation of electricity in a gas turbine

Biogas can be converted to current via gas turbines of medium and large capacity (20 MW_{el} and more) at a maximum temperature of ca. 1200 °C. The tendency is to go to even higher temperatures and pressures, whereby the electrical capacity and thus the efficiency can be increased.

The main parts of a gas turbine are the compressor, the combustion chamber, and the turbine.[48]

Ambient air is compressed in the compressor and transmitted to the combustion chamber, where biogas is introduced and combustion takes place. The flue gas that is so formed is passed to a turbine, where it expands and transfers its energy to the turbine. The turbine propels on the one hand the compressor and on the other hand the power generator. The exhaust gas leaves the turbine at a temperature of approximately 400–600 °C. The heat can be used for driving a steam turbine downstream, for heating purposes, or for preheating the air that is sucked in (see Figure 5.12).

The gas turbine is regulated by changing the biogas supply into the combustion chamber.

Gas turbines are characterized by very low emission values. When feeding decontaminated biogas, the NO_x value in the exhaust gas is ca. 25 ppm. The CO content can be considerably reduced by a catalyst downstream.

Higher efficiencies can be obtained by higher turbine inlet temperatures, which presupposes particularly temperature-resistant materials and complicated technologies for blade cooling. Therefore gas turbines of the highest efficiency are relatively maintenance-intensive.

48) Cp. BRO 13

5.3.1.5 Generation of electricity in a micro gas turbine

Micro gas turbines are small high-speed gas turbines with low combustion chamber pressures and temperatures. They are designed to deliver up to ca. 200 kW_{el} electrical power.

Nearly all micro gas turbine manufacturers offer turbines of radial design with combustion air compressor, combustion chamber, generator, and heat exchanger (Figure 5.13). Micro gas turbines are characterized by a single shaft on which the compressor, the turbine, and the generator are fixed. The turbine propels the compressor, which compresses the combustion air and at the same time the generator. Thus radial forces to the bearings and to the shaft are avoided, which allows a simple design; e.g., the bearings can be "gas-lubricated" because of the low load. The gas lubrication can be accomplished by passing compressed air through the bearings. Oil changes as required for normal turbines are not necessary because of the oil-free running of the micro gas turbine.

For normal operation, the turbine sucks in the combustion air. The fuel is normally supplied to the combustion air in the combustion chamber. When biogas with a low calorific value is used it can also be mixed with the combustion air before the turbine. In the latter case, a little biogas has only to be supplied directly to the combustion chamber for fine adjustment.

The up to 100 000 rpm rotating generator produces high-frequency alternating current, which is converted in an electronic device (Figure 5.14), so that it can be fed synchronously into the power network.

The electrical efficiency of 15–25% of today's micro gas turbines is still unsatisfactorily low. An attempt to increase the efficiency has been made by preheating the combustion air in heat exchange with the hot turbine exhaust gases. But great improvements are still necessary before micro gas turbines will penetrate the market of industrial biogas plants. However, already today the coupling of a micro gas turbine with a micro steam turbine to form a micro gas and steam turbine seems interesting and economical because of its high electrical efficiency (ca. 50%).

Micro gas-turbines are regulated only by varying the fuel supply.

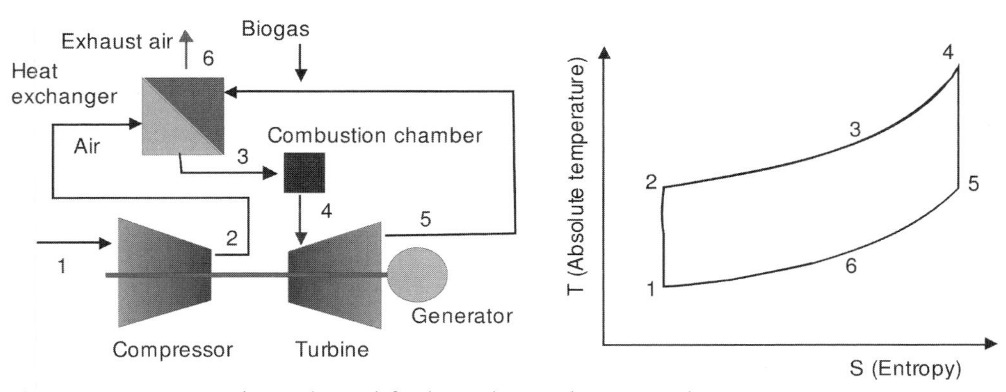

Figure 5.13 Micro gas turbine: scheme (left), thermodynamical process (right).

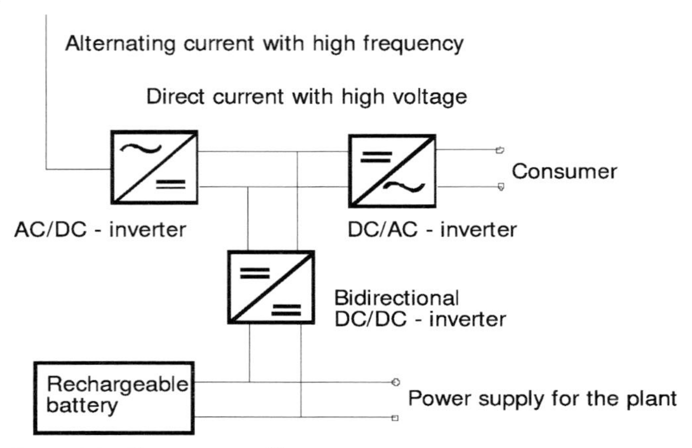

Figure 5.14 Electronic gear.[49]

Table 5.4 Some worldwide suppliers of micro gas turbines.

Name	*Country*	*Electrical capacity [kW]*	*Type of turbine*
Allied Signal	USA	75	Radial one-stage
Bowman/GE/Elliott	USA	45, 60, 80	Radial one-stage
Capstone	USA	30, 60	Radial one-stage
Honeywell Power Systems	USA	75	
NREC 4	USA	70	Radial two-stage
Turbec	Sweden	Ca. 100	Radial one-stage

The maintenance interval can be as long as 2000–8000 h. It can be longer when the operating temperature is increased above the normal operating temperature of ca. 10 °C.

In the United States, many micro gas turbines are in operation for the generation of electricity from natural gas. The machines are well approved for low capacities. Some experiments on the supply of biogas to micro gas-turbines have been made in Sweden (Table 5.4).

Turbec[50]

Combined heat and power generation can be achieved by micro gas turbines ready for installation like the turbine module Turbec T100. In this, the combustion air bypasses the generator in order to provide heat exchange between exhaust gas and combustion air.

The Turbec T100 turbine runs at 70 000 rpm. The shaft with the centrifugal compressor, the radial-flow turbine, and the four-pole permanent magnet of the generator are pivoted on both sides of the permanent magnet. As compressor, a Scroll compressor is installed with a suction head of 0.02–1.0 bar. In the combus-

49) Cp. WEB 57

50) Cp. WEB 63

tion chamber, a pressure of 4.5 bar is maintained. The Turbec T100 is controlled by a fully automatic module controller (power module controller PMC), which also regulates the start-up.

When fueling with biogas, a gas pressure of 6–8.5 bar, a gas temperature of 0–60 °C before compression, and a calorific value of the gas of 38–50 MJ kg^{-1} are required.

The fast-rotating generator supplies alternating current of high frequency. This is transformed in a static frequency changer into normal current of 50 cycles per second. Thus, the number of revolutions of the turbine is decoupled from the frequency. The micro gas turbine can therefore be operated with a variable rotation rate nearly without any loss of efficiency.

The performance data of the Turbec T100 are:

Electrical power capacity:	100 KW
Nominal electrical efficiency:	33%
Overall efficiency:	77%

Downsteam from the Turbec micro gas-turbine, a cross-flow heat exchanger transfers the residual heat of the exhaust gas to water, which can be used anywhere in the plant for heating purposes. With it, an exhaust gas temperature of 85 °C can be reached at an exhaust gas flow rate of 0.80 kg s^{-1}.

The limiting values of the CARB (Californian Air Resource Board) can be achieved.

Capstone

Another type of micro gas turbine is the Capstone micro turbine, which could perhaps also be used for biogas supply. Its generator is cooled by the air that is sucked in. The compressor works radially.

5.3.2 Controlling the CHP

All operational data of the CHP,e.g., capacity, temperatures, and pressures are constantly supervised, controlled, and monitored. Deviations from given or prescribed values result in warnings or switching off the entire plant. A modern SPS (programmable controller) enables a completely independent operation of the CHP nowadays. The SPS automatically performs the following tasks:

- Start and stop sequencing program
- Monitoring the analog values of the generator as well as evaluation of digital breakdown reporting data
- Power adjustment and rpm
- Synchronizing control
- Heating water temperature control
- Lambda regulation.

A modern CHP is controlled on the basis either of the heat requirements or of requirements of the consumers for electric current.

The heat requirement-dependent control registers the temperature of the returning heating water and selects and opens heat flows from different aggregates connected in parallel, and/or connects a peak load boiler.

The electric current-dependent control selects and switches equipment on and off depending on its power. With power stations working in parallel, the controller calculates the necessary CHP capacity, in order to adjust the net consumption to a desired value (usually zero-load).

Number of revolutions per minute and electric power are regulated by adjustment of the injection pump of Diesel engines or by the butterfly valve of four-stroke cycle engines according to the power generation and the demand for electricity. If the engine delivers an energy surplus, the number of revolutions per minute or the electrical power rise presuming the generator runs network-synchronously.

When the engine starts or when the plant is working in isolation,[51] the controller smoothly adjusts the rpm, continuously measuring the actual value by the starter ring of the engine. For a four-pole generator, this should be aimed at 1500 rpm when the current is provided at 50 cycles per second.

After connecting the CHP to the power network, the controller works as the power control. The actual value of the generator's power is sensed by a transducer. The aimed-at value depends on the mode of operation of the biogas plant. The value can be set at a standard value or be adjusted to the electric current requirements of the object to be supplied.

5.3.3 Emission control[52]

During combustion, several chemical reactions occur and develop different reaction products depending on pressure, temperature, and oxygen and nitrogen concentration. These products are to be found in the exhaust gas, as can be seen from the Table 5.5 for five different biogas-driven engines.

The principal contaminants are nitrogen oxides (NO_x), carbon monoxide (CO), and particles (dust, unburnt carbon).

The concentrations of these main compounds vary for engines with a combustion heat performance less than 1 MW in continuous operation, as can be seen from Figure 5.15.[53] The NO_x content of the exhaust increases with the methane content of the biogas. The other impurities in the exhaust like CO, formaldehyde, and unburnt carbon decrease with increasing methane content.

5.3.3.1 Regulations

The exhaust gas emissions of engines are restricted in Europe. The limits even differ among the European countries. In Table 5.6, the limiting values for Austria and Germany are given as an example. They refer to 0 °C, 1013 mbar, 5% O_2 and regular running at nominal capacity.

51) i.e. without feeding current into the public power network

52) Cp. BOK 53

53) Cp. BOK 42

Table 5.5 Average amounts of contamination in the exhaust gases of combustion engines driven with biogas (* above measuring range, GM = four-stroke cycle engine, ZS = ignition oil Diesel engine).

Contamination	*Unit*	*Plant 1*	*Plant 2*	*Plant 3*	*Plant 4*	*Plant 5*
		13 kWel GM	*22 kWel ZS*	*50 kWel ZS*	*80 kWel ZS*	*132 kWel ZS*
CO	mg m^{-3}	>3354*	>3549*	2511	3231	1142
NO	mg m^{-3}	618	708	1083	301	1062
NO_2	mg m^{-3}	29	4	62	227	48
NO_x (given as NO_2)	mg m^{-3}	973	1086	1719	687	1672
SO_2	mg m^{-3}	2	7	42	0	26
O_2	% by volume	4.4	4.2	9.1	10.4	9.1
CO_2	% by volume	15.8	14.2	10.2	8.8	10.1
Temperature of the exhaust gas	°C	50.6	255.6	216.7	183.1	307.7
λ	–	1.27	1.26	1.79	2.04	1.80

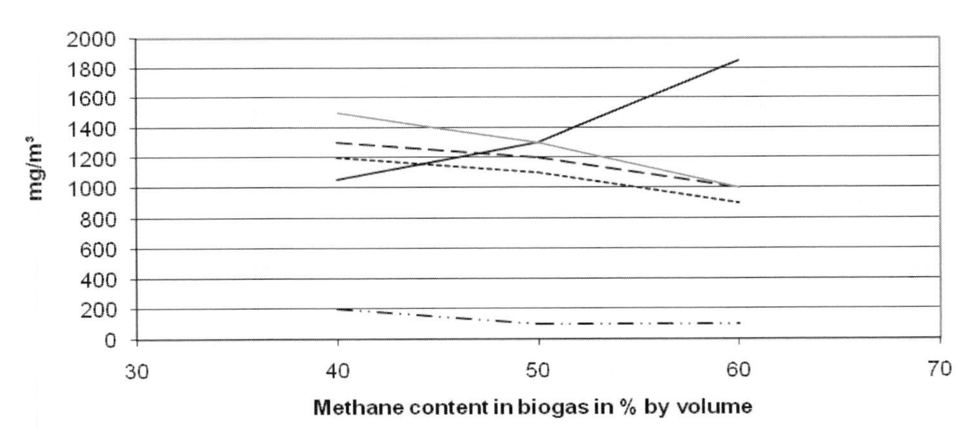

Figure 5.15 Impurities in the exhaust gas of a CHP at different methane contents (—) NO_X; (-··-) Formaldehyde; (—) Total-carbon; (······) Methane loss; (- - -) CO.

The measurement and recording of the dust load can be omitted for all combustion engine systems, since even in the exhaust gas of ignition oil Diesel engines only concentrations below 8 mg m^{-3} could be detected.

5.3.3.2 Measures for the reduction of emissions[54)]

Three-way catalyst Three-way catalysts work satisfactorily for emission control with motorcar engines. A catalyst consists of a metallic or ceramic carrier on which the catalyst materials platinum and rhodium are vapor-deposited.

54) Cp. WEB 62

Table 5.6 Prescribed limits in Austria and Germany (FWL = total produced energy).

Contamination	*Limits of emissions in Austria*		*Limits of emissions in Germany*
	<250 kW FWL	*≥250 kW FWL*	
Dust	n.a.	n.a.	20 mg m^{-3} in general 50 mg m^{-3} for ignition oil Diesel engines
Nitrogen oxide (NO_x)	n.a.	400 mg/m^3 and 500 mg/m^3 for lean-burn engines	500 mg m^{-3} for gas four-stroke engines 1500 mg m^{-3} resp. 1000 mg/m^3 with gas compression-ignition engines or ignition oil Diesel engines
Carbon monoxide (CO)	650 mg/m^3	650 mg m^{-3} and 400 mg m^{-3} for lean-burn engines	1000 mg m^{-3} with gas four-stroke engines 650 mg m^{-3} (at >3 MW FWL) 2000[55] mg m^{-3} with gas compression-ignition engines or ignition oil Diesel engines
Non-methane hydrocarbons	n.a.	150 mg m^{-3}	n.a.
Hydrogen sulfide (H_2S)	n.a.	5 mg m^{-3}	n.a.
SO_2	n.a.	n.a.	350 mg m^{-3}
Formaldehyde	n.a.	n.a.	60 mg m^{-3}

At the surface of a three-way catalyst, nitrogen oxides (NO and NO_2) are reduced, and at the same time hydrocarbons (C_mH_n) and carbon monoxide (CO) are oxidized. Carbon dioxide (CO_2), nitrogen (N_2) and water (H_2O) are formed.

$$NO + CO \rightarrow \frac{1}{2} N_2 + CO_2$$

$$2(m + n/4)NO + C_mH_n \rightarrow (m + n/4)N_2 + m\ CO_2 + n/2H_2O$$

In the three-way catalyst, the oxygen content is measured by a lambda probe. The measured values are transferred to the programmable controller (SPS) of the CHP. The SPS regulates the gas feed automatically in such a way that the engine is operated at a fuel air ratio of $\lambda = 1$. Thus, no oxygen is present in the exhaust gas.

The catalyst has to be individually designed according to the engine data such as quantity of exhaust gas, waste gas emissions, and exhaust gas temperature. Depending on the design, the catalyst requires a temperature of 400–500 °C in the upstream gas.

55) These requirements have to be considered dynamically.

Table 5.7 Characteristic figures for a CHP running at its optimum.

	Electrical efficiency	*Carbon monoxide emissions (CO)*	*Nitrogen oxide emissions (NO_x)*
Diesel engine	42	n.a.	2500 mg/Nm^3
Four-stroke engine in lean-burn range, without exhaust recycling, with precompression, without catalyst	38	1.000 mg/Nm^3	250 mg/Nm^3
Four-stroke engine in $\lambda = 1$ operation, without exhaust recycling, without precompression, with 3-way catalyst	35	1.000 mg/Nm^3	30 mg/Nm^3
Four-stroke engine in $\lambda = 1$ operation, with mixture cooling, with exhaust recycling, with precompression, with 3-way catalyst	41	100 mg/Nm^3	1 mg/Nm^3 Related to a oxigene content of 5% in the exhaust

By the reactions in the catalyst the exhaust gas temperature increases, but normally does not exceed 650 °C. The temperature is regulated by the SPS of the CHP.

By working with a three-way catalyst at a fuel air ratio of $\lambda = 1$ +/– 0.5% and recycling the exhaust gases, the best operation of the CHP is achieved,[56),57),58)] as can be seen from Table 5.7.

SCR catalyst

Another catalyst type is the SCR catalyst. It works by selective catalytical reduction (SCR) of the exhaust gases. Depending on the capacity of the engine, urea solution is injected into the exhaust just behind the engine. The urea is converted directly to ammonia and carbon dioxide. The nitrogen (NO_x) oxides react with ammonia (NH_3) and form molecular nitrogen (N_2) and water vapor (H_2O).

Oxidation catalyst

With Diesel engines, emissions of CO and unburnt hydrocarbons (HC) can also be reduced by oxidation catalysts.

However, the application of an oxidation catalyst is frequently out of question, because of the sulfur content in the exhaust gas. Sulfur poisons the catalyst, and SO_3 is formed, which reacts with the water formed to give sulfuric acid. The acid corrodes the exhaust system at the low temperatures at which the acid condenses.

Filter for unburnt carbon

In a mesh filter, up to 90% of particles of unburnt carbon are removed. If the filters are fully loaded, they must be exchanged. At temperatures over 600 °C, their service life extends, because the carbon particles burn to CO_2 and CO. The temperature at which the particles are burnt can be lowered by ca. 150 °C if a catalyst layer is fixed on the filter surface.

56) Cp. WEB 83
57) Cp. WEB 45
58) Cp. WEB 75

With partial load, the exhaust gas temperature is too low for filter regeneration. Therefore the CHP control supervises the exhaust back-pressure and automatically raises the equipment to its full operating load and thus provides a sufficiently high exhaust gas temperature, when measured values indicate a possible contamination of the filter.

5.4 Lessons learnt from experience

The bioreactor should continuously supply biogas of consistent quality

- Substrate should be fed into the bioreactor at a constant rate. In the partial load operation, when too little substrate is supplied and therefore too little biogas is available, the NO_x value remains low, but the CO value exceeds the limit value.
- The bioreactor should be adjusted by regularly measuring the pH value and the organic dry matter as COD value in order to obtain a high and consistent biogas yield with high methane content.

The heat of the exhaust gas from the biogas combustion engine should be used. At least the possibility of fitting a heat exchanger in the exhaust pipe should be considered.

Modern engines should be equipped with special devices

- for the achievement of a low NO_x value. But is has to be kept in mind that adjustments for this purpose decrease the efficiency of the engine.
- to keep the NO_x value low at a high efficiency. Such a device could be a denitrifying device in the exhaust gas stream (SCR catalyst), recirculation of a part of the exhaust gas, integrated filters for unburnt carbon, or integrated fuel regulation.

The settings of the engine should be carefully done with special reference to the biogas composition.

- When operating the engine as an ignition oil Diesel engine, the NO_x concentration in the exhaust gas is similar to that of a normal Diesel engine. But the CO concentration behaves differently: at partial load operation the CO concentration increases for $\lambda > 2$. The reaction from CO to CO_2 occurs so slowly that a large part of the burned intermediate product CO reaches the exhaust pipe. Therefore, ignition oil Diesel engines should be preferably used for operation under full load. A low CO concentration could be reached only by recycling exhaust gases, so that the engine is run at $\lambda < 1.9$.

- Modern power generation plants by default are equipped with an engine management. With variable fuel air regulation, e.g., λ regulation, the mode of operation can be easily adapted to short-term variations of the gas quality.
- If the fuel/air ratio is too high, unacceptably high NO_x emissions are released. High ignition oil fractions of more than 10% cause high NO_x emissions.

Regular engine maintenance and emission measurements are necessary and prescribed by authorities. Only in this way can the safe operation of the power generator be ensured.

- An emission measurement is to be performed in Germany within a period of, e.g., 3–12 months after start-up (acceptance test).
- The emission measurements have to be regularly repeated every three years (recurring measurements).
- The measurements may be performed only by a certified organization.[59)]
- The execution of the measurements and/or the measuring reports is to be made according to the regulations:[60)] The substrates and their rates are to be indicated in the measuring report. At least three measurements are to be accomplished with unimpaired mode of operation with highest emissions and one measurement under regular operating conditions. Each measurement must last minimum of half an hour; the result of each measurement has to be indicated as a half-hour average value. The measuring schedule has to be communicated to the responsible authorizing agency promptly; e.g., at least eight days before beginning. For the proper execution of the measurements, a suitable sampling device has to be installed in the exhaust pipe and recommended measuring technique have to be taken in consideration.
- The CO emissions can be reduced by regular cleaning of the spark plugs with a four-stroke gas engine.
- The exhaust gases have to be discharged perpendicularly upward into the free air. Plants which have to be approved must have a chimney of a minimum height, which has to be calculated according to prescriptions.[61)] The chimney must be at least 10 m high above the soil and/or 3 m high above the ridging.
- Chimneys and exhaust nozzles may not have covers. As protection against rain, fall deflectors are recommended.

59) Cp. LAW 13
60) Cp. LAW 8
61) Cp. LAW 8

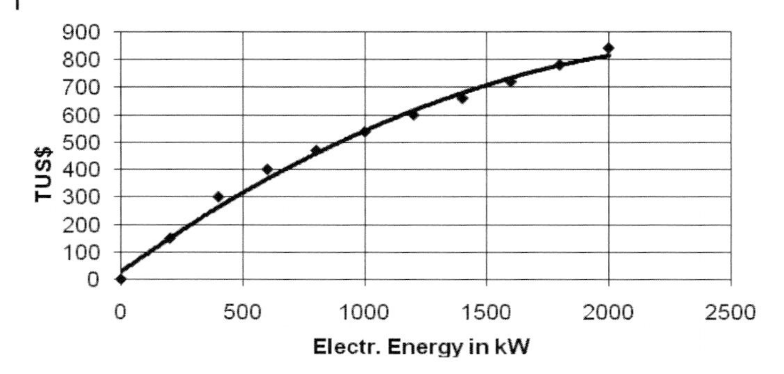

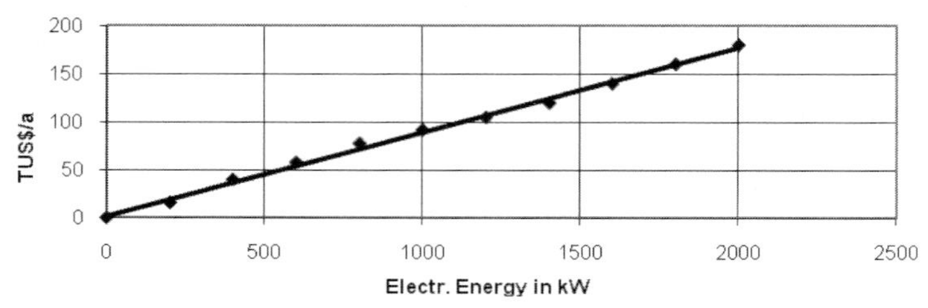

Figure 5.16 Investment costs (above) and operational costs (personnel, maintenance, repairs, insurance) (below) in TUS$ for a CHP depending on the capacity.

The plants and engine manufacturers must provide regular training periods for all operational personnel in order to communicate their experience.

5.5 Economy

The dependency of the electrical power and the price of a CHP on its capacity is shown in the Figure 5.16.

If the CHP is fed with decontaminated biogas equivalent to natural gas, it could have a lifetime of 15 years. In agricultural biogas plants, where roughly desulfurized biogas is fed, a maximum lifetime of eight years is to be expected.

5.6 CHP manufacturers

There are many CHP manufacturers,[62] who design and construct CHPs, particularly for the generation of electricity from biogas. The features of CHP plants available on the market in Europe vary widely.

62) Cp. WEB 62

6
Biogas for feeding into the natural gas network

Apart from the local direct conversion of biogas to current and heat, there is the possibility to clean the biogas, to separate methane and carbon dioxide,[63] and to feed the methane into the low-pressure natural gas network (see also Section 2.1.1.). Carbon dioxide can be compressed and filled in bottles for sale.

Before feeding the biogas into the natural gas network, the following features must be adjusted:[64]

- Pressure
- Density
- Total sulfur
- Oxygen and humidity content
- Wobbe index.

Unlimited feed of sewage gas/biogas

Sewage gas/biogas can be added to natural gas H without restrictions, if it is prepared and fulfills the following requirements:

- CH_4 content: at least 96% by volume
- CO_2 content: no special definition
- O_2 content: less than 0.5% by volume
- H_2S content: max. $5\,mg\,Nm^{-3}$
- Water vapor dewpoint: saturation temperature below the average soil temperature of the region where the gas is supplied
- Relative humidity: max. 60%.

Limited feed of sewage gas/biogas

A maximum of 5% by volume of sewage gas/biogas is allowed to be added to the natural gas H flow rate if it is partially prepared as follows:

- CH_4 content: natural CH_4 content of sewage gas/biogas
- CO_2 content: natural CO_2 content of the sewage gas/biogas

63) Cp. JOU 11

64) Cp. BOK 39

Biogas from Waste and Renewable Resources. An Introduction.
Dieter Deublein and Angelika Steinhauser

ISBN: 978-3-527-31841-4

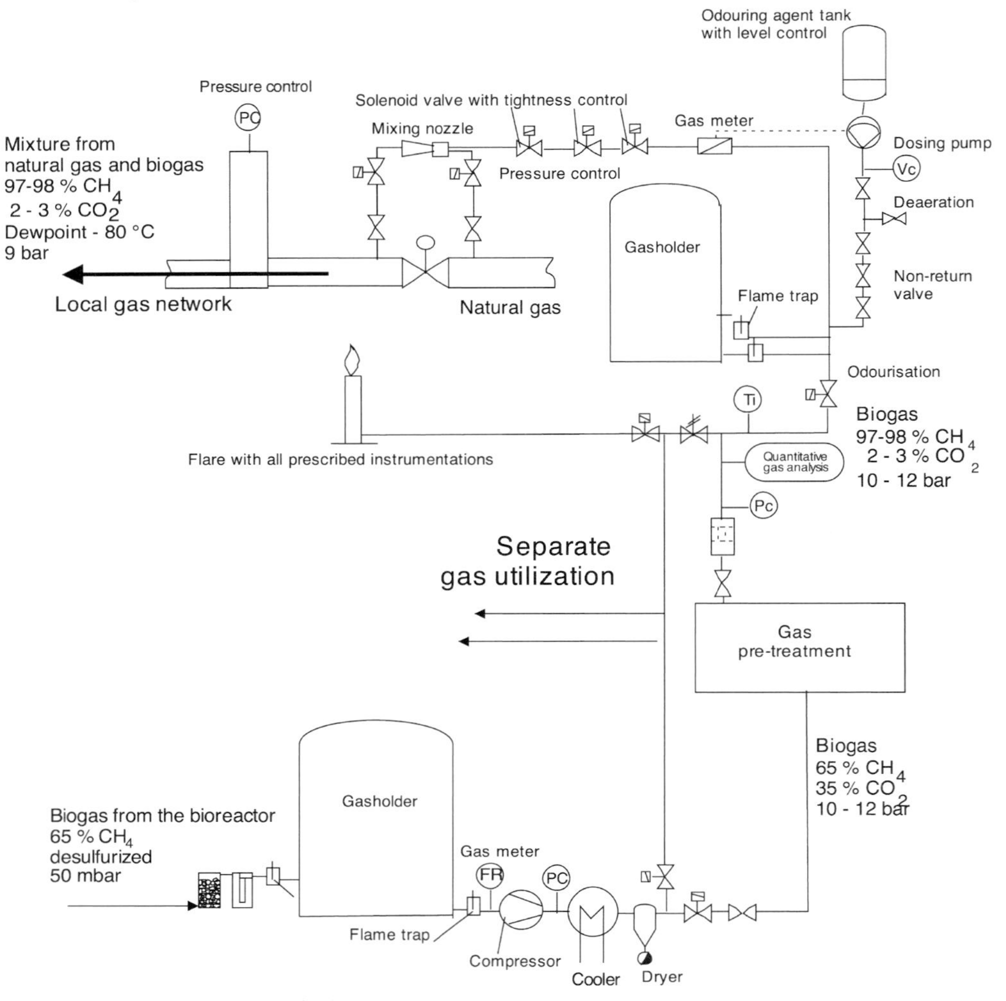

Figure 6.1 Plant for feeding biogas into the natural gas network – all pressures given in bar gauge.

- O_2 content: less than 0.5% by volume
- H_2S content: max. 5 mg Nm^{-3}
- Water vapor dewpoint: saturation temperature below the average soil temperature in the region where the gas is supplied
- Relative humidity: max. 60%.

In some countries, the composition of the gas mixture at the feed point to the customers is decisive for pricing. In order to maintain the gas quality at the feed point, gas mixing is permitted (Figure 6.1), e.g., L-gas may be mixed with H-gas in order to raise the quality. Mixing devices are controlled by the calorific value or

Wobbe index using the variables pressure, natural gas flow rate, and gas flow rate.[65)] In order to ensure a homogeneous mixture in the downstream network, the mixing devices have to be driven continuously. The addable rate of CO_2 depends on the CO_2 rate in the natural gas stream (normally not more than 2%) and the quantity of nitrogen in the natural gas. Wide variations in the composition of the biogas cause large expenditure on the regulation system and need a system which can give "attenuation" in the summer as well as raising the quality in the winter.

A minimum calorific value at the feed point is a result of the use of natural gas only for heating and steam boiling in houses. The standard is difficult to keep up regarding techniques, even more in H-gas networks than in L-gas networks. But the minimum calorific value nowadays is less important than other features of the gas; e.g., for gas engines the methane number, which indicates the knocking behavior of gases. The methane number considerably influences the lifetime of the engine. In respect to the engine's lifetime, biogas, with its methane number of 130, is much more suitable for power generators than natural gas, with the methane number of 85–90. Natural gas does not really cause problems associated with the low methane number when fed to power stations. However, piston seizures were observed, when feeding natural gas mixed with liquid gas – the butane in liquid gas has a methane number of 10, propane of 35. Thus, when the calorific value of biogas is increased by adding other gases, the modification might just go in the wrong direction.

Biogas could be more easily fed into the natural gas network if the quality of the transported gas (natural gas + biogas) was what decided the price. Than the costumer has to pay for the energy taken and not for the gas flow rate of a certain quality.

If the biogas is not completely compatible with the natural gas distributed in the gas network, then the maximum feedable biogas rate is in general determined by the Wobbe index of the resulting gas mixture. In Denmark up to 25% biogas with a methane content of 90% may be fed to the natural gas network. Without any preparation, up to 8% by volume may be mixed with the natural gas if the biogas has a methane content of at least 60%.

There are special stations for the automatical supply of biogas into the natural gas network, depending on the quantity and quality of the gas. Such stations consist of

- gas filters for the protection of the controlling system
- compressors, in order to safeguard the gas pressure in the network
- gas quality measuring instruments, e.g., a calibratable process gas chromatograph
- shut off device with ball valves in case of unexpected situations
- gas pressure controller to maintain the initial gas pressure
- sometimes a special gas mixing device.

65) Cp. WEB 68

Before connecting to the gas network, the following have to be installed:
- Two automatic shut-off valves with an intermediate tightness-checking device. In case of interruptions of the operation, incidents, and power failures, these valves must automatically close.
- One valve to be operated manually. This valve may only be opened by personnel responsible to the owner of the gas network for safety reasons.

Before feeding the gas into the network, it must be odorized. Because odorizing plants are very expensive, they are only installed in large biogas plants. For feeding of biogas from small plants it is recommended to odorize the main natural gas stream a little further upstream. The odorization can be done either by injection of the odorizing agents directly into the natural gas stream or by passing part of the gas flow through a container in which the odorizing agent evaporates.

Recently, tertiary butylmercaptan (TBM) was recommended as an odorizing agent. In comparison to tetrahydrothiophen (THT), it has the following advantages:
- high smell intensity
- low absorption rate at pipe walls
- easier disposability.

The lower the biogas pressure at the feeding point, the more advisable is the installation of a compressor plant.

6.1
Biogas for feeding into the natural gas network in Switzerland

In the year 1995, a biogas preparation plant[66] (Kompo-mobil I) was developed to supply gas to vehicles powered by retooled gasoline and/or Diesel engines. Components of that biogas preparation plant were
- Absorber of hydrogen sulfide on activated charcoal
- Pressurized water scrubber for the removal of the CO_2
- Molecular sieve dryer for the removal of the water vapor.

With this preparation plant, biogas was obtained with a methane content of 93% and a very low content (5 ppm) of water vapor and hydrogen sulfide. The power requirement of the entire preparation plant was however much too high, approaching 30% of the calorific value of the cleaned biogas. In the subsequent project (Kompo-mobil II), molecular sieve technology in form of pressure swing adsorption was applied, and more favorable values were reached:

66) Cp. BOK 56

- Approximately 30% lower investment costs than those of Kompo-mobil I
- Lower energy consumption –8% of the calorific value of the prepared biogas
- Increase in the methane content in the prepared gas to 96–98%
- Lowering of the hydrogen sulfide content to below the limit of detection.

The cost of the entire preparation plant amounted to about 450000 US$ (start-up 1995). Calculating on this basis and assuming an annual operating period of 4500 h, specific costs were 0.068 USct/kWh for biogas. These specific costs would correspond to 0.61 US$ per liter gasoline.

Despite these improvements, the preparation of biogas and feeding it into the local natural gas network or using it directly as fuel for vehicles were not competitive with the generation of electricity in a CHP.

Since the end of the 1990s, a local gas network has been built up around lake Zurich, and a considerable percentage of this is supplied with biogas out of particular established biogas plants.

Each of these is designed to produce more than 50 m^3 biogas per hour. The plants receive organic wastes from households, from industry, and from agriculture as well as industrial waste water, particularly from the food, animal feed, and paper industries. The biogas preparation includes the following process steps:

- desulfurization
- gas drying
- pressure swing adsorption for CO_2 separation.

Since the heat produced during the generation of electricity cannot be used economically in the local networks, and since the biogas plant works continuously and cannot produce biogas only if a vehicle is to be refuelled, the decision was made to feed all prepared biogas into the natural gas network. This gas mixture is sold under the labelling “Naturgas” in the area around Zurich, because it does not completely meet the specifications of natural gas.

6.2 Biogas for feeding into the natural gas network in Sweden

In the year 1992 in Laholm a biogas plant was constructed, mainly to contribute to the reduction of the eutrophication of Laholm bay, in which problems due to pollution by ferzilizers were increasing. Other objectives of the project were the production of biogas for Laholm and the production of a biological fertilizer for the surrounding agriculture.

The plant processes 25 000 Mg liquid manure per year as well as 10 000 Mg other waste materials, in particular from fifteen different food manufacturers.

Biogas production reaches about 20–30 GWh annually, with a methane content of ca. 70%. Up to the year 2000 this biogas was used in a CHP, and about 300 dwellings were supplied with heat. When not enough biogas was available, the CHP could be operated with natural gas.

A substantial disadvantage of the system was that in periods with low heat requirements almost 40% of the produced biogas had to be burnt in the flare.

In the year 2001, therefore, a biogas preparation plant was constructed, which is able to prepare 250 $m^3 h^{-1}$ biogas with 60–65% methane content to natural gas quality. In detail, the following steps are accomplished:

- desulfurization
- carbon dioxide separation by means of chemical absorption (Selexol)
- adaptation of the Wobbe index by addition of 5–10% propane gas.

The prepared biogas is used in a CHP as in earlier times. When the heat consumption decreases, the prepared biogas is fed into the local low-pressure natural gas network and distributed in the city of Laholm.

6.3 Biogas for feeding into the natural gas network in Germany

A project in Stuttgart Mühlhausen[67] was promoted by the European Union in the 1990s with the object of gathering experience in the preparation of sewage gas to natural gas quality. Before the project started, a basic calculation had shown that the preparation of biogas and its feeding into the natural gas network could be economical under the given circumstances.

For the preparation of the sewage gas, the procedure of chemical absorption (scrubbing under pressure) in an MEA reactor (monoethanolamine) was chosen. The experiences with the plant were quite positive. Crude sewage gas was prepared at the rate of 400 $m^3 h^{-1}$ and was afterwards fed into the natural gas network for several years without problems. As can be seen from Table 6.1, the plant actually operated profitably.

In Germany, the natural gas distributing companies have given up their initially discouraging attitude to biogas. Today it is assumed, according to a careful estimation, that in future 100 billion KWh biogas per year will be fed into the natural gas network, which is equal to 10% of the today's natural gas demand.

In Plienigen, near Munich, starting from the year 2006, an agricultural biogas plant is feeding 3.9 Mio m^3 biogas per year into the natural gas network, equivalent to ca. 2.9 MW. The plant is supplied with 32 000–35 000 Mg corn and energy-affording plants per year from surrounding farmers, who cultivate land of ca. 500 ha area. The residue is taken back to the farmers as fertilizer.

67) Cp. BOK 52

Table 6.1 Economics of the sewage gas pre-treatment part of the sewage water treatment plant in Stuttgart-Mühlhausen.

Pretreatment plant	***US$***
Investment costs	
0.75 Mio US$; 10 years amortization; 7% interest rate, annuity 14.24%	106 800
Operational costs	
Personnel	18 000
Power consumption	61 334
Other costs	34 730
Maintenance costs	
3.5% of the investment costs	26 250
Costs per year for the pretreatment plant	247 114
Biogas plant	
Investment costs	
0.9 Mio US$; 20 years amortization; 7% interest rate, annuity 9.44%	84 960
Maintenance costs	
2.0% of the investment costs	18 000
Costs per year for the biogas plant	102 960
Total costs	350 074
Income from the sales of biogas ($2.2 \times 10^6\,m^3 \times 0.2035$ US$/m^3)	–447 700
Totel revenue per year at 80% work load	97 625

Three horizontal bioreactors with a capacity of 1000 m^3 each are installed as well as three vertical reactors with a volume of 2700 m^3 each for after-fermentation. Two containers serve for residue disposal, with 10 000 m^3 volume in total.

The output is 920 m^{3}h^{-1} biogas. This biogas is first desulfurized, then dehumidified, and afterwards enriched to a methane content of approximately 96% by pressure swing adsorption (PSA). The PSA works with carbon molecular sieves. By the PSA procedure, the biogas volume rate is reduced to 485 m^{3}h^{-1} with natural gas quality. A conventional natural gas compressor compresses the cleaned biomethane to 40 bar. At this pressure the biogas is fed into the high-pressure gas network of the urban natural gas supply.

Also in the Niederrhein region, two biogas plants, and, near Bremen, another plant, have started operation and are feeding biogas into the natural gas network nearly simultaneously.

7
Biogas as fuel for vehicles

With respect to economy and technology, the utilization of biogas as fuel looks interesting, since compressors are already integrated in service stations and the ecological aspect could be marketed plausibly. In comparison to feeding the biogas into the natural gas network, its utilization as fuel is in general less problematic and can be realized much more cheaply. Four-stroke engines adapted to the combustion of natural gas are suitable for this purpose.

The features, which biogas should have in order to be used as fuel, are internationally recommended in the ISO/DIS 15403, and, since the year 2004 in Switzerland, nationally by the guideline 13G of the SVGW.[68] In Sweden, the standards required for biogas fuel are listed in Table 7.1.

The calorific value of the biogas/air mixture is ca. 15% lower than that of a gasoline air mixture for vehicles. Therefore the power would be expected to be15% lower at the same compression ratio of the engine. The power of the engine is further reduced by the low rate of combustion of biogas. This disadvantage is compensated by the high anti-knock quality of the biogas, so that a higher compression ratio of the engine can be used. By using a four-stroke engine instead of a gas Diesel engine, the efficency could be improved.

In both cases, an appropriate pressure (200–250 bar) tank is needed in the vehicle, of course for biogas as well as for natural gas. This tank must have a volume of 200 L, five times as much as a gasoline tank for the same cruising range.

7.1
Example project: "chain of restaurants in Switzerland"[69]

In the food processing companies and in the restaurants of a Swiss cooperative, about 2500 Mg bio wastes arise annually. Until the end of the year 2000, most of the bio wastes were composted (1500 Mg/year), and the rest were taken for pig fattening. Both utilizations were under discussion because of their odor emissions. Today, all bio wastes from all food processing companies and restaurants, even leftovers, are supplied to a biogas plant and converted to biogas and fertilizer.

68) Cp. JOU 31

69) Cp. BOK 55

Biogas from Waste and Renewable Resources. An Introduction.
Dieter Deublein and Angelika Steinhauser

ISBN: 978-3-527-31841-4

Table 7.1 Prescriptions in Sweden for biogas as fuel according to Swedish standard SS 15 54 38 (t_S = dewpoint temperature).

	Normal four-stroke engines	*Lean-burn engines*
Wobbe Index	12.2–13.1 kWh/Nm^3	12.4–12.9 kWh/Nm^3
CH4 content	97 ± 2% by volume	97 ± 1% by volume
Methane number	>130	>130
O_2 content	<1% by volume	<1% by volume
$CO_2 + O_2 + N_2$ content	<5% by volume	<4% by volume
Total nitrogen (without N_2)	<20 mg/Nm^3	<20 mg/Nm^3
H_2S content	<23 mg/Nm^3	<23 mg/Nm^3
H_2O content	<32 mg/Nm^3	<32 mg/Nm^3
Dewpoint temperature for the highest pressure in the gasholder	t_S − 5 °C	t_S − 5 °C

The biogas drives a CHP and/or is fed to the natural gas network after appropriate preparation. The cooperative owns a gas station and eight trucks with engines refitted for gas burning, which are run with biogas. The trucks supply the restaurants close to the city daily.

The cooperative has had such good experience with this system that they are planning to introduce it at other locations also. The system depends heavily on features of the actual infrastructure, such as the technology of the biogas plant, the gas piping connections, the organization, and of course the goodwill of the employees.

The 2500 Mg per year bio wastes of the cooperative correspond to 200 000 L Diesel oil with respect to the produceable usable energy. If a consumption of 33 L Diesel oil per 100 km for each truck is assumed, the 8 trucks can drive 606 000 kilometers per year with biogas.

7.2 Example projects in Sweden

The Swedish government has set the goal that Sweden shall be independent of imported oil and natural gas within the foreseeable future. In many Swedish cities it is usual that "green" cars, fed with renewable fuels, are exempt from road duty and parking fees.

In Linkoeping,[70),71)] ca. 200 km south of Stockholm, as early as 1990, refitting of some urban buses for the combustion of biogas was started. Today, the entire fleet of city buses as well as taxis, transporters, refuse collectors, private vehicles, and even the train (Figure 6.28), which moves regularly between Linkoeping and the city of Vestervik, situated 80 km away, are driven by biogas. Great efforts had to

70) Cp. BOK 54

71) Cp. WEB 85

be made to make it possible to drive the train with biogas. For example, the Diesel engine had to be replaced with two biogas engines. Eleven pressure bottles containing biogas are now enough for driving the train ca. 600 km.

In the biogas plant, 50 000 $Mg\,a^{-1}$ of slaughterhouse wastes, human excrement, alcohol confiscated by the customs administration, and liquid manure are first sanitized at 70 °C and then fermented mesophilically for 30 days.

The crude biogas is prepared in five preparation plants. One of it operates by the PSA procedure and four by the wet pressure-scrubbing process. The overall capacity of the preparation plant amounts to 1800 $Nm^3\,h^{-1}$ crude biogas. In the plants, the methane content is increased from 65–70% to 96–97%.

The residues from the fermentation are distributed on agriculturally used land as high-quality fertilizer.

In Sweden, similar projects exist for the preparation of sewage gas in Stockholm Henriksdahl, in Norrkoeping, and in Helsingborg.[72)]

72) Cp. WEB 32

Part VII
Residues and waste water

Biogas from Waste and Renewable Resources. An Introduction.
Dieter Deublein and Angelika Steinhauser

ISBN: 978-3-527-31841-4

1
Residues

In most biogas plants the retention time of the biomass is ca. 20 days in the anaerobic reactor. As mentioned above, only part of the organic matter is degraded within this time. A certain amount of residue remains (Table 1.1).

This residue contains more or less water depending on the applied technology. Residues from a wet thermophilic fermentation have a higher water content than those from mesophilic or dry processes.

Before the residue can be distributed as valuable fertilizer on agricultural land, it must be sufficiently dehydrated, and in many cases sanitized, deodorized, and rotted.

If high temperatures are applied during fermentation a hygienization can be avoided since the residue is discharged in a sufficiently sanitized condition. In the most commonly used wet fermentation processes, additional procedures for hygienization of the residue have to be provided, especially if the residue is discharged after the second phase of degradation at a temperature no higher than 30 °C.

The residue looses its typical smell after 12–24 h of anaerobic fermentation and remains without smell afterwards. Therefore, when the fermentation process is adequate, additional deodorization is not necessary.

Rotting is the aerobic decomposition of the residue. This takes ca. 2–6 weeks, depending on how long the matter was subjected to the anaerobic fermentation and the desired degree of decomposition.

The aerobic rotting usually takes place in a totally enclosed or merely roofed composting plant consisting of rotting tunnels or compost hills. The figures given in Table 1.2 are achievable.

Instead of composting, residues from biogas production can also be gasified, carbonized, and/or pyrolyzed. All procedures work with reduced oxygen supply, but differ with respect to the temperature and retention time of the residues in respective reactors.

Compost generated in an anaerobic process is of only slightly inferior quality to compost produced completely aerobically, because it contains only a little more salt and nutrients. But anaerobic compost has a strongly differing nitrogen reaction dynamic, the reason for which still has to be investigated however. High

Biogas from Waste and Renewable Resources. An Introduction.
Dieter Deublein and Angelika Steinhauser

ISBN: 978-3-527-31841-4

Table 1.1 Some components of the residue of bio wastes (kg/Mg).

Dry matter	480
Organic dry matter	240
Total nitrogen (N)	6.2
Phosphate (P_2O_5)	3.2
Potassium (K_2O)	4.6
Calcium (Ca)	20
Magnesium (Mg)	3.6

Table 1.2 Characteristic figures for aerobically rotted residue after 3 weeks anaerobic degradation in one stage under quasi-dry and mesophilic conditions compared with prescribed limits.[1)]

Parameter[2)]		***Unit***	***Limit value according to regulations***[3)]	***Value after 7 weeks rotting***	***Value after 10 weeks rotting***
Breathing activity	AT_4	mg O_2/g DM	5	3.0	2.6
Gas formation	GB_{21}	Nl/kg DM	20	n.a.	9.2
TOC in the eluate		mg/L	250	240	220
Upper calorific value	H_o	MJ/kg DM	6000	9700	9700[4)]
TOC in the residue		% DM	18.0	15.2	14.8

ammonia content, which occasionally occurs, can be reduced by the same methods as those used to separate ammonia from substrate.

Usually, anaerobic compost is of versatile application in garden and landscape gardening.

If residues are to be used as landfill, they should coincide with the values given in Table 1.2.

1) Cp. LAW 15

2) Either the values for AT_4 or GB_{21} respectively either Ho or TOC in the residue are to be guaranteed

3) without considering the possible exceedance

4) in fraction <20 mm after 10 weeks 8800 MJ/kg DM

2
Waste water

The waste water from the plant varies greatly in quality according to whether the plant is operated under mesophilic or thermophilic conditions. Waste water from thermophilic plants is much more contaminated. In general all waste water from a biogas plant must be biologically treated before it is discharged into a watercourse. The easiest option is to feed it into the communal sewage treatment plant.

The water can also be sprayed as liquid fertilizer onto agricultural land. This procedure is recommended when communal sewage treatment plants are too far away to consider building a connecting pipeline. But it must be remembered that, waste water may not be sprayed onto agricultural land when the soil is frozen, which means during the winter period of at least 3 months. This is to protect the rivers from receiving excessive amounts of plant nutrients.

Other possibilities are the precipitation of magnesium ammonium phosphate out of the waste water or crystallization of the nitrogen in the form of ammonium hydrogen carbonate, or membrane filtration of the impurities and recycling of the water. This can be recommended in plants where dry biomass is fed.

Biogas from Waste and Renewable Resources. An Introduction.
Dieter Deublein and Angelika Steinhauser

ISBN: 978-3-527-31841-4

Attachment I
Typical design calculation for an agricultural biogas plant

The size of an agricultural plant should be suitable for the number of domestic animals and the area available for cultivating co-ferments. From these values, the daily biogas yield can be calculated.

Biogas from Waste and Renewable Resources. An Introduction.
Dieter Deublein and Angelika Steinhauser

ISBN: 978-3-527-31841-4

Table A.1 Products and biogas yields from an average substrate.

Description		*Basic substrate*						*Biomass*		*Feedback from the residue storage tank*	*Total*
	Unit	*Liquid manure from cattle*	*Muck from cattle*	*Liquid manure from pigs*	*Muck from pigs*	*Liquid manure from poultry*	*Muck from poultry*	*Silo maize*	*Grass Silage*		
Animals in GVE*	GVE	100	0	0	0	0	0	–	–	–	100
Liquid manure /GVE.day	$m^3/(GVE.d)$	0.050	0.050	0.050	0.050	0.050	0.050	–	–	–	
Area for cultivation (A)	ha/a	–	–	–	–	–	–	15.00	15.00	–	30.00
Yield per hectare	Mg/ha/a	–	–	–	–	–	–	50.0	50.0	–	
Yield per day ($\dot{M}_G$)	Mg/d	5.00	0.00	0.00	0.00	0.00	0.00	2.00	2.00	2.50	11.5
DM-content	%	7.0	20.0	5.0	27.5	15.0	75.0	32.0	35.0	0.0	14.7
Yield of DM per day	kg DM/d	350	0	0	0	0	0	640	700	0	1690
oDM in DM	%	85.0	90.0	90.0	90.0	75.0	75.0	95.0	95.0	0.0	13.7
Yield of oDM (DM_{BR})	kg oDM/d	298	0	0	0	0	0	608	665	0	1571
Spec. yield of biogas	$m^3/(kgoDM.d)$	0.50	0.40	0.50	0.40	0.50	0.45	0.65	0.65	0.00	0.62
Yield of biogas ($\dot{V}_{BR}$)	m^3/d	149	0	0	0	0	0	395	432	0	976

* One animal unit (GVE) corresponds to the liquid manure from one full-grown cow, 5 calves, 6 beef cattle, or 250 hens.

Based on the daily rate of biogas and some additional assumptions, the equipment of a complete biogas plant can be designed as follows:

Preparation tank

The preparation tank shall be a vertical cylindrical container of concrete.

In the preparation tank, only the daily produced liquid manure shall be stored for $t_{PT}=10$ days in order to carry out cleaning and maintenance work at the bioreactor. The density of the liquid manure ρ_G can be set equal to the density of water $\rho_G=\rho_W$.

A factor $f_{VPT}=1.25$ shall be assumed to take into consideration the volume for air and fixtures.

The relationship between height and diameter of the tank shall be $H_{PT}/D_{PT}\approx 2$.

Design

Volume: $V_{PT}=\dot{M}_G\cdot t_{PT}/\rho_G\cdot f_{VPT}=5\,Mg/d\cdot 10\,d/1000\,kg/m^3\cdot 1.25=62.5\,m^3$,
Height: $H_{PT}=6.8\,m$,
Diameter: $D_{PT}=3.4\,m$,

Pump of the preparation tank

In the preparation tank a centrifugal pump with a wide chamber and a submerged motor shall be installed.

The pump of the preparation tank shall be able to deliver $\dot{V}_{VP}=10\,m^3/h$ liquid manure to the bioreactor or to pump the complete volume of the bioreactor ($V_{BR}=431\,m^3$ as calculated below) within $t_{BRI}=5\,h$ (given below). Its efficiency is assumed to be $\eta_{VP}=0.5$. Its pressure head shall be 1 bar.

Design

Throughput: $(\dot{V}_{VP})_1=10\,m^3/h$ or
$(\dot{V}_{VP})_2=V_{BR}/t_{PT}=431\,m^3/5\,h=86\,m^3/h$
Pressure head: $\Delta P_{VP}=1$ bar,
Capacity of the motor: actual: $P_{VP}=(\dot{V}_{VP})_1\cdot\Delta P_{VP}/\eta_{VP}\approx 0{,}6\,kW$;
nominal: $P_{VP}=(\dot{V}_{VP})_2\cdot\Delta P_{VP}/\eta_{VP}\approx 5{,}0\,kW$

Silo for maize and/or grass

The complete harvest ($\dot{M}_S=750\,Mg/a$) of maize (density $\rho_S=0.7$ Mg/m^3) of one year shall be stored in a concrete silo, which is accessible to traffic. The dimensions can be freely chosen. The gras is dried and stored on fields.

Design

Volume: $V_S=\dot{M}_S/\rho_S=750\,Mg/a/0{,}7\,Mg/m^3=1100\,m^3$
Breadth: $B_S=10\,m$,
Height: $H_S=3.5\,m$,
Length: $L_S=32\,m$

Silo conveyors

Two screw conveyors in series with nominal capacity $\dot{V}_{SC} = 1\,m^3/h$ each and motor capacity $P_{SC} = 5\,kW$ each are driven between silo and preparation tank twice a day for $t_{SC} = 1\,h/d$.

Design

Total power consumption of the two screw conveyors: $(P_{SC})_{tot} = 2\ P_{SC} \cdot 2 \cdot t_{SC}/24\,h = 2\ \cdot 5\,kW \cdot 2 \cdot 1\,h/24\,h = 0.8\,kW$

Bioreactor

A vertical cylindrical tank of concrete shall be used as bioreactor.

The residence time of the substrate in the bioreactor shall be $t_{BR} = 30$ days. A factor $f_{VBR} = 1.25$ has to be chosen to take into consideration the volume for air and fixtures in the bioreactor.

The relation between height and diameter of the bioreactor shall be $H_{BR}/D_{BR} \approx 1/2$.

The completely filled bioreactor shall be emptied within $t_{BRI} = 5\,h$ at a flow rate of $v_{BRI} = 0.5\,m/s$.

Two propeller-agitators (diameter $D_{BRR} = 0.5\,m$, Newton number $Ne_{BRR} = 0.5$, revolution $n_{BRR} = 150\,rpm$) shall be installed for intermittent mixing and breaking off the floating layer with a working period of $t_{BRR} = 5\,min/h$. Both agitators are equipped with submerged motors and their height shall be adjustable by chains.

Design

Volume: $V_{BR} = \dot{M}_G/\rho_G \cdot t_{BR} \cdot f_{VBR} = 11.5\,Mg/d/1000\,kg/m^3 \cdot 30\,d \cdot 1.25 = 431\,m^3$

Height: $H_{BR} = 5.5\,m$,

Diameter: $D_{BR} = 10\,m$,

The volume load of this medium-sized bioreactor is then $B_{BR} = DM_{BR}/V_{BR} = 1571\,kg\ oDM/d/431\,m^3 = 3.64\,kg\ oDM/m^3 \cdot d$.

The average volume load for small plants is $B_{BR} = 1.5\,kg\ oDM/m^3 \cdot d$ and of large plants is $B_{BR} = 5\,kg\ oDM/m^3 \cdot d$.

Diameter of the discharge pipe:

$$D_{BRI} = \sqrt{(V_{BR}/t_{BRI}/v_{BRI} \cdot 4/\pi)} = \sqrt{(431\,m^3/5\,h/0.5\,m/s \cdot 4/\pi)} \approx 0.3\,m$$

Capacity per agitator drive:

$$P_{BRR} = 1.3 \cdot Ne_{BRR} \cdot \rho_G \cdot n_{BRR}{}^3 \cdot D_{BRR}{}^5 = 1.3 \cdot 0.5 \cdot 1000\,kg/m^3 \cdot (150 \cdot \pi/30)^3 \cdot 0.5^5\,m^5 = 78.7\,kW \approx 80\,kW$$

Power consumption of both agitators:

$$(P_{BRR})_{tot} = 2 \cdot P_{BRR} \cdot t_{BRR} = 2 \cdot 80\,kW \cdot 5\,min/h = 13.4\,kW$$

Heating (pipes)

For fermentation, mesophilic temperatures shall be chosen at $\vartheta_{BR}=50\,°C$. The lowest outside temperature in winter is $\vartheta_A=-20\,°C$ (humid soil)). The substrate with its specific heat capacity of $c_{SU}=4.2\,kJ/kg\cdot°C$ has thus to be heated from 20 °C to 50 °C, i.e. ca.. $\Delta\vartheta_{SU}=30\,°C$.

The bioreactor walls shall be insulated with a layer ($s_{BR}=0.1\,m$ thick) of polystyrene. The heat transmission coefficient of polystyrene is $\lambda_{BR}=0.05\,W/m\cdot K$. Although the substrate surface does not reach the ceiling of the bioreactor, the complete wall shall be taken into consideration when calculating the heat losses; heat losses through the ceiling are negligibly low, because the ceiling is in contact with gas and/or air inside and outside.

The heat transfer coefficients inside at the wet bioreactor wall shall be assumed to be $(\alpha_{BR})_i=4000\,W/m^2\cdot°C$ for agitated liquid and outside to $(\alpha_{BR})_a=400\,W/m^2\cdot°C$ for humid soil; Then the k-factor can be calculated:

$$k_{BR}=1/((\alpha_{BR})_i+s_{BR}/\lambda_{BR}+(\alpha_{BR})_a)=1/(1/4000+0.1/0.5+1/400)$$
$$=0.5\,W/m^2\cdot°C$$

The maximum temperature difference between substrate and environment is $\Delta\vartheta_{BR}=\vartheta_{BR}-\vartheta_A=(50\,°C)-(-20\,°C)=70\,°C$

The heating medium (warm water) shall cool down from $\vartheta_{HE}=70\,°C$ to $\vartheta_{HA}=60\,°C$ and the temperature difference can be calculated

$$\Delta\vartheta_H=\vartheta_{HE}-\vartheta_{HA}=10°C$$

The flow rate of the heating medium in the heating pipe shall be $v_H=1\,m/s$.

The heat transfer coefficient inside and outside the heating pipes shall be asumed to be the same $(\alpha_H)_i=(\alpha_H)_a=400\,W/m^2\cdot°C$ for slowly flowing liquid; The heating pipe is well heat-conducting and therefore negligible in the calculation; then the k-factor for the heating pipe wall can be calculated

$$k_H=1/(1/((\alpha_H)_i+(\alpha_H)_a)=1/(1/400+1/400)=200\,W/m^2\cdot°C$$

The average temperature difference between heating medium and substrate in the bioreactor is

$$\Delta\vartheta_{BH}=(\vartheta_{HE}+\vartheta_{HA})/2-\vartheta_{BR}=15°C$$

Design

Heat for heating the substrate:

$$Q_{SU}=\dot{M}_G\cdot c_{SU}\cdot\Delta\vartheta_{SU}=11.5\,Mg/d\cdot4.2\,kJ/kg\cdot°C\cdot30\,K=17\,kW$$

Surface area of the bioreactor, which conducts heat:

$$A_{BR}=\pi\cdot D_{BR}{}^2/4+\pi\cdot D_{BR}\cdot H_{BR}=250\,m^2$$

Heat losses of the reactor:

$Q_{BR} = k_{BR} \cdot A_{BR} \cdot \Delta\vartheta_{BR} = 0.5\,W/m^2.°C \cdot 250\,m^2 \cdot 70\,°C = 8.8\,kW$

Necessary heat: $Q_V = Q_{SU} + Q_{BR} = 17\,kW + 8.8\,kW = 25.8\,kW$

Necessary heating liquid $\dot{V}_w$, for heat supply to the bioreactor:

$\dot{V}_w = Q_V/(c_w \cdot \rho_w \cdot \Delta\vartheta_H) = 25.8\,kW/$
$(4.2\,kJ/kg.°C \cdot 1000\,kg/m^3 \cdot 10\,°C) = 6.14\,m^3/h$

Diameter of the heating pipe:

$D_{HR} = \sqrt{(\dot{V}_w/v_H \cdot 4/\pi} = \sqrt{(6.14\,m^3/h/1\,m/s \cdot 4/\pi)} = 0.046 \approx 0.05\,m$

Length of the heating pipe: $L_{HR} = Q_V/(k_H \cdot \Delta\vartheta_{BH} \cdot \pi \cdot D_{HR}) = 25.8\,kW/(200\,W/m^2.°C \cdot 15\,°C \cdot \pi \cdot 0.05\,m = 55\,m$

The result of the calculation is, that a pipe with three windings of a diameter of $D_W = 8\,m$ is sufficient. Because sinking layers and a disturbance of the heat transfer must be taken into consideration, a higher number of windings should be chosen.

Aeration

Aeration with an air flow rate $\dot{V}_L$ referring to the biogas flow rate $\dot{V}_{BR}$ of $\dot{V}_L / \dot{V}_{BR} = 0.04$ is sufficient for desulfurization. Oil-free compressed air shall be blown in at a pressure of $p_{k2} = 6$ bar. The velocity of the air in the air pipe shall be $v_L = 2\,m/s$.

Design

Blown in air:

$\dot{V}_L = \dot{V}_L/\dot{V}_{BR} \cdot \dot{V}_{BR} = 0.04.976\,m^3/d = 1.63\,Nm^3/h$

Diameter of the air pipe:

$D_L = \sqrt{(\dot{V}_L/v_L \cdot 4/\pi)} = \sqrt{(1.63\,Nm^3/h/2\,m/s \cdot 4/\pi)} \approx 0.02\,m$

Compressor with pressure vessel of volume $V_K = 0.05\,m^3$

Design

Volume rate of the compressor:

$\dot{V}_K = 1.7\,Nm^3/h$

Pressure head

from $p_{K1} = 1$ bar to $p_{K2} = 6$ bar (there is to be assumed friction in the pipe)

Capacity of the compressor:

$P_K = 0.5\,kW$

Gasholder

A low-pressure gasholder of plastic foil shall be used

The relationship of the volume of the bioreactor to the volume of the gasholder shall be $(V_{BR}/V_{GS}) = 1:2$. The ratio V_{BR}/V_{GS} is the usual value (between 1:1 and 1:3).

Design

Volume of the gasholder: $V_{GS} = V_{BR}/(V_{BR}/V_{GS}) = 431/(1/2) = 862\,m^3$.

The gasholder enables the storage of the biogas production of nearly 1 day.

Engine

The biogas-energy content shall be $\dot{E}_{spec} = 6\,kWh/m^3$.

The biogas plant shall be equipped with an ignition oil Diesel engine in the CHP. 9% by weight of ignition oil shall be added to the biogas in a ratio of $\dot{M}_{OIL}/\dot{M}_{BR} = 0.09$. The energy content of the ignition oil is $E_{OILspec} = 10\,kWh/kg$.

Efficiency of the engine:

Electrical efficiency: $\eta_{el} = 30\%$

Thermal efficiency: $\eta_{th} = 50\%$

Design

Consumption of ignition oil assuming a density of the biogas of $\delta^* = 1.11\,kg/m^3$:

$$\dot{M}_{BR} = \dot{V}_{BR} \cdot \delta^* = 976\,m^3/d \cdot 1.11\,kg/m^3 = 1083\,kg/d$$

$$\dot{M}_{OIL} = \dot{M}_{BR} \cdot (\dot{M}_O / \dot{M}_{BR}) = 1083\,kg/d \cdot 0.09 = 97.6\,kg/d$$

Yield of energy:

$$E_{tot} = E_{spec} \cdot \dot{V}_{BR} + E_{OILspec} \cdot \dot{M}_{OIL}$$

$$= (6\,kWh/m^3 \cdot 976\,m^3/d + 10\,kWh/kg \cdot 97.6\,kg/d)/24\,h/d = 284\,kW$$

$$E_{el} = E_{tot} \cdot \eta_{el} = 284 \cdot 0.3 = 85.4\,kW$$

$$E_{th} = E_{tot} \cdot \eta_{th} = 284 \cdot 0.5 = 142.0\,kW$$

Nominal capacity of the engine with a reserve of 30%:

$$E = 111\,kW$$

Residue storage tank

A vertical cylindrical tank of concrete shall be used as residue storage tank.

The residence time of all residue in the storage tank shall be $t_E = 100\,d$, according to the period in which the soil is frozen, when the residue could not penetrate into the soil, would possibly flow into rivers, and would deteriorate the water quality. It has to be taken into consideration that some water ($\dot{V}_E = 2.5\,m^3/d$) from the residue tank is fed back into the bioreactor.

A factor $f_{VB} = 1.1$ is chosen to take into consideration the volume for air and fixtures in the storage tank. The storage tank shall have the same height as the bioreactor.

Design

Volume: $V_E = (\dot{M}_G/\rho - \dot{V}_E) \cdot t_E \cdot f_{VE} = (11.5\,Mg/d/1\,Mg/m^3 - 2.5\,m^3/d) \cdot 100\,d \cdot 1.1 = 990\,m^3$.

Height: $H_E = 5.5\,m$
Diameter: $D_E = 15\,m$

Power and heat consumption of the entire plant

Table A.2 Calculated energy consumption of the plant.

Energy consumer	*Abbreviation*	*Energy*
Agitators (2 pieces)	$(P_{BRR})_{tot}$	13.4 kW
Pump	P_{VP}	0.6 kW
Screw conveyors for maize and grass	$(P_{SC})_{tot}$	0.8 kW
Air compressor	P_K	0.5 kW
Total power consumption	$\mathbf{E_{Eel}}$	**15.3 kW**
Heat losses via the bioreactor wall	Q_{BR}	8.8 kW
Heat for heating the substrate	Q_{SU}	17.0 kW
Total heat consumption	$\mathbf{Q_V}$	**25.8 kW**

Attachment II
Economy of biogas plants for the year 2007 (Calculation on the basis of the example of Attachment I)

From experience, the complete investment costs for a biogas plant can be estimated at $KA_{spec}=300.00–500.00\,US\$/m^3$ volume of the bioreactor. The smaller number stands for large plants, the higher number for small plants. Depending on the degree of automation and the spectrum of planning and supervision work, the given range can be exceeded. A simple plant for power generation (CHP) costs additionally $KK_{spec}=650.00\,US\$/kW_{el}$ from experience.

The bioreactor was calculated in the example in Attachment I to have a volume of $V_{BR}=431\,m^3$ and the CHP to have a nominal capacity of $E=111\,kW_{el}$.

The economic calculation shall be carried out as an example based on investment costs of $KA=200\,000.00\,US\$$. To be added are the costs for the CHP accumulated $K_{CHP}=72\,000.00\,US\$$. Included in the costs for the CHP is the electrical connection between plant and power network according to the technical and legal possibilities for the shortest connection. The total investment costs are thus $K=272\,000.00\,US\$$.

The costs are given below in differentiated form as is generally usual.

Capital-bound costs per year in US$

For the concrete works shall be taken $x_B=KB/KA=0.63$ of the investment costs. These costs can be amortized within $t_B=20$ years.

For technical equipment shall be taken $x_T=KT/KA=0.37$ of the investment costs. These costs can be amortized within $t_T=10$ years.

The complete costs of the CHP of $K_{CHP}=72\,000\,US\$$ can be amortized within $t_K=4$ years.

The interest rate shall be 6% ($Z_R/K=0.06/a$) of the total investment costs.

Calculation:

Costs for the concrete works:

$KB=x_B \cdot KA/t_B=0.63 \cdot 200\,000.00\,US\$/20\,a=6300.00\,US\$/a$

Costs for technical equippment:

$KT=x_T \cdot KA/t_T=0.37 \cdot 200\,000.00\,US\$/10\,a=7400.00\,US\$/a$

Biogas from Waste and Renewable Resources. An Introduction.
Dieter Deublein and Angelika Steinhauser

ISBN: 978-3-527-31841-4

Costs for the CHP:

$KK = K_{CHP}/t_K = 72\,000.00\ US\$/4\,a = 18\,000.00\ US\$/a$

Costs for interest:

$Z_R = Z_R/K \cdot K = 0.06/a \cdot 272\,000.00\ US\$ = 16\,300.00\ US\$/a$

Total capital bound costs:

48 000.00 US$/a

Consumption-bound costs per year ($t_s = 8640\,h/a$)

The substrate and the co-ferments are available free of charge. Costs for transportation of substrate and residue do not arise.

The power consumption of the plant shall be $E_{Eel} = 15.3\,kW$, the ignition oil consumption $\dot{M}_{OIL} = 97.6\,kg/d$, the heat consumption $Q_V = 25.8\,kW$.

Costs for electrical power shall be $KS_{spec} = 0.07\ US\$/kWh$, costs for ignition oil shall be $K_{OILspec} = 0.50\ US\$/kg$, and costs for heat shall be $KW_{spec} = 0.05\ US\$/kWh$.

The costs for cultivation of renewable resources shall be $KR_{spec} = 1500\ US\$/ha.a$. For cultivating maize shall be available an area of $A_M = 15\,ha$.

Costs for electrical power:

$KS = E_{Eel} \cdot KS_{spec} \cdot t_s = 15.3\,kW \cdot 0.07\ US\$/kWh \cdot 8640\,h/a = 9253.00\ US\$/a$

Costs for ignition oil:

$KO = \dot{M}_{OIL} \cdot K_{OILspec} \cdot t_s = 97.6\,kg/d/24\,h/d \cdot 0.50\,US\$/I \cdot 8640\,h/a$
$= 17568.00\,US\$/a$

Costs for heat:

$KW = Q_V \cdot KW_{spec} \cdot t_s = 25.8\,kW \cdot 0.05\ US\$/kWh \cdot 8640\,h/a$
$= 11\,146.00\ US\$/a$

Costs for cultivation:

$KR = A_M \cdot KR_{spec} = 15\,ha \cdot 1500\ US\$/ha.a = 22\,500.00\ US\$/a$

Total consumption-bound costs: 60 467.00 US$/a

Operation-bound costs per year ($t_s = 8640\,h/a$)

The specific annual costs for maintenance shall be for concrete works 0.5% ($y_B = 0.005/a$) of the investment costs ($x_B \cdot KA$). for technical equipment 3% ($y_T = 0.03/a$) of the investment costs ($x_T \cdot KA$) and for the CHP 4% ($y_{CHP} = 0.04/a$) of the investment costs for the CHP.

The personnel costs for operators shall be $KP_{spec} = 20.00\ US\$/h$. An annual working time of $t_P = 500\,h/a$ is assumed.

Costs for maintenance for concrete works:

$KX = y_B \cdot x_B \cdot KA = 0.005/a \cdot 0.63 \cdot 200\,000.00\ US\$ = 630.00\ US\$/a$

Costs for maintenance for techniques:

$KY = y_T \cdot x_T \cdot KA = 0.03/a \cdot 0.37 \cdot 200\,000.\ 00\ US\$ = 2220.00\ US\$/a$

Costs for maintenance of the CHP

$KZ = y_{CHP} \cdot K_{CHP} = 0.04/a \cdot 72\,000.00\ US\$ = 2880.00\ US\$/a$

Personnel costs:

$KP = KP_{spez} \cdot t_P = 20.00\,US\$ \cdot 500\,h/a = 10\,000.00\,US\$/a$

Total operation bound costs:

15 730.00 US$/a

Other costs per year (t_s = 8640 h/a)

The costs per year KV for insurance of the plant shall be 0.5% (z = 0.005/a) of the investment costs.

Costs for insurance:

$KV = z \cdot K = 0.005/a \cdot 272\,000.00\,US\$ = 1360.00\,US\$/a$

Total other costs:

1360.00 US$/a

Total costs

125 557.00 US$/a

These costs are balanced by incomes, calculated below. The high income from the sales of electrical power is due to the governmental subsidy.

By the fermentation, the quality of the residue which is used as fertilizer is improved, which can only roughly be taken in consideration in the balancing.

Income per year (t_s = 8640 h/a)

A subsidy of KB_{spec} = 0.24 US$/kWh shall be assumed (actual subsidy in Germany).

Sales of electrical power:

$G = E_{el} \cdot t_s \cdot KB_{spec} = 85.4\,kW \cdot 8640\,h/a \cdot 0.24\,US\$/kWh$
$= 177\,085.00\,US\$/a$

Sales of heat:

$W = E_{th} \cdot t_s \cdot KW_{spec} = 142.0\,kW \cdot 8640\,h/a \cdot 0.05\,US\$/kWh$
$= 61\,366.00\,US\$/a$

Sales of fertilizer:

D = 1000.00 US$/a

Total income:

239 451.00 US$/a

Annual revenue of the biogas plant: 239 557.00 US$/a – 125 557.00 US$/a = 113 894.00 US$/a

Literature

BOK Books/proceedings

BOK 1. Publication of the European enquete-commission 1988
BOK 2. Publication of the European enquete-commission 1994 and 1995
BOK 3. BMWi, VDEW, Statistic of the coal economy, DIW 2002
BOK 4. Bickel et al.: Natura. Themenband Stoffwechsel. Klett-Verlag. Stuttgart 1995
BOK 5. VDI-Report 1620, 2001
BOK 6. Schattner,S., Gronauer, A., Mitterleitner, H., Reuss, M.: Bewertung der Biogastechnologie:, publication of the Bayerische Landesanstalt für Landtechnik Weihenstephan
BOK 7. Publication of the German Bundesministerium für Verbraucherschutz, Ernährung und Landwirtschaft, 2003
BOK 8. C.A.R.M.E.N. Almanac, Energetische Verwertung nachwachsender Rohstoffe: Biogas 2002
BOK 9. Hoppenheidt, K., Mücke, W.: Hygieneanforderungen an Verfahren und Produkte bei der Bioabfallverwertung; LfU-Fachseminar "Vollzug der Bioabfall-VO" in Augsburg, 2000
BOK 10. Energieagentur NRW: "Biogas: Strom und Wärme aus Gülle" 2002
BOK 11. Maishanu, S.M., Muse, A., Sambo, A.S.: Energy and Environment Vol. *3*, 1990
BOK 12. Mudrak, K., Kunst, S.: Biologie der Abwasserreinigung, 4. Auflage, G. Fischer-Verlag, Stuttgart, 1994
BOK 13. Öko-Institut – Institut für angewandte Ökologie e. V.: Gesamt-Emissions-Modell integrierter Systeme (GEMIS), Version 4, 2001
BOK 14. Puls, J., Poutanen, K., Körner, H.U., Viikari, L., Biotechnical utilization of wood carbohydrates after steaming pre-treatment, Springer Verlag 1985
BOK 15. Reitberger, F.: Emissionsminderungsmöglichkeiten bei Biogasanlagen – entlang der Prozesskette, Fachtagung Biogasanlagen Anforderungen zur Luftreinhaltung 2002
BOK 16. Kaltschmitt, M., Hartmann, H.: Energie aus Biomasse; Springer-Verlag, 2001
BOK 17. Knoche, J., Biogasanlagen: Genehmigungspflichten, Fachtagung Biogasanlagen: Anforderungen zur Luftreinhaltung, 2002
BOK 18. Savado, T., Nakamura, Y., Kobayashi, F.: Effects of Fungal Pretreatment and Steam Explosion Pretreatment an enzymatic Saccharification of Plant Biomass. Biotechnology and Bioengineering, Vol. *48*, p. 719–724, 1995.
BOK 19. Schlegel, H.G.: Allgemeine Mikrobiologie, Thieme-Verlag, 6. Auflage 1985
BOK 20. Schön, M.: Verfahren zur Vergärung organischer Rückstande in der Abfallwirtschaft, Abfallwirtschaft in Forschung und Praxis, 66, Schmidt-Verlag, 1994
BOK 21. Toussaint, B., Excoffier, mG., Vignon, M.R.,: Saccharification of steam-exploded poplar wood, Biotechnology and bioengineering, *38*, 11, p. 1308–1317, 1991
BOK 22. Publication of the Intergovernmental Panel on Climate Change (IPCC), 2003
BOK 23. Wichmann, P.: Können landwirtschaftliche Biogasanlagen eine Alternative zum Kanalanschlusszwang

Biogas from Waste and Renewable Resources. An Introduction.
Dieter Deublein and Angelika Steinhauser

ISBN: 978-3-527-31841-4

darstellen? Oberfränkische Energietage 1999

BOK 24. Wiemer, Kern: Witzenhausen-Institut – Neues aus Forschung und Praxis: Bio- und Restabfallbehandlung VII, Fachbuchreihe Abfall-Wirtschaft des Witzenhausen-Instituts für Abfall, Umwelt und Energie, 2003

BOK 25. Baserga, U.: Vergärung von Energiegras zur Energiegewinnung,. Bundesamt für Energiewirtschaft Bern, Jahresbericht 1994

BOK 26. Bergey's manual of systematic bacteriology 2005

BOK 27. Biogashandbuch Bayern, published by Bayerisches Staatsministerium für Umwelt, Gesundheit und Verbraucherschutz (StMUGV), 2006

BOK 28. Böhnke, B., Bischofsberger, W., Seyfried, C.F.: Anaerobtechnik, Springer Verlag, 1993

BOK 29. Brock Mikrobiologie, Spektrum-Verlag Gustav Fischer, 2000

BOK 30. Publication of the German Bundesministerium für Umwelt, Naturschutz und Reaktorsicherheit; Erneuerbare Energien, Referat Öffentlichkeitsarbeit, 2003

BOK 31. Edelmann, W., Engeli, H.: Biogas aus festen Abfällen und Indutrieabwässern – Eckdaten für PlanerInnen. Published by Swiss Bundesamt für Konjunkturfragen, 1996

BOK 32. Gross, F. u. Riebe, K.: Gärfutter. Stuttgart: Eugen Ulmer Verlag, p. 130, 141, 1974.

BOK 33. Schulz, H.: Biogas – Praxis, Ökobuch, 1. Auflage, ISBN 3-922964-59-1, 1996

BOK 34. Huber, S.; Mair, K.: Untersuchung zur Biogaszusammensetzung bei Anlagen aus der Landwirtschaft. Ergebnisbericht des Bayerischen Landesamts für Umweltschutz, 1997

BOK 35. 2. Arbeitsbericht des ATV-Fachausschusses 7.5 – Anaerobe Verfahren zur Behandlung von Industrieabwässern; Technologische Beurteilungskriterien zur anaeroben Abwasserbehandlung; Zeitschrift Korrespondenz, Abwasser, Ausgabe 2, p. 217, 1993

BOK 36. 3. Arbeitsbericht des ATV Fachausschusses 7.5: Geschwindigkeitsbestimmende Schritte beim anaeroben Abbau von organischen Verbindungen in Abwässern; Zeitschrift Korrespondenz Abwasser, Ausgabe 1, p. 101, 1994

BOK 37. Oechsner, A., Weckenmann, D., Buchenau, C.: Erhebung von Daten an landwirtschaftlichen Biogasanlagen in Baden-Württemberg. Agrartechnische Berichte, *H28*, 1999

BOK 38. Möbius, C.H., Helble, A.: Stand der Technik in Konzeption und Sanierung von Abwasserbehandlungsanlagen für Zellstoff- und Papierfabriken in Mitteleuropa; 28. Internationale Jahresfachtagung, lecture held in Bled (SI), 2001

BOK 39. Act Energy – Aktionsgemeinschaft Regenerative Energie e.V.: Biogaseinspeisung ins Erdgasnetz – Technik, Wirtschaftlichkeit und CO_2-Einsparungen, lecture 2002

BOK 40. Mudrack, K.: Biochemische und mikrobielle Gegebenheiten bei der anaeroben Abwasser- und Schlammbehandlung, Anaerobe Abwasser- und Schlammbehandlung – Biogastechnologie. München, Wien, Oldenbourg, 1983

BOK 41. Boback, R.: Gasaufbereitung mittels Tieftemperaturrektifikation", lecture held at the FNR-Workshops "Aufbereitung von Biogas" in Braunschweig, 2003

BOK 42. Gronauer, A., Effenberger, M., Kaiser, F., Schlattmann, M.: Biogasanlagen-Monitoring und Emissionsverhalten von Blockheizkraftwerken, Abschlussbericht der bayerischen Landesanstalt für Landtechnik, Arbeitsgruppe Umwelttechnik der Landnutzung, an das bayerische Staatsministerium für Landesentwicklung und Umweltfragen, 2002

BOK 43. Jansson, M.: Water scrubber technique for biogas purification to vehicle fuel, Vortrag auf der Tagung "Biogas und Energielandwirtschaft – Potenzial, Nutzung, Grünes Gas, Ökonomie, Ökologie", lecture held in Potsdam, 2002

BOK 44. Keicher, K., Krampe, J., Rott, U.: Systemintegration von Brennstoffzellen in Kläranlagen – Potenzialabschätzung für Baden-Württemberg, BWI 22006, 2004

BOK 45. Kopp, J., Dichtl, N.: Influence of surface charge and exopolysaccharides on the conditioning characteristics of sewage sludge; Chemical water and wastewater treatment V, Proceedings of the 8th Gothenburg Symposium in Prag, 1998

BOK 46. Köttner, M., Kaiser, A.: Übersicht über die Verfahren der Trockenvergärung, lecture held in Bad Hersfeld on the 4. congress about gras, 2001

BOK 47. Langhans, G.: Some aspects of rheology in bioorganic suspensions and anaerobic fermentation broth, Paper presentation at Conference, Microbiology of Composting, Innsbruck, 2000

BOK 48. Möller, U.: Einführung zum 3. Bochumer Workshop, Schriftenreihe Siedlungswasserwirtschaft 7, 1985

BOK 49. Muche, H. et al: Biogas-Entschwefelung – Möglichkeiten und Grenzen in "Technik anaerober Prozesse", lecture given at the technical university Hamburg-Harburg in cooperation with the DECHEMA e. V., Hamburg, ISBN 3-926959, 1998

BOK 50. Reher, S.: "Kraftstoffe aus Biogas – Technik, Qualität. Praxisbeispiele", lecture given at the FNR-Workshops "Aufbereitung von Biogas" in Braunschweig, 2003

BOK 51. Tiehm, A., Neis, U.: Technical University of Hamburg-Harburg Reports on sanitary and environmental engineering 25, GFEU-Verlag, Hamburg, ISSN 0724-0783; ISBN 3-930400-23-5, 2002

BOK 52. Publication of the Landeshauptstadt Stuttgart, Tiefbauamt, Eigenbetrieb Stadtentwässerung, 2003

BOK 53. Zell, B.: Emissionen von Biogasblockheizkraftwerken, Fachtagung Biogasanlagen Anforderungen zur Luftreinhaltung, 2002

BOK 54. Bertil Carlsson, public services Linköping: "Upgrading and biogas utilisation in Linköping", Bornimer Agrartechnische Berichte 32, p. 83, 2002

BOK 55. Publication oft he Swiss Bundesamt für Energiewirtschaft, Salat im Tank – Migros-Lastwagen fahren mit Biogas. Und Sie? Broschüre der Informationsstelle Biomasse, 2002

BOK 56. Publication oft he Swiss Bundesamt für Energiewirtschaft, Kompo-Mobil – Biogasnutzung in Fahrzeugen, written by Nova Energie, Tänikon, 1997

BOK 57. Tentscher, Wolfgang: Gasaufbereitung mittels nasser Gaswäsche in Schweden", lecture held at the FNR-Workshops "Aufbereitung von Biogas" in Braunschweig, 2003

BOK 58. MC Carty, P.L.: One hundred Years of anaerobic treatment, Second international Symposium of anaerobic digestion Travemünde 1957

BOK 59. Kroiss, H.: Anaerobe Abwasserreinigung, Wiener Mitteilungen Wasser-Abwasser-Gewässer, 62, 1986

BOK 60. Kuhn, E.: working paper 219, Kuratorium für Technik und Bauwesen in der Landwirtschaft, KTBL-Schriften, Landeswirtschaftsverlag Münster Hiltrup 1995

BOK 61. Römpp Chemie Lexikon a-Z, Thieme Verlag, 1999

BOK 62. Zeilinger, J.: presentation about, "costs of the supply of energetical wood, generation and thinning", lecture held on the conference on "energetical woods" at the LUFA in Thüringen/Germany 1993

BOK 63. Klinski, St.: Einspeisung von Biogas in das Erdgasnetz; studies edited by the agency of renewable resources (FNR), 2006

BOK 64. Bartz, M.: research cooperation in the development of the chinese energy sector, TU INTERNATIONAL 55, 2004

BOK 65. Zell, B.: Emissionen von Biogasblockheizkraftwerken, lecture held on Fachtagung des BayLfU, 2002

BOK 66. KTBL-Taschenbuch Landwirtschaft; VDI-Richtlinie 3472, 1998/1999

BOK 67. Zeschmar-Lahl, B.: Die Klimarelevanz der Abfallwirtschaft im Freistaat Sachsen, Zwischenbericht an das Sächsisches Staatsministerium für Umwelt und Landwirtschaft, 2002

BOK 68. Widmann, R.: Anaerobe Verfahren der biologischen Abfallverwertung, Biologische Abfallverwertung, p. 144, Ulmer Verlag Stuttgart, 2000

BOK 69. Schwedes, J.: Verbesserter Abbau von Klärschlämmen durch Zellaufschluss – Teilprojekt 4a; DFG-Geschäftszeichen Schw 233/19-2, Schw 233/19-3, Schw 233/19-4 und Schw 233/19-5, 2000

BOK 70. Moser, D.: Anaerobe Abwasserreinigung: Beeinflussende Faktoren der Versäuerung eines Zitronensäurefabrikabwassers. Wiener

Mitteilungen, 173, ISBN 3-85234-064-0, 2002

BOK 71. Hartmann, L., Jourdan, R. (1987), Untersuchungen zur Reduzierung der Schwefelwasserstoff-Bildung beim anaeroben Abbau durch Luftzugabe. BMFT-Forschungsbericht 02W5404, Institut für Ingenieurbiologie und Biotechnologie des Abwassers, University of Karlsruhe, 1987

BOK 72. Wolters, D.: Bioenergie aus ökologischem Landbau, Wuppertal Papers, Nr. 91, ISSN 0949-5266, 1999

BOK 73. KA-Wasserwirtschaft, Abwasser, Abfall 49, 12, 2002

BOK 74. Austermann-Haun, U.: Seyfried, C F.; Rosenwinkel, K-H. (1997), Full scale experiences with anaerobic pre-treatment of wastewater in the food and beverage industry in Germany. Water Science & Technology, *36*, 2–3, p. 321–328, 1997

BOK 75. Austermann-Haun, U.: Inbetriebnahme anaerober Festbettreaktoren. Publication of the Institut für Siedlungswasserwirtschaft und Abfalltechnik of the university of Hanover, H. 93, 1997

BOK 76. Soyez, K., Plickert, S., Koller, M.: Von der Abfall- zur Rohstoff- und Energiewirtschaft- Umsetzung der Ziele Nachhaltigkeit, Klimaschutz und Ressourceneffizienz in der MBA-Technologie, lecture at the Wetzlarer Abfalltage 2001

LAW laws, prescriptions, official guidelines

LAW 1. Germany: Abwasserverordnung 2004

LAW 2. Germany: BGR 104 “Explosionsschutz-Regel”

LAW 3. Germany: DVGW-Merkblatt G 262/ Nutzung von Deponie-, Klär- und Biogasen 1991

LAW 4. Germany: Technische Regeln – Arbeitsblätter G 260/G 685

LAW 5. Germany: DIN 4102

LAW 6. Germany: law about energy supply

LAW 7. Germany: EnWG, § 16 Anforderungen an Energieanlagen, 1998

LAW 8. Germany: TA-Luft

LAW 9. Germany: Richtlinien zur Förderung von Maßnahmen zur Nutzung erneuerbarer Energien, Bundesanzeiger Nr.58, 2002

LAW 10. Switzerland: SVGW

LAW 11. German DIN 1045 and DIN 1046 and DIN 11622

LAW 12. German VDI/VDE-Richtlinie 2180, DIN V 19 250

LAW 13. Germany: Bundesimmissionsschutz-Gesetz

LAW 14. Germany: Verbändevereinbarung zum Netzzugang bei Erdgas (VV Erdgas II) zwischen den Verbänden BDI (Berlin), VIK (Essen), BGW (Berlin) and VKU (Köln), 2002

LAW 15. Germany: Abfallablagerungsverordnung (AbfAblVO), 2001

LAW 16. Germany: Sicherheitsregeln für landwirtschaftliche Biogasanlagen der landwirtschaftlichen Berufsgenossenschaft

LAW 17. Germany: ATV-DVWK Merkblätter

LAW 18. Germany: Die Gülleverordnung des Landes NRW, Herausgeber: Der Minister für Umwelt, Raumordnung und Landwirtschaft NRW, 1986

LAW 19. Germany: Bayerisches Staatsministerium für Ernährung, Landwirtschaft und Forsten/Oberste Baubehörde im Bayerischen Staatsministerium des Innern, 2, 1986

LAW 20. Germany: Beschluss des “Hessisches Ministerium für Landwirtschaft, Forsten und Naturschutz” 1988

LAW 21. Germany: (Anforderungskatalog JGS-Anlagen) der Obersten Baubehörde im Bayerischen Staatsministerium des Innern, 1992

LAW 22. Germany: Ministeriums für Umwelt und des Ministeriums für Ländlichen Raum, Ernährung, Landwirtschaft und Forsten des Landes Baden-Württemberg, 1992

LAW 23. Germany: Umweltministeriums des Landes Mecklenburg-Vorpommern (Verwaltungsvorschrift JGS-Anlagen – VVJGSA) 1993

LAW 24. Germany: GUV 19.8, Richtlinien für die Vermeidung der Gefahren durch explosionsfähige Atmosphäre mit Beispielsammlung-Explosionsschutz-Richtlinien – (EX-RL), Bundesverband der Unfallkassen, Munich, 2000

LAW 25. Germany: BetrSichV and European guideline 1999/92/EG, article 8 (ATEX 137) 1999
LAW 26. Germany: TRBA 466
LAW 27. Germany: Wasserhaushaltsgesetz (WHG § 19)
LAW 28. German VDI 3478 (1996), Biologische Abgasreinigung- Biowäscher und Rieselbettreaktoren. Beuth, Berlin, 1996
LAW 29. German VDI 3477 (1991). Biologische Abgas-/Abtuftreinigung Biofilter, Beuth, Berlin, 1991
LAW 30. Germany: Bioabfallverordnung (BioAbfVO)

DIS Dissertations/theses/research reports

DIS 1. Schmidt, K.: Thesis: Konzeption und Aufbau einer kleintechnischen Biogasanlage mit nachfolgender Vergärung von Restmüll, 2002
DIS 2. Betzl, H.: Thesis at the university of applied scences in Munich, 2002
DIS 3. Lucke, I.: Thesis at the Hochschule Vechta:, Biogas, die regenerative Energie der Zukunft, 2002
DIS 4. Wenzel, W.: Thesis at the university of Berlin: Mikrobiologische Charakterisierung eines Anaerobreaktors zur Behandlung von Rübenmelasseschlempe, 2002
DIS 5. Harnack, M.: Thesis at the university of Dresden: Untersuchungen zur Leistungssteigerung der anaeroben Schlammbehandlung unter Berücksichtigung der landwirtschaftlichen Verwertung, 1992
DIS 6. Jander, F.: Dissertation at the university of Kiel: Massenkultur von Mikroalgen mit pharmazeutisch nutzbaren Inhaltsstoffen unter Verwendung von CO_2 und $NaHCO_3$, gewonnen aus den Abgasen eines BHKWs, 2001
DIS 7. Yin, X.: Biogas; Thesis at the University of Applied Sciences in Munich, 2005
DIS 8. Kunz, P., Wörne, D.: Institut für Siedlungswasserwirtschaft of the university Braunschweig: Nachweis der biologischen Verfügbarkeit von Klärschlämmen nach Desintegration: Rührwerkskugelmühle im Rahmen einer gezielten Denitrifikation; Klärschlammdesintegration, H. 61, p. 209–214, 1998
DIS 9. Jourdan, R.: researches at the Institut für Ingenieurbiologie und Biotechnologie des Abwassers of the university Kiel: Untersuchungen zur Reduzierung der Schwefelwasserstoffbildung beim anaeroben Abbau durch Luftzugabe, 1987
DIS 10. Falk, O., Sutor, G., Wiegland, S.: researches at the Lehrstuhl für Energie- und Umwelttechnik der Lebensmittelindustrie oft he technical university of Munich: Altspeisefette Aufkommen und Verwertung, 2001
DIS 11. Ochsner, H.: researches at the university Hohenheim: Desulfuration of biogas with Fe-(II)-SO_4, Publikation 01/99–01/00

JOU learned journals/magazines

JOU 1. ATV-report: Verfahren und Anwendungsgebiete der mechanischen Klärschlammdesintegration, Korrespondenz Abwasser, *47*, 4, p. 570, 2000
JOU 2. Binswanger, S., Siegrist, H., Lais, P.: Simultane Nitrifikation/Denitrifikation von stark ammoniumbelasteten Abwässern ohne organische Kohlenstoffquellen, Korrespondenz Abwasser, 44, p. 1573–1580, 1997.
JOU 3. Braun, R.: Neue Entwicklungen im Bereich der Biogaserzeugung, lecture held at the Symposium Biogas – Brennstoffzellensysteme am Museum Arbeitswelt in Steyr (Österreich), 2001
JOU 4. Braun, R.: Verwertung und Entsorgung organischer Nebenprodukte und Reststoffe der Industrie, Müll und Abfall 12, 1992
JOU 5. Bryant, M.P.: Microbial methane production – theoretical aspects, Journ. Anim. Sci. 48, p. 193–201, 1979
JOU 6. Bryant, M.P.; Wolin, E.A.; Wolfe, R.S.: Methanobacillus omelanskii, a symbiotic association of two spezies of bacteria, Arch. Microbiology 59, 1967
JOU 7. Davis, R. D., Hall, J. E.: Production, treatment and disposal of wastewater sludge in Europe from a UK perspective. Water Pollution Control *7*, 2, p. 9–17, 1997.

JOU 8. Englert, G.: Wirtschaftlichkeit von Siloanstrichen. Bauen für die Landwirtschaft 27, 2, p. 23–25, 1990.

JOU 9. Ernst, C.: Biologische Arbeitsstoffe in abwassertechnischen Anlagen – Gefährdungsbeurteilung und Schutzmaßnahmen, Entsorgung 1, 2001.

JOU 10. Paesler, L., Lepers, J., Röper, R.: Energetische Verwertung von holzartiger Biomasse, Umwelt-Magazin 4/5, 2004

JOU 11. Friedrichs, G., Hartmann, U., Kaesler, H., Zingrefe, H.: Biogas – Möglichkeiten und Voraussetzungen der Einspeisung in die Netze der öffentlichen Gasversorgung, Gas, Erdgas 144, p. 59–65, 2003

JOU 12. Benzinger, S., Dammann, E., Wichmann, K.: Nutzung von Eisenhydroxidschlamm aus der Grundwasseraufbereitung in kommunalen Abwasseranlagen. Korrespondenz Abwasser, 43, 9, p. 1552–1560, 1996

JOU 13. Greenpeace Magazin 6, p.32, 2003

JOU 14. Grüneklee, C.E., Artzt, G.: Energie aus Müll, Zeitschrift Chemie Technik 6, p. 56 ff, 1996

JOU 15. Imhoff, K.: Zum Wert der Schlammfaulung und Hinweise zum Faulverfahren, Korrespondenz Abwasser 27, 1980

JOU 16. Kaltschmitt, M.: Einsatzmöglichkeiten von Biomasse in Deutschland – Potenziale und Nutzung. Blickpunkt Energiewirtschaft 1, p. 1, 2003

JOU 17. KA-Wasserwirtschaft, Abwasser 49, 12, 2002

JOU 18. Kesten, E.: Energiefarming – Neue Aufgaben für die Pflanzenzüchtung; Aus der landwirtschaftlichen Praxis, p. 89–93, 2003

JOU 19. Krüger, J.: Prozessabwässer behandeln lassen, Umweltmagazin 2, 2004

JOU 20. Krull, R., Diekmann R., Lindert, M.: Behandlung von Bioabfall in Vergärungsanlagen – eine Marktübersicht, Bioabfall 5, p. 22–25, 1995

JOU 21. Martin, P., Ellersdorfer, E., Zeman, A.: Auswirkungen flüchtiger Siloxane in Abwasser und Klärgas auf Verbrennungsmotoren, Korrespondenz Abwasser, *43*, 9, p. 1574–1578, 1996

JOU 22. Nature 16, 1999

JOU 23. Neis, Uwe: Ultrasound in water, wastewater and sludge treatment; WATER 21, 2000

JOU 24. Palmowski, L., Müller, J.: Einfluß der Zerkleinerung biogener Stoffe auf deren Bioverfügbarkeit, Müll und Abfall, 31, 6, p. 368–372, 1999

JOU 25. Sahm, H.: Biologie der Methanbildung, Chemie-Ingenieur-Technik 53, p. 854, 1981

JOU 26. Scheffer, K.: Ökonomische und ökologische Optimierung des Anbaus und der energetischen Nutzung von landwirtschaftlichen Kulturpflanzen; Aus der landwirtschaftlichen Praxis, p. 89–93, 2003

JOU 27. Einsatz von Lamellenabscheidern in Belebungsbecken, KA-Abwasser, Abfall 1, p. 82–87, 2003

JOU 28. Burbaum, H., Dickmann, T., Kéry, K., Pascik, I., Radermacher, H.: Biokatalytische Verbesserung der Klärschlammfaulung durch Enzyme, KA 49, 8, p. 1110–1119, 2002

JOU 29. Winter, J.: Mikrobiologische Grundlagen der anaeroben Schlammfaulung GWF 126, 2, p. 51, 1985

JOU 30. Eder, B., Eder J., Gronauer, A.: Mehr Gas als aus der Gülle; WOCHENBLATT-Serie "Biogas", Teil 3: Welche Einsatzstoffe Gas liefern, BLW 47, 2004

JOU 31. KA-Wasserwirtschaft, Abwasser, Abfall, 49, 12, 2002

JOU 32. Mielke, H., Schöber-Butin; B.: Pflanzenschutz bei Nachwachsenden Rohstoffen, Zuckerrübe, Öl- und Faserpflanzen; Mitteilungen aus der Biologischen Bundesanstalt für Land- und Forstwirtschaft Berlin-Dahlem, H. 3491, 2002

JOU 33. Landtechnik, H. 4, 1986

JOU 34. Mitteilungen des Instituts für Bautechnik (Berlin), p. 35, 1988

JOU 35. Zeitschrift Bauen für die Landwirtschaft 1, 25, Beton-Verlag GmbH, Düsseldorf, 1998

JOU 36. U.S. Publ. Health Service, Bull, 173, 1927

BRO brochures/handouts

Bro 1. BEKON Energy Technologies GmbH, 2002

Bro 2. Bioferm GmbH, 2002
Bro 3. C.A.R.M.E.N. e.V. Straubing, 2003
Bro 4. Dywidag Umweltschutztechnik GmbH, 1997
Bro 5. ENTEC Umwelttechnik GmbH, Schilfweg 1, A-6972 Fußach/Österreich
Bro 6. Fachverband Biogas e.V.: Biogas-Potentiale in Deutschland, Journal 1. 2001
Bro 7. Börner, F., Blechert, T.: Gülle Entsorgungskonzepte, Veröffentlichung der Firma GEA Wiegand GmbH
Bro 8. Bayerische Asphaltmischwerke
Bro 9. Limus GmbH
Bro 10. Langhans, G.: Vergärung und Co–Fermentation; unser Know-how der Linde-KCA-Dresden GmbH, 1996
Bro 11. Langhans, G.: Wachsende Akzeptanz für Vergärungsanlagen – Vorteile und Risiken einer interessanten Entsorgungstechnolgie, Linde-KCA-Dresden GmbH, 1999
Bro 12. Passavant-Werke AG, 1990
Bro 13. Siemens 2003
Bro 14. Uhde GmbH, Dortmund
Bro 15. Wolf, Osterhofen
Bro 16. KTBL-Arbeitspapier 235 "Energieversorgung und Landwirtschaft", 1996

PAT Patents

PAT 1. DE 10116144
PAT 2. DE 10254309A1
PAT 3. DE 19721243C2
PAT 4. DE 10260968

WEB Internet-publications

WEB 1. www.VTA-Technologie GmbH: Abschlussbericht_Miltenberg_gsd.pdf
WEB 2. www.userpage.fu-berlin.de/~voelker/Vorlesung_Meeresgeologie/sediment4.htm
WEB 3. www.1998_EU_Solarenergiegenerator.pdf: Schwickart, G., Schimetzek, V., Ihrig, D.F.: Verknüpfung von Biomasseproduktion durch Algen mit anaerober Energiegewinnung – Ein vollbiologischer Solarenergiegenerator; 1998
WEB 4. www.agrikomp.de/lang_de/agrikomp/anlagenkomponenten_biolene.htm
WEB 5. www.agrikomp.de/lang_de/agrikomp/anlagenkomponenten_paddelgigant.htm
WEB 6. www.agrikomp.de/lang_de/agrikomp/anlagenkomponenten_vielfrass.htm
WEB 7. www.bayerisches-energie-forum.de/portal/news_detail,2123,27373,detail.html
WEB 8. www.bgr.de/bt_klima/klima_kohlenstoff_ergebnisse.htm
WEB 9. www.bhkw-infozentrum.de/statement/biomassevo.html
WEB 10. www.biogas.ch/f+e/arasieg.htm
WEB 11. www.biogas2004[1].pdf, 2004
WEB 12. www.biogas.ch/f+e/oekobil.htm: Edelmann, W., Ilg, M., Joss, A., Schleiss, K., Steiger, H.: Ökologischer, energetischer und ökonomischer Vergleich von Vergärung, Kompostierung und Verbrennung fester biogener Abfallstoffe
WEB 13. www.biogas-zentrum.de
WEB 14. www.Biologie_der_Gaserzeugung.pdf
WEB 15. www.biomasse-info.net/Energie_aus_Biomasse/Basisdaten
WEB 16. www.biothane.com
WEB 17. www.bmwa.gv.at/.../989E24A3-326E-4A70-BB9D-837A7CBBD883/10415/BeurteilungvonBiogasanlagen2bAnhang1bis6.pdf
WEB 18. www.boxer99.de/biogas_biogaspotentiale.htm
WEB 19. www.boxer99.de/biogas_fachvortrag_01.htm
WEB 20. www.boxer99.de/subs/Tabellen/primaerenergieverbrauch_deutschland.htm
WEB 21. www.bussundhempel.de
WEB 22. www.ceno-tec.de
WEB 23. www.eeg-aktuell.de/ezfilemanager/downloadtemp/BMU_EE_zahlen.pdf
WEB 24. www.eisele.de
WEB 25. www.elib.uni-stuttgart.de/opus/volltexte/2004/2149/pdf/Schlussbericht_BWI22006.pdf
WEB 26. www.energetik-leipzig.de/Ausg_4_02.html
WEB 27. www.energiekonsens.de/aktivitaeten/energiewirtschaft/download/gutachten09-07.pdf
WEB 28. www.energieland.nrw.de/leitprojekte/leitprojekte/deutsch/leit_bio_imk.html

WEB 29. www.erneuerbareenergie.at/inhalt.htm

WEB 30. www.erneuerbareenergien.de/0604/s_59-61.pdf: Schmack, D. Nusko, R.: Biogas in Brennstoffzellen – in fünf Jahren serienreif?

WEB 31. www.fachberatung-biologie.de/Themen/dunkelreakt/seitendunkel/calvinzyk.htm

WEB 32. daf.zadi.de/download/PPT_Scholwin.pdf, 2006

WEB 33. www.fat.admin.ch/pdf/FAT_Bericht_546_D.pdf

WEB 34. www.fh-muenster.de/fb4/lehrgebiete/abwasser/diplomarbeiten/kast/kast.htm

WEB 35. www.fiz-agrar.de/SEARCH/ELFEN/FORMBIB/DDD/70-96-21060.pdf

WEB 36. www.home.landtag.nrw.de/.../Lathen.jpg

WEB 37. www.geocities.com/exobiologie/Abstracts/Abstract_11.html

WEB 38. www.glymes.com/images/Genosorb 2002.pdf

WEB 39. www.gns-halle.de/pdf/ANAStripV.pdf; Verfahren patentiert durch GNS – Gesellschaft für Nachhaltige Stoffnutzung mbH, Weinbergweg 23, 06120 Halle

WEB 40. www.graskraft.de/resonanz/taspo.html

WEB 41. www.gts-oekotech.de/docs/Beitrag_Wetzlarer_Abfalltage_2001.PDF

WEB 42. www.biogasleitfaden_band_4.pdf: Wetter, C., Brügging, E.: Leitfaden zum Bau einer Biogasanlage – Band IV, herausgegeben vom Umweltamt des Kreises Steinfurt

WEB 43. www.icdp-online.de/sites/mallik_de/objectives/physical_properties.html

WEB 44. www.idw-online.de/public/pmid-56564/zeige_pm.html

WEB 45. www.12-20-05_rich-burn_engine_control_briefing.pdf; 2005

WEB 46. www.ifa-tulln.ac.at: Braun, R.: Prinzipien und Systeme von Vergärungsanlagen für biogene Kommunalabfälle

WEB 47. www.igb.fraunhofer.de/WWW/GF/Wasser/dt/GFWM_222_KA-Schaum.dt.html

WEB 48. www.inelektro.de/Applikationen/Hz_ap_02.pdf

WEB 49. www.initiative-brennstoffzelle.de/ibz/live/nachrichten/show.php3?id=99&p=0&nodeid=0

WEB 50. www.ivaa.de/verzeichnis

WEB 51. www.jyu.fi/bio/ymp/alisivut/ymp32sl01LI1.pdf: Rintala, J.A., Jain, V.K., Kettunen, R.H.: Comparative status of the world-wide commercially available anaerobic technologies adopted for biomethanation of pulp and paper mills effluents

WEB 52. www.klaerschlammdesintegration.de/Bericht2.pd

WEB 53. www.koehler-ziegler.de/de/programm/zugehoerige_links/entschwefelung.html

WEB 54. www.limnotec.de/fault.htm

WEB 55. www.linde-kca.com

WEB 56. www.lipp-system.de

WEB 57. www.minibhkw.de/Tagung_MiniBHKW/Mikro-Gasturbinen.pdf: Dielmann, K.; Mikrogasturbinen, Aufbau und Anwendung, 2005

WEB 58. www.mlur.brandenburg.de/i/biogas04.htm

WEB 59. www.rhvmauthausen.at

WEB 60. www.roevac.de/html/german/supply-sub5.htm

WEB 61. www.rux.de

WEB 62. www.schmitt-enertec.de

WEB 63. www.turbec.com/products/techspecific.htm

WEB 64. www.scienceticker.info/news/EpZEupuVZpxhKFdQdk.shtml http://www.uni-protokolle.de/nachrichten/id/27100/

WEB 65. www.shell-wollishofen.ch/motorsport/weltenergieverbrauch.htm

WEB 66. www.solargeneration.de/solar_studie.pdf

WEB 67. www. mtu-cfc.com

WEB 68. www.svgw.ch/deutsch/filesPR/MB_TISG_013.pdf

WEB 69. www.US$solar.org/download/Memorandum_LaWi.pdf, 2006

WEB 70. www.tu-berlin.de/forschung/IFV/wasser/schrift/band6/6-ott.pdf

WEB 71. www.tu-harburg.de/forschung/fobe/1995-2000/a1998.2-09/w.43.945152987761.html

WEB 72. www.ubka.uni-karlsruhe.de

WEB 73. www.umwelt-fuks.de/dokuprechtl.html

WEB 74. www.uni-giessen.de/~gf1265/GROUPS/KLUG/Methanogene.html
WEB 75. www.energie.ch/themen/haustechnik/blockheizkraftwerke
WEB 76. www.w amonco_ferreira.pdf
WEB 77. www.utec-bremen.de/biogas/bsp_gas_labor.htm
WEB 78. www.utec-bremen.de/biogas/bsp_gas_vfzhh.htm
WEB 79. www.volker-quaschning.de/downloads/sonnenforum2000_1.pdf
WEB 80. www.vorspanntechnik.com
WEB 81. www.science.orf.at/science/news/77026
WEB 82. www.wind-energie.de/zeitschrift/neue-energie/jahr-2000/inhalte/ne0011/novemeber3.htm
WEB 83. www.wtz.de/3DGMK/Sektion3Vortrag1.htm: Prochaska, U., Doczyck, W.: Hocheffiziente Klärgasnutzung durch Gasmotoren mit Abgasrückführung und Lambda = 1 Betrieb
WEB 84. www.wzv-malchin-stavenhagen.de/html/sites/i_pfanni.htm
WEB 85. www.energie-cites.org/db/linkoping_113_en.pdf
WEB 86. www.fachverband-biogas.de/
WEB 87. www.wdr.de/tv/q21/typo3temp/pics/3de193e8b9.jpg
WEB 88. www.atb-potsdam.de/Hauptseite-deutsch/ATB-Schriften/Sonstige/Kap_4_bis_11.pdf
WEB 89. www.fnr-server.de/pdf/literatur/pdf_41folien4c_end.pdf, 2006
WEB 90. Kromus, S, Narodoslawsky, M, Krotschek, C.: Grüne Bioraffinerie, Integrierte Grasnutzung als Eckstein einer nachhaltigen Kulturlandschaftsnutzung, www.nachhaltigwirtschaften.at/nw_pdf/0218_gruene_bioraffinerie.pdf, 2002
WEB 91. www.vdi.de; published in VDI-Nachrichten on 16.11.1999
WEB 92. www.igv-gmbh.de/referenz3.htm: Puls, O., Mikroalgenproduktionsanlage in Klötze (Altmark)
WEB 93. www.innovations-report.de/html/berichte/energie_elektrotechnik/bericht-59246.html, 2006
WEB 94. www.greencarcongress.com/2005/06/sweden_to_launc.html#trackback 2006
WEB 95. www.erneuerbare-energien.de: Extrapolation of data about electricity generation from bio resources, report for the task force "Renewable Energies – Statistics" (AGEE-Stat), 2004
WEB 96. www.treehugger.com/files/2006/05/biogas_generati_1.php, 2006
WEB 97. Yapp, J., Rijk, A.: CDM Potential for the commercialization of the Integrated Biomass System, www.potentiale in Asien.pdf, 2006
WEB 98. www.i-sis.org.uk/BiogasChina.php
WEB 99. Glaser, N.: China auf neuem Weg; Interview mit. Knecht; www.inwent.org/E+Z/content/archiv-ger/05-2006/inw_art1.html
WEB 100. www.snisd.org.cn/enhtm/four-in-one%20biogas%20technology.doc, 2006
WEB 101. www.sleekfreak.ath.cx:81/3wdev/VITAHTML/SUBLEV/ES1/BIOGAS3M.HTM, 2006
WEB 102. www.zeit.de/online/2006/01/russland_ukraine
WEB 103. www.energieforum.ru/ru, 2006
WEB 104. www.inforse.dk/europe/pdfs/VisionUA_ppt.pdf, 2006
WEB 105. www.biogasnutzung_polen_03_d.pdf
WEB 106. www.biogasertraege.pdf
WEB 107. Plöchl, M.: Technische Nutzung von Biogas, www.kap_4_bis_11.pdf, 2004
WEB 108. www.tu-berlin.de/fb6/energieseminar/veroeffentlichung/Rottetrommel_ES04_Bericht.pdf, 2006
WEB 109. www.iswa.uni-stuttgart.de/alr/sso/Kap5/kap5.html, 2006
WEB 110. www.checklistebiogas.pdf, 2006
WEB 111. www.alr_110706_vanbergen.pdf
WEB 112. Konzepte der Fermentiertechnik bei der Nassvergärung: http://energytech.at/biogas/portrait_kapitel-6.html, 2006
WEB 113. www.pius-info.de/dokumente/docdir/weiHer/praxis_info/WH_0105_pdf_102aid_biogasanlagen.html, 2006
WEB 114. www.carmen-ev.de/dt/hintergrund/biogas/monits00s.pdf, 2006
WEB 115. www.lrz-neukirchen.de
WEB 116. www.Nawaros_Gruber.pdf, 2006

WEB 117. www.gasmotoren_analyse.pdf: Herdin, R.: Standesanalyse des Gasmotors im Vergleich zu den Zukunftstechniken (Brennstoffzellen und Mikroturbine) bei der Nutzung von aus Biomasse gewonnenen Kraftstoffen

WEB 118. www.gtp-merichem.com/downloads/lo-cat.pdf, 2006

WEB 119. www.linde-anlagenbau.de/anlagenbau/gasaufbereitungsanlagen/rectisol_waesche.php

Index

Biogas from Waste and Renewable Resources. An Introduction.
Dieter Deublein and Angelika Steinhauser

ISBN: 978-3-527-31841-4

k

l

m